Water Quality and the Early Life Stages of Fishes

Major support for preparation of this volume
was provided by the

U. S. Environmental Protection Agency
16th Annual Larval Fish Conference
and the
Early Life History Section, American Fisheries Society

Conference Sponsor

University of Rhode Island

Publication Sponsor

Early Life History Section, American Fisheries Society

Editorial Board

Lee A. Fuiman, Editor

David A. Bengtson   Lawrence J. Buckley

Grace Klein-MacPhee

Water Quality and the Early Life Stages of Fishes

Edited by

Lee A. Fuiman
Marine Science Institute
University of Texas at Austin
Port Aransas, Texas USA

Selected Papers from a Conference Held in
Kingston, Rhode Island, USA
June 16-20, 1992

American Fisheries Society Symposium 14

Bethesda, Maryland
1993

The American Fisheries Society Symposium series is a registered serial.
Suggested citation formats follow.

Entire book

Fuiman, L. A., editor. 1993. Water quality and the early life stages of fishes. American Fisheries Society Symposium 14.

Article within the book

Hall, L. W., Jr. 1993. A review of in situ and on-site striped bass contaminant and water-quality studies in Maryland waters of the Chesapeake Bay watershed. American Fisheries Society Symposium 14:3-15.

© Copyright by the American Fisheries Society, 1993

Library of Congress Catalog Card Number: 93-73537
ISSN 0892-2284   ISBN 0-913235-86-5

Address orders to

American Fisheries Society
5410 Grosvenor Lane, Suite 110
Bethesda, Maryland 20814, USA

CONTENTS

Preface vii

Introduction viii

Laboratory and Field Studies

A Review of In situ and On-Site Striped Bass Contaminant and Water-Quality Studies in Maryland Waters of the Chesapeake Bay Watershed
Lenwood W. Hall, Jr., Susan E. Finger, and Michael C. Ziegenfuss 3

Sublethal Effects of Methyl Parathion, Carbofuran, and Molinate on Larval Striped Bass
Alan G. Heath, Joseph J. Cech, Joseph G. Zinkl, Brian Finlayson,
and Robert Fujimura 17

Production, Mortality, and Transport of Striped Bass Eggs in Congaree and Wateree Rivers, South Carolina
James S. Bulak, Noel M. Hurley, Jr., and John S. Crane 29

Effects of Changes in Age-0 Survival and Fishing Mortality on Egg Production of Winter Flounder in Cape Cod Bay
John Boreman, Steven J. Correia, and David B. Witherell 39

Pathological Conditions of Narragansett Bay Young-of-the-Year Winter Flounder
Sharon A. MacLean 47

Effects of Chemical Stresses on Behavior of Larval and Juvenile Fishes and Amphibians
W. J. Birge, R. D. Hoyt, J. A. Black, M. D. Kercher, and W. A. Robison 55

Methodology

Behavioral Methods for Assessing Impacts of Contaminants on Early Life Stage Fishes
Edward E. Little, James F. Fairchild, and Aaron J. DeLonay 67

The Teratological and Pathological Effects of Contaminants on Embryonic and Larval Fishes Exposed as Embryos: a Brief Review
Joel E. Bodammer 77

Calibrating Starvation-Induced Stress in Larval Fish Using Flow Cytometry
Gail H. Theilacker and W. Shen 85

Use of Mesocosm Studies to Examine Direct and Indirect Impacts of Water Quality on Early Life Stages of Fishes
    James F. Fairchild and Edward E. Little 95

Using Mesocosms to Assess the Influence of Food Resources and Toxic Materials on Larval Fish Growth and Survival
    Grace Klein-MacPhee, Barbara K. Sullivan, and Aimee A. Keller 105

Contaminant Exposure and Population Growth of English Sole in Puget Sound, Washington: the Need for Better Early Life-History Data
    John T. Landahl and Lyndal L. Johnson 117

Individual-based Modelling of Environmental Quality Effects on Early Life Stages of Fishes: a Case Study Using Striped Bass
    Kenneth A. Rose, James H. Cowan, Jr., Edward D. Houde, and
    Charles C. Coutant 125

Management Issues

The Importance of Habitat to the Early Life History of Estuarine Dependent Fishes
    Donald E. Hoss and Gordon W. Thayer 147

Certain Decisions with Uncertain Data: Early Life-History Data and Resource Management
    Penny Howell 159

The Potential Role of Larval Fish Culture in Alleviating Population and Habitat Losses
    G. Joan Holt 167

Preface

The annual Larval Fish Conferences have been associated with the Early Life History Section of the American Fisheries Society since the founding of that section. In many respects, the Section is a product of the conferences. The leadership of the Early Life History Section provides a measure of continuity and oversight by lending support and helping arrange future conferences. The Section also promotes publication of papers from the conferences. Toward that end, the Early Life History Section appoints a Publications Editor to provide technical editing and to manage the publication process. The editor selects an editorial board to conduct the rigorous peer-review process for each manuscript submitted. As Publications Editor, I have elected to assemble volumes that address one or more themes, rather than compile a volume of proceedings for each conference. This is intended to make each volume a more valuable resource, albeit for a narrower audience.

I wish to express my sincere gratitude to Drs. David A. Bengtson, Lawrence J. Buckley, and Grace Klein-MacPhee who conceived and arranged a very interesting conference and for their work as associate editors during the production of this volume. I also appreciate the assistance of Ms. Patty Baker of the University of Texas Marine Science Institute who prepared the final typescript and Mr. Greg Whitledge who assisted in some of the technical editing chores. The entire editorial board thanks the people named below for providing thoughtful scientific reviews of the original manuscripts.

LEE A. FUIMAN
Marine Science Institute
University of Texas at Austin

Kenneth Able
David Bengtson
Walter Berry
Dianne Black
John Boreman
Joseph Brown
Larry Buckley
Chris Chambers
Sue Cielinski
James Cowan
Dena Gadomski

Mark Gibson
Jeffrey Giddings
Tim Gleason
Len Hall
Karsten Hartel
W. Huntting Howell
John Hunter
Cynthia Jones
Grace Klein-MacPhe
William Krueger
Ed Little
Donald Miller

Thomas Monroe
Dean Perry
Sherry Poucher
Chris Powell
Charles Roman
Tom Savoy
John Stegeman
Saran Twombly
Richard Voyer
Patrick Walsh
Judith Weis

Introduction

The critical importance of the early life stages of fishes, defined here as embryos, larvae, and early juvniles, to fish population dynamics has long been recognized by both fishery biologists and environmental scientists. Larvae in particular have been shown to be extremely sensitive to biological and physical fluctuations in the environment (resulting in variable recruitment) and to anthropogenic inputs to the environment (based on laboratory toxicity tests). The early life stages of fishes suffer a variety of adverse interactions with humans, ranging from exposure to toxic chemicals and habitat degradation to power-plant entrainment. In addition, overfishing of stocks can reduce the numbers of reproducing adults.

Given the international interest in environmental quality and the specific interests of the membership of the American Fisheries Society Early Life History Section, we the organizers of the 16th Annual Larval Fish Conference chose "Environmental Quality and the Early Life Stages of Fishes" as the conference theme. The conference was divided into three major areas: Problem Identification (9 papers presented), Research Approaches (24 papers), and Mitigation/Management Approaches (7 papers). This volume represents a selection of those presentations, all of which have passed a rigorous peer-review process.

The study of environmental effects on fish early life stages is conducted at several biological levels of organization (subcellular to populations) with a variety of methods (e.g., histopathology, behavior, population simulations) at several spatial scales (beaker to mesocosm to lakes and oceans). The research is conducted by scientists from academia, government, and the private sector. Some of that research is used to make policy and to make management decisions. Our goal was to bring together scientists and managers to talk about as many of the biological levels, methods, and scales as possible, in order to provide a comprehensive overview of this rapidly expanding field. This volume of papers faithfully reflects the diversity of the presentations at the conference. We hope that it will stimulate workers in this field, especially those now entering.

DAVID A. BENGTSON
Department of Zoology
University of Rhode Island

LAWRENCE J. BUCKLEY
Graduate School of Oceanography
University of Rhode Island

GRACE KLEIN-MACPHEE
Graduate School of Oceanography
University of Rhode Island

Laboratory and Field Studies

A Review of In situ and On-Site Striped Bass Contaminant and Water-Quality Studies in Maryland Waters of the Chesapeake Bay Watershed

Lenwood W. Hall, Jr.

*The University of Maryland System, Maryland Institute
for Agriculture and Natural Resources, Agricultural Experiment Station,
Wye Research and Education Center, Queenstown, Maryland 21658, USA*

Susan E. Finger

*U. S. Fish and Wildlife Service, National Fisheries
Contaminants Research Center, Columbia, Missouri 65201, USA*

Michael C. Ziegenfuss

*The University of Maryland System, Maryland Institute
for Agriculture and Natural Resources, Agricultural Experiment Station,
Wye Research and Education Center, Queenstown, Maryland 21658, USA*

Abstract.—The objective of this review is to compare survival data from in situ and on-site striped bass, *(Morone saxatilis)* prolarva tests conducted in the Nanticoke River, Choptank River, upper Chesapeake Bay, and Potomac River of the Chesapeake Bay watershed from 1984 to 1990 and to discuss the possible effects of contaminants and water-quality conditions on survival of striped bass prolarvae in these habitats. Acidic conditions (low pH, monomeric aluminum and other metals) were reported to reduce survival of prolarvae in the Nanticoke River, although these conditions were not present in every year. Acidic conditions and trace metals (Al, Cd, Cu, and Zn) in the Choptank River were also suspected as factors contributing to mortality of prolarvae during in situ tests. Survival of striped bass prolarvae was generally greater in the upper Chesapeake Bay when compared with the other habitats. Low salinity, high conductivity, and lack of toxic contaminants were suspected as contributing to the high survival. Low survival of prolarvae was reported during all 4 years of testing in the Potomac River. Sudden reductions in temperature and presence of various trace metals (Al, Cd, Cu, Pb, and Zn) were suspected as contributing to the low survival in this system. Results from this 7-year study suggested that environmentally realistic acidic conditions, contaminants (primarily trace metals), and low temperatures can reduce survival of striped bass prolarvae. Each of these factors was site-specific to the four spawning areas.

The majority of the striped bass *(Morone saxatilis)* migrating along the Atlantic coast are spawned in tributaries of the Chesapeake Bay. Kohlenstein (1980) has reported that the four major spawning areas for the Maryland waters of the Chesapeake Bay are located in the Nanticoke River, Choptank River, Potomac River, and the upper Chesapeake Bay. With the exception of 1989, striped bass stocks have generally declined in these habitats in recent years due to the low production of juveniles (Boreman and Austin, 1985). Numerous factors, such as habitat modification, disease, predation, food availability, stream flow, entrainment and impingement by power plants, fishing pressure, and contaminants, have been postulated as possible causes of striped bass decline. The two most commonly postulated reasons for striped bass decline are fishing pressure and contaminants, including adverse water quality (Goodyear 1985). The focus of this review is contaminants.

Various pathways exist for contaminants to affect striped bass reproductive success due to the spatial extent of the various life stages. Striped bass is an anadromous species and its early life stages (egg, prolarval, larval, and young juvenile stages) are found in freshwater or low-salinity areas of spawning rivers. Older juveniles are in the estuary and adults live in the bay areas or the ocean near the mouths of spawning rivers. The most susceptible life stage (prolarvae) is found in areas (freshwater rivers) where exposure to contaminants is most likely due to greater human activity and anthropogenic inputs. Contaminants may cause adverse effects on early life stages by previous accumulation in the parent stock, environmental exposure of egg and sperm to contaminants during fertilization, exposure of sensitive early life stages to adverse contaminants or a combination of all factors. Mehrle et al. (1982) isolated numerous organic and inorganic contaminants from striped bass tissues and the environment in both the Chesapeake Bay and other Atlantic coast estuaries. Laboratory studies were conducted to determine the effects of multiple organic and inorganic contaminants, representing concentrations in east coast tributaries, on various life stages of striped bass to assess the impact of a mixture of chemical contaminants on growth, swimming stamina, and response to natural environmental challenges such as prey-predator relationships and water-quality variations (CNFRL 1983; Buckler et al. 1987). Major conclusions from these early laboratory studies were that larvae were most susceptible to contaminant mixtures in freshwater (when compared to low-salinity conditions), pH < 6.5 in combination with 100 µg/L aluminum caused significant mortality to young larvae, and pH sensitivity of striped bass declined after 50 to 80 days of age. These laboratory studies provided valuable baseline information necessary for assessing the role of contaminants on declining stocks of striped bass.

The next phase of contaminants research with striped bass was to conduct in situ and on-site field studies where susceptible young life stages were exposed to ambient water-quality conditions during the spring spawning season. In situ and on-site field studies were conducted by the University of Maryland's Agricultural Experiment Station (formerly Johns Hopkins University) and the U. S. Fish and Wildlife Service in the four major striped bass spawning areas in Maryland waters of Chesapeake Bay from 1984 through 1990. These field studies were designed to assess the survival of striped bass prolarvae after 96-h exposures to ambient environmental conditions. A suite of inorganic and organic contaminant and water-quality conditions were reported concurrently with these studies to evaluate the possible effects of these factors on survival of young larvae.

The objective of this review is to compare and summarize survival data from previously published in situ and on-site striped bass prolarva survival tests conducted in Maryland's four major striped bass spawning areas from 1984 through 1990 and to discuss the possible adverse effects of contaminants and water quality on survival of young larvae in these habitats.

Methods

Study Areas

Ninety-six-hour in situ and/or on-site studies with striped bass prolarvae were conducted in the following spawning areas from 1984 to 1990: Nanticoke River (1984 to 1990); Choptank River (1984, 1987 to 1990), Potomac River (1986, 1988 to 1990), and upper Chesapeake Bay (1985 to 1990) (Figure 1). Results from the in situ and on-site tests were presented together because these methods produced similar results (Hall et al., 1988). In situ and on-site studies in the upper Chesapeake Bay were conducted in the Chesapeake and Delaware (C & D) Canal from 1985 to 1987; open water areas of the upper Chesapeake Bay (near Betterton, Maryland) were tested in 1989 and 1990. All tests were conducted during April and May for each respective year. In situ tests were generally conducted in three locations in each habitat except the Choptank River where two locations were tested.

In Situ and On-Site Test Procedures

All tests were conducted with young prolarvae, 24 to 48 h old, obtained from one of the following hatcheries: Dennis Wildlife Center, Bonneau, South Carolina; Delmarva Ecological Laboratory, Middletown, Delaware; Richloam Fish Hatchery, Webster, Florida; or Manning Hatchery, Brandywine, Maryland. Most of the prolarvae used for testing were Chesapeake Bay stock obtained from the Manning Hatchery.

In situ test chambers and testing procedures have been described in detail by Hall (1984), Ziegenfuss et al. (1990), and Hall et al. (1991a). One to four 96-h tests were generally conducted at one to three stations in each habitat during the spawning period, when water temperatures ranged from 13 to 20°C. Field studies from 1988 to 1990 were generally more comprehensive on both a spatial and temporal scale. Controls for these in situ tests were housed in chambers in situ (identical to those in the field) located in 945-L circular tanks in an on-site laboratory. Approximately 50% of the circular tanks volume was renewed daily. Five hundred prolarvae

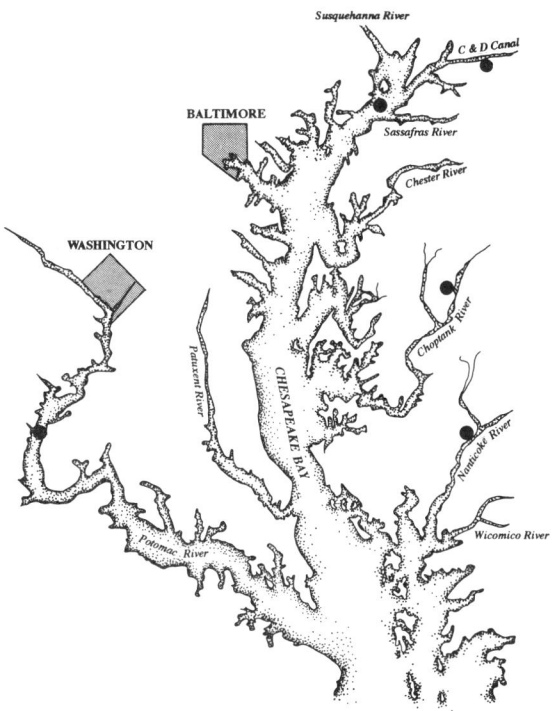

FIGURE 1.—Location of spawning areas for in situ and on-site studies in Maryland waters of Chesapeake Bay (•).

were tested in each chamber; replicate chambers were used for ambient habitat and control exposures. Survival of prolarvae was evaluated after 96 h using subsampling procedures described in detail in Hall et al. (1985).

Two types of on-site tests were conducted using procedures described in detail by Mehrle et al. (1984) and Hall et al. (1989). The Mehrle et al. (1984) flow-through method was designed to test the toxicity of a series of 24-h composite samples collected from each habitat station and renewed daily for 4 d. A series of treatments containing 100, 50, 25, and 12.5 % habitat water and two controls (contaminant-free water) were prepared in 250-mL beakers. Ten beakers with one prolarva in each were used for each treatment. Survival was evaluated every 24 h for 96 h.

The on-site test described in detail by Hall et al. (1988) involved the use of a mobile laboratory (2.5 × 7.7 m trailer) housed on the shore of the spawning area. Habitat water was pumped from a Gould submersible pump into head tanks on top of the mobile laboratory. The water then entered a flow-through, solenoid-valve proportional dilutor system calibrated to deliver 100, 50, and 25% habitat water and control water (contaminant-free) conditions to 56-L aquaria. Ten prolarvae were tested in 1-L polycarbonate chambers with 202-μm Nitex windows. Four chambers (replicates) were used for each treatment. Survival of prolarvae was evaluated every 24 h during the 96-h test.

Contaminant and Water-Quality Procedures

Aqueous samples were collected at all field locations during the tests. Composite samplers were used to collect water at 24- to 48-h intervals for analyses of inorganic and organic contaminants listed in Table 1. Detailed procedures for preparation and analyses of samples and quality-assurance/quality-control procedures were described by Hall et al. (1989) and Hall et al. (1991a). Metals were analyzed by furnace atomic absorption spectrophotometry and, in the case of aluminum, by direct aspiration spectrophotometry (USEPA 1979). Halogenated hydrocarbon pesticides and polychlorinated biphenyls were measured by electron-capture gas chromatography and acid-extractable organic compounds were analyzed by gas chromatography and mass spectrometry (USEPA 1982).

Temperature, salinity, dissolved oxygen, pH, conductivity, and hardness were measured daily on grab samples from all field locations and the controls. Ammonia, nitrate, nitrite, total phosphorous, and turbidity were measured twice a week. Precipitation volume and pH were also evaluated after major rainfall events in each study area. For experiments conducted from 1988 to 1990, Datasonde units suspended from in situ rafts were used to collect the following water-quality conditions at 0.5- to 1-h intervals at each station: temperature, pH, conductivity, salinity, and dissolved oxygen.

Results

The volumes of data on prolarva survival, contaminants, and water-quality conditions generated from this 7-year research effort can not be presented in detail in this paper. Therefore, we have summarized the data for each of the four habitats by year in Tables 2 through 5. Mean percent survival was the mean of values usually represented by nine, 96-h survival values from three stations for three experiments. Survival was classified as low (< 20%), medium (20 to 40%) and high (> 40%) based on comparison with natural mortality rates (Dahlberg 1979; Dey 1981). Control survival of > 80% in 96-h was generally used to define an acceptable test. Potentially toxic water-quality and/or contaminant conditions were identified based on a comparison of ambient conditions with results from laboratory toxicity studies or water-quality criteria.

Mean survival of striped bass prolarvae tested in situ in spawning areas of the Nanticoke River was classified as low during 5 of the 7 years (Table 2). Low survival in 1984, 1987, 1988, 1989, and 1990 was consistently associated with the presence of pH conditions less than 6.5 with concurrent elevated concentrations of aluminum and other metals (Hall et al. 1985; Buckler et al. 1987). The highest prolarva survival (70%) was reported in 1985 when pH conditions were near neutral and toxic metals were not identified. Salinity conditions of 2.1 to 3.2‰ reported in the spawning area of the Nanticoke River in 1985 were also considered as important factors contributing to survival of prolarvae (Hall 1991).

Prolarva survival data by year from the Choptank River was similar to the Nanticoke River (Table 3). Low survival of prolarvae was reported in 1984, 1988, 1989, and 1990. Acidic conditions (pH < 6.5), aluminum, and other metals were reported at potentially toxic levels simultaneously with the low prolarva survival in 1984, 1989, and 1990. The low survival of prolarvae in 1988 was

TABLE 1.—Desired detection limits for contaminants assayed at all field locations. Stated values were not detected in all samples over the 7-year study period.

Contaminant	Detection limit (µg/L)
Aroclor 1248	0.014
Aroclor 1254	0.024
Aroclor 1260	0.028
DDE	0.002
Toxaphene	0.06
Chlordane	0.01
Perylene	0.04
Fluorene	0.04
Phenanthrene	0.04
Anthracene	0.04
Fluoranthrene	0.04
Pyrene	0.04
Benz(a)anthracene	0.04
Chrysene	0.04
Aluminum	1
Arsenic	3
Cadmium	0.5
Chromium, total	1
Copper	1
Lead	10
Nickel	5
Selenium	2
Tin	1
Zinc	10

likely related to cadmium concentrations of 4 to 15 µg/L, copper concentrations of 4 to 25 µg/L, zinc concentrations of 32 to 331 µg/L, and aluminum concentrations of 50 to 280 µg/L (Finger 1989). The moderate survival of prolarvae in 1987 occurred in the presence of potentially toxic concentrations of monomeric aluminum (150 µg/L), cadmium (3 µg/L), and copper (40 µg/L).

Survival of striped bass prolarvae in the upper Chesapeake Bay in all 5 years from 1985 to 1990 was classified as high, based on the natural rates of mortality for this species (Table 4). Mean survival of prolarvae ranged from 40 to 80%. Potentially toxic water-quality and contaminant conditions were generally not found in this spawning area. The presence of low-salinity conditions in the upper Chesapeake Bay was suspected as a contributing factor for enhancing survival of striped bass prolarvae during many of the experiments. However, there were cases of high prolarva survival without the presence of salinity during the 1990 experiments (Hall et al. 1992).

Survival of striped bass prolarvae in the Potomac River was consistently low during all 4 years of testing (Table 5). Low survival of prolarvae in 1986, 1987, and 1989 was reported concurrently with the presence of potentially toxic metals (Al, Cd, Cu, and As in various years) and sudden reductions in temperature. Low survival in 1990 was

TABLE 2.—Results from 96-h in situ and on-site toxicity tests with striped bass prolarvae in the Nanticoke River from 1984 to 1990.

Year	Type of test	Number of locations	Number of tests	Mean survival (%) and classification	Results	Reference
1984	In situ	3	2	4, low	> 90% mortality after 4-d exposure; dissolved aluminum ($\approx$0.12 mg/L), pH (mean = 6.3), and soft water were suspected causes of mortality.	Hall et al. (1985)
	On-site	3	2	0, low	100% mortality after 48-h exposures to full strength river water; aluminum, pH, and soft water as reported by Hall et al. (1985) were suspected causes of mortality.	Mehrle et al. (1984)
1985	On-site	1	1	70, high	High survival of prolarvae; adverse water quality (i.e., pH) and contaminants not reported.	Mehrle et al. (1986)
1986	On-site	1	1	30, moderate	70% mortality concurrent with pH < 6.5.	Mehrle et al. (1986)
1987	In situ On-site	1	1	20, low	$\approx$ 70% mortality during test; pH as low as 6.39 suspected stressful.	Finger (1987)
1988	In situ	3	3	14, low	57-100% mortality during 3, 96-h in situ and on-site tests; zinc (0.030 mg/L), aluminum (0.090-0.120 mg/L), copper (0.003-0.018 mg/L), and low pH ($\approx$ 6.1) suspected causes of mortality.	Rago (1988)
1989	In situ	3	4	11, low	63-100% mortality after 4, 96-h tests; pH conditions < 6.5 suspected stressful.	Finger et al. (in press)
1990	In situ	3	3	8, low	> 95% mortality during first 2 tests with pH < 6.5; during test 3 mortality was lower and pH increased.	Rago (1990)

attributed to the presence of Cr, Zn, and As; drastic temperature reductions did not occur during the 1990 experiments.

Discussion

Nanticoke River

In situ studies in the Nanticoke River in 1984 and various other years through 1990 suggested that acidification (low pH, monomeric aluminum and other metals) in this Chesapeake Bay eastern shore tributary was responsible for mortality of striped bass prolarvae. This finding was significant in 1984 as acidification was not previously identified as a potential problem for anadromous fishes in Chesapeake Bay. Chesapeake Bay is generally considered to be resistant to acidification due to the presence of saline conditions and alkalinity that create buffering capacity sufficient to neutralize acidic inputs; however, the capacity of this system to resist change in pH conditions decreases in tributaries leading into the bay. Bowman (1984) reported that 23 tributaries in Maryland's coastal plain (freshwater streams) had mean pH values below 6.0 for 4 consecutive weeks during a very "wet spring" in April and May of 1983. Precipitation in Maryland is as acidic as many regions of the northeastern United States as annual average pH values have been reported in the range of 3.5 to 4.5 (Hall 1987). The soils of the coastal plain may also create potential acidification problems resulting from runoff during precipitation events because soil pH levels below 5 have been reported in many non-agricultural locations in Maryland (Miller 1976). The thickness of coastal plain soils precludes interaction between acid deposition and bedrock. The combination of acidic precipitation, acidic soils, and low buffering capacity of surface waters resulted in acidic conditions in spawning areas (freshwater streams) of various anadromous fishes.

Despite the data available on acidic conditions in first- and second-order Maryland coastal plain streams, potential acidification effects were generally dismissed (before 1984) in mainstream rivers where striped bass spawn. During our 1984 studies, we reported hardness values in the Nanticoke River as low as 23 mg/L $CaCO_3$, thus suggesting that this river may have low buffering capacity and may be susceptible to acidification. Our studies in 1984 also showed that short-lived pH depressions (ca. 6.0 to 6.4) in the Nanticoke River can occur during precipitation events with rapid changes in river

TABLE 3.—Results from 96-h in situ and on-site toxicity tests with striped bass prolarvae in the Choptank River from 1984 to 1990.

Year	Type of test	Number of locations	Number of tests	Mean survival (%) and classification	Results	Reference
1984	On-site	1	2	0, low	100% mortality after 48h in river water; aluminum and other metals, acidic conditions, and soft freshwater suspected stressful.	Mehrle et al. (1984)
1987	In situ On-site	1	2	25, moderate	55-70% mortality in 2, 96-h on-site and in situ tests; stressful factors: low hardness (36-48 mg/L $CaCO_3$), monomeric aluminum (0.150 mg/L), cadmium (0.003 mg/L), copper (0.040 mg/L).	Hall et al. (1988)
1988	In situ On-site	2	3	0, low	100% mortality in 3, 96-h in situ and on-site tests; cadmium (0.004-0.015 mg/L), zinc (0.032-0.331 mg/L), aluminum (0.050-0.280), and copper (0.004-0.025 mg/L) suspected in causing mortality.	Finger (1989)
1989	In situ	2	4	3, low	87-100% mortality; pH < 6.5, cadmium (0.005-0.023 mg/L), zinc (0.019-0.071 mg/L), copper (0.009-0.023 mg/L) and monomeric aluminum (0.009-0.056 mg/L) suspected in causing mortality.	Finger et al. (in press)
1990	In situ	2	3	8. low	100% mortality in first 2 tests with pH < 6.5; 64-88% mortality in test 3 with pH ≈ 7 and salinity ≈ 0.5 ‰.	Rago (1990)

stage, turbidity, and dissolved aluminum levels. Although pH conditions between 6.0 and 6.4 are not extremely low, these conditions were reported to be lethal to striped bass prolarvae (CNFRL 1983; Buckler et al. 1987).

The "acidification hypothesis" was rapidly adopted as an explanation to account for declining stocks of striped bass along the Atlantic coast of the United States. While our data from the Nanticoke and Choptank Rivers provide support for this hypothesis, it is important to note that the two other spawning areas for striped bass in Maryland waters of Chesapeake Bay, the Upper Bay and the Potomac River (discussed later), also experienced low production of juvenile striped bass in recent years, in the absence of acidic conditions. Since these two spawning habitats are not susceptible to acidification, the "acidification hypothesis" does not apply to all spawning areas in Maryland waters of Chesapeake Bay.

In situ studies with striped bass prolarvae in the Nanticoke River provided valuable data on water-quality and contaminant conditions present during a critical period of development (prolarval stage) for this anadromous species. These data are important for determining survival of young striped bass prolarvae in their spawning area during the spawning season. The presence of older life stages of striped bass (juveniles) is documented in the Nanticoke River (and the other three habitats) by the State of Maryland. In Maryland waters of the Chesapeake Bay, a juvenile index of young of the year is determined each summer to provide an indication of good or poor survival of the previous spring spawn. The index is determined by using a series of seine hauls in various locations within each habitat (Boone 1980). An index of 10 or greater may be necessary to provide a strong year class of striped bass. For example, dominant year classes in 1958, 1964, and 1970 had juvenile indices ranging from 19.2 to 30.4 (Maryland Department of Natural Resources 1986).

In situ and on-site tests were not designed to predict the juvenile index for striped bass. Many biotic and abiotic factors can influence the survival of striped bass during the time this species requires to develop from prolarvae to juveniles. However, since both endpoints are used to gauge reproductive success of striped bass we have compared these data for each habitat by year (Figure 2). Survival of striped bass prolarvae from in situ and on-site studies in the Nanticoke River from 1984 to 1990 were compared with the State of Maryland's juvenile index for young-of-the-year striped bass (Figure 2). Based on a comparison of in situ and/or on-site prolarva survival data and the juvenile index there is close agreement between these two endpoints for 5 of the 7 years. The only year when a large differ-

TABLE 4.—Results from 96-h in situ and on-site toxicity tests with striped bass prolarvae in the upper Chesapeake Bay from 1985 to 1990. Experiments were conducted in the Chesapeake and Delaware Canal (C&D Canal) in 1985, 1986, and 1987. Open water areas of the Upper Bay (near Betterton, Maryland) were tested in 1989 and 1990.

Year	Type of test	Number of locations	Number of tests	Mean survival (%) and classification	Results	Reference
1985	In situ	3	2	50, high	Significant mortaility not reported; adverse water-quality and contaminant conditions not identified.	Hall et al. (1987b)
	On-site	3	2	60, high	Significant mortality not reported; adverse water-quality and contaminant conditions not identified.	Finger (1985)
1986	On-site	1	1	80, high	Sigificant mortality not reported; adverse water-quality and contaminant conditions not identified.	Mehrle et al. (1986)
1987	On-site In situ	1	4	60, high	Significant mortality not reported.	Hall et al. (1988)
1989	In situ Low saline	3	3	40, high	48-60% mortality during 2 tests; significantly higher mortality during third test; low temperatures (10 -11°C) stressful.	Hall et al. (1991a)
1990	In situ	3	3	45, high	Mortality not significant; salinity not detected.	Hall et al. (1992)

ence was reported between these endpoints was 1985. Survival of prolarvae from in situ tests was high while the juvenile index was low. An explanation for this difference is that the temporal and spatial coverage from toxicity tests during 1985 was limited to one 96-h on-site test in only one location in the Nanticoke River (Finger 1985). Adverse water-quality or contaminant conditions or other adverse conditions may have occurred later in the spawning season or in different locations in the spawning area.

Choptank River

One of the conclusions from this 7-year study is that each of the four spawning areas in Maryland waters of Chesapeake Bay has unique water-quality and contaminant conditions that may influence survival of young striped bass prolarvae. However, the chemical characteristics of the Choptank and Nanticoke River are similar because each of these neighboring eastern shore Chesapeake Bay rivers is susceptible to acidification. As with the Nanticoke River, the combination of acidic precipitation, acidic soils, and low buffering capacity of the Choptank River create potentially acidic conditions (pH depressions) in the striped bass spawning area. When significant precipitation events occurred concurrently with the presence of young prolarvae in the river during the spring, high mortality of this life stage was reported.

Concentrations of some trace metals in the Choptank River were high. Maximum values of cadmium (15 µg/L), copper (40 µg/L), and zinc (301 µg/L) have been reported in this river during various striped bass spawning seasons. These concentrations exceeded the U. S. Environmental Protection Agency's acute water-quality criteria for freshwater (USEPA 1987). Environmental concentrations of both cadmium and copper reported in the Choptank River were reported toxic to young striped bass in laboratory toxicity studies (CNFRL 1983). The source for these metals was not known but potentially toxic concentrations have been reported during the last several years of the study. Finger et al. (in press) have reported that the highest concentrations of these metals in the Choptank River in 1989 corresponded with periods of highest precipitation. Precipitation may therefore be a direct source of metals (atmospheric deposition) or an indirect source due to increased runoff (nonpoint source input) from the terrestrial environment.

A comparison of prolarva survival data from in situ and on-site tests with the State of Maryland's juvenile index demonstrated close agreement between these two endpoints during 4 of the 5 years of testing (Figure 2). The major difference occurred in 1989 as low survival of prolarvae was reported from four in situ tests but the juvenile index for striped bass was extremely high (97.8). Possible

TABLE 5.—Results from 96-h in situ and on-site toxicity tests with striped bass prolarvae in the Potomac River from 1986 to 1990.

Year	Type of test	Number of locations	Number of tests	Mean survival (%) and classification	Results	Reference
1986	In situ	3	3	12, low	77.5–95.5% mortality; stressful conditions: monomeric aluminum (0.09 mg/L), cadmium (0.007 mg/L), copper (0.072 mg/L), and sudden decreases in temperature below 11°C.	Hall et al. (1987a)
1988	In situ On-site	1	3	11, low	85–95% mortality; stressful conditions; cadmium (0.004 mg/L), lead (0.012 mg/L), chlordane (0.000152 mg/L), and sudden drops in temperature.	Hall et al. (1988)
1989	In situ	3	3	16, low	67–97% mortality; stressful conditions: low temperatures, chromium, and arsenic.	Hall et al. (1991a)
1990	In situ	3	3	8, low	82.5–98% mortality; potentially stressful concentrations of chromium, zinc, and arsenic.	Hall et al. (1992)

reasons for the difference in these two endpoints are: (1) Low pH (< 6.6) occurred continuously during the spawning season due to persistent precipitation. Therefore, prolarvae were not subjected to rapid pH depressions. (2) The 11 to 13% survival of prolarvae reported during experiment 3 was sufficient to produce a high number of juveniles. Reason 1 is a plausible explanation because pH conditions reported from continuous monitoring with Datasonde units showed that critical pH values below 6.6 were consistently reported throughout the spawning season. One of the most detrimental effects of low pH on striped bass prolarvae is the rapid "rate of change" that occurs during a rain event. Doroshev (1970) reported that sudden shifts in pH of 0.8 to 1.0 unit were toxic to striped bass larvae. Since pH was consistently low during the spawning season, prolarvae were not subjected to rapid depressions and they may have acclimated to these low-pH conditions. The low-pH conditions (< 6.6) were present during adult spawning, fertilization of the eggs, embryo development, hatching of eggs, and development of prolarvae. Sager et al. (1986) reported that 19-d-old striped bass larvae continuously exposed to pH ≤ 6 in low-salinity water suffered no apparent mortality in a hatchery. However, the low-salinity conditions present in these acidic waters may have also enhanced the survival of young striped bass and mitigated effects from acidic conditions. Although the low salinity conditions may have partially mitigated adverse effects of acidic conditions, the data reported by Sager et al. (1986) demonstrate that striped bass larvae can survive pH conditions < 6.5 if continuously exposed during their development.

The other plausible explanation for the high juvenile index in 1989 is that the prolarva survival reported during the third experiment (11 to 13%) was sufficient to establish high numbers of juveniles in the Choptank River. Anadromous fishes such as striped bass experience natural mortalities as high as 99% during the first year of life (Dahlberg 1979). Therefore, it is conceivable that the survival reported during the third experiment in 1989 would be sufficient to produce large numbers of juveniles if habitat conditions were unusually favorable during the remainder of their developmental period.

Upper Chesapeake Bay

The upper Chesapeake Bay is the largest and most important spawning area for striped bass in Maryland waters of the Chesapeake Bay (Kohlenstein 1980). Since striped bass spawning habitat in the Upper Bay covers such a large area, attempting to conduct in situ and on-site tests to represent this spawning area with only three locations has spatial limitations. Based on our data, the major conclusion from 5 years of field studies in this habitat is that survival of prolarvae was generally high or within the natural range as reported by other investigators (Polgar 1977; Dahlberg 1979; Dey 1981). The high survival of prolarvae in the Upper Bay was generally related to the presence of low salinity in the spawning areas and lack of potentially toxic contaminants or water-quality conditions (i.e., low pH). Various investigators have reported that the presence of low salinity enhances the survival of striped bass larvae (Lal et al. 1977; Burton 1982; Geiger and Parker 1985). Most of our

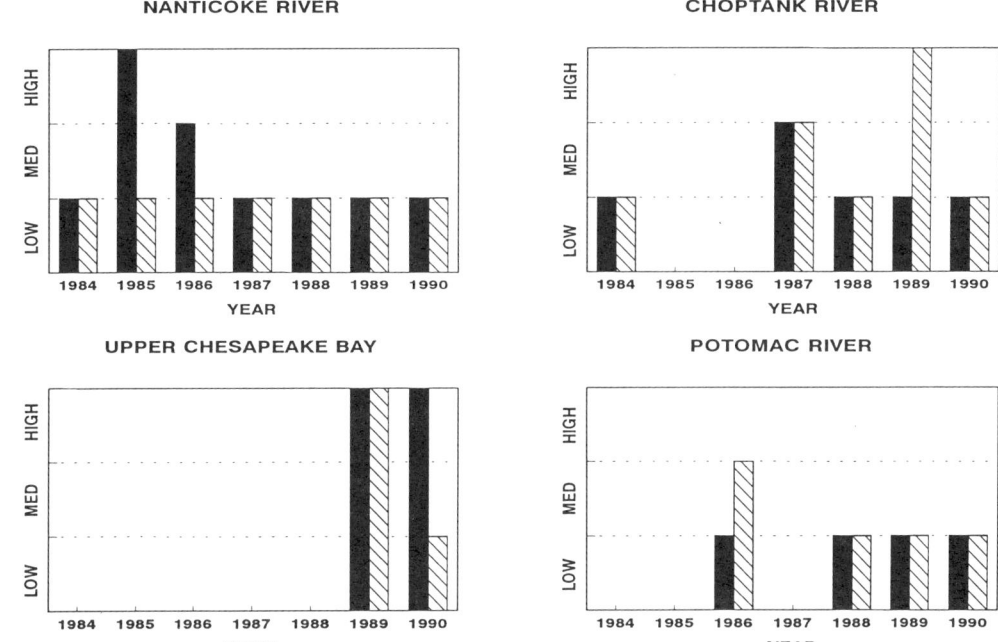

FIGURE 2.—A comparison of in situ larva survival (solid bars) in each of the four habitats with the State of Maryland's juvenile index.

prolarva data in the Upper Bay support this finding, with the exception of the 1990 studies (Hall et al. 1992). In 1990, we reported high survival of prolarvae during in situ tests in low-conductivity water without the presence of salinity conditions. Rago (1990) also reported similar high survival of striped bass prolarvae in low-conductivity waters of southern Chesapeake Bay tributaries during 96-h in situ studies.

Although the general conclusion from the Upper Bay field studies was that prolarva survival was high or adequate during all years of testing, there was an example of low prolarva survival during the third experiment in 1989 (Hall et al. 1991a). Survival of prolarvae during this test ranged from 6 to 22%, thus suggesting a potential problem in habitat quality (water quality, contaminants, or other factors). Temperature reductions occurred during this experiment and resulted in a final temperature of 10 to 11°C. This temperature range was likely stressful because it was in the lower lethal range reported for striped bass prolarvae (Davies 1973; Doroshev 1970). Lack of salinity at the in situ stations during this experiment was also likely a factor contributing to mortality of young striped bass.

A comparison of our prolarva survival data from in situ tests with the State of Maryland's juvenile index is presented in Figure 2. Only data from 1989 and 1990 were presented because the spatial coverage of the Upper Bay during these years was more representative than data from previous years. A comparison of these data showed good agreement in 1989 but poor agreement in 1990 as survival from in situ tests was high while the juvenile index was low. The high survival of the prolarval stage in 1990 does not necessarily ensure high numbers of juveniles in the spawning areas since numerous biotic and abiotic factors influence survival during the 2- to 3-month developmental period.

Potomac River

Survival of striped bass prolarvae from field experiments in the Potomac River was consistently low during all 4 years of testing. Factors suspected in causing mortality were inorganic contaminants and temperature reductions. Concentrations of various metals were consistently higher than would be expected in this ambient environment and on numerous occasions the concentrations exceeded the EPA water-quality criteria. As recently as 1990, we reported the following concentrations of metals at our station on the Virginia side of the river: arsenic = 8.8 µg/L, cadmium = 4.7 µg/L, chromium

= 150 µg/L, copper = 37 µg/L, nickel = 281 µg/L, lead = 14 µg/L, and zinc = 270 µg/L. All of these concentrations were in excess of their respective water-quality criteria for freshwater (USEPA 1987). Concentrations of cadmium, mercury, nickel, lead, and zinc were also reported at potentially stressful concentrations in sediment at this station in 1990 (Hall et al. 1991b). Reported concentrations of these metals exceeded the "Effects Low-Range" values defined by Long and Morgan (1990) as concentrations where toxic effects are possible. The spatial extent of potentially toxic metals in sediment of this spawning area was also reported in 1990 as possibly toxic concentrations of cadmium, mercury, nickel, lead, and zinc were reported at a nearby site near Indian Head (Hall et al. 1991b). All of the above metals except cadmium were also reported at possibly toxic concentrations in sediment at another location in the spawning area near Freestone Point (Hall et al. 1991b).

In addition to the stressful conditions caused by metals in the Potomac River striped bass spawning area, naturally occurring stressful reductions in temperature (cold shock) were also reported periodically in 1986, 1988, and 1989. In various cases during all 3 years, temperatures were reduced rapidly in the spawning area to lower lethal levels of 10 to 11°C (Davies 1973; Doroshev 1970). It is highly likely that temperature and metals acted synergistically to create stressful conditions in the striped bass spawning area during the first 3 years of testing. However, temperature reductions were not suspected in causing the low survival of prolarvae in the 1990 field experiments (Hall et al. 1992).

Other indicators of toxic conditions were also reported in the Potomac River striped bass spawning area. In 1986, we reported significantly higher mortality of striped bass juveniles during two successive 7-d in situ studies at our up-river station when compared with the other two down-river stations (Hall et al. 1987a). The temporal pattern of yearling mortality was similar during both 7-d studies as spikes in mortality occurred at the end of the weekend (Sunday and Monday). Yearling mortality during both 7-d experiments was reported concurrently with pH conditions > 9. The high pH conditions were likely related to a point source and not biological activity because water temperatures were less that 15°C. An inspection of NPDES permits showed that two sewage treatment plants discharging into a creek above our station (Nebesco Creek) had been previously cited for fecal coliform and high pH values (State Water Control Board 1984).

We speculated that the weekend discharges of these sewage treatment plants may have been responsible for the toxic conditions reported at this station. It is noteworthy that prolarva experiments were not in progress when drastic yearling mortality was reported, but naturally occurring early life stages of striped bass were likely present in the river during this spring period (April) in 1986.

Another example of toxic conditions in the Potomac River striped bass spawning area occurred in 1990. During our third prolarva experiment, a large "fish kill" was reported in our testing area (Hall et al. 1992). Mortality of various resident species such as adult catfish were reported during the "fish kill." Conditions severe enough to cause mortality of adult catfish would be more than detrimental to sensitive larval stages of striped bass even if a moderate amount of dilution occurred. Specific factors responsible for the "fish kill" were not conclusively identified. Potentially stressful concentrations of chromium (14 µg/L) and zinc (119 µg/L) were reported during our third experiment and the "fish kill" (Hall et al. 1992).

A comparison of prolarva survival from in situ and on-site field tests with the State of Maryland's juvenile index demonstrated close agreement between these two endpoints for all years of testing except 1986 (Figure 2). The prolarva survival reported from our experiments in 1986 was low while the juvenile index was moderate (9.9). Possible reasons for the differences are: (1) a spawn occurring after our in situ studies under more suitable water-quality conditions produced enough young juveniles to achieve the index; (2) prolarva survival of 12% (mean survival for the experiments) was sufficient to provide the reported index; or (3) the three stations used for in situ tests were all located on the Virginia side of the river and thus represented a limited spatial scale not representative of the middle of the river or the Maryland side of the river due to the width of this spawning area (ca. 3 km). The better correlation between results from prolarva tests and the juvenile index during the last 3 years of testing may be related to the testing procedures that provided a more environmentally realistic scale for spatial and temporal testing. Locations of Potomac River stations in 1989 and 1990 were on the Virginia side of the river, middle of the river, and Maryland side of the river. The temporal scale used during the last 3 years of testing was also improved as the testing period covered a larger part of the spawning season.

Acknowledgments

We would like to acknowledge the U. S. Fish and Wildlife Service for sponsoring this research. Special thanks are extended to Paul Rago for his suggestions on the study design.

References

Boone, J. G. 1980. Estuarine fish recruitment survey — July, 1980 through June 30, 1981. Maryland Department of Natural Resources, Federal Aid in Fish Restoration Performance Report, F-27-R-6, Annapolis.

Boreman, J., and H. M. Austin. 1985. Production and harvest of anadromous striped bass stocks along the Atlantic coast. Transactions of the American Fisheries Society 144:3-7.

Bowman, M. 1984. Chesapeake Bay freshwater feeder stream chemistry. Pages 73-79 in Working group on acid deposition, editors. The potential effects of acid deposition in Maryland. Office of Environmental Programs, Baltimore, Maryland.

Buckler, D. R., P. M. Mehrle, L. Cleveland, and F. J. Dwyer. 1987. Influence of pH on the toxicity of aluminum and other inorganic contaminants to east coast striped bass. Water, Air, and Soil Pollution 35:97-106.

Burton, D. T. 1982. An evaluation of the potential toxicity of treated bleached kraft mill effluent to the early life stages of striped bass *(Morone saxatilis)*. Johns Hopkins University, Applied Physics Laboratory Report CPE-8204, Laurel, Maryland.

CNFRL (Columbia National Fisheries Research Laboratory). 1983. Impacts of contaminants on early life stages of striped bass. U. S. Fish and Wildlife Service, Progress Report 1980-1983. CNFRL, Columbia, Missouri.

Dahlberg, M. D. 1979. A review of survival rates of fish eggs and larvae in relation to impact assessments. Marine Fisheries Review 41(3):12.

Davies, W. D. 1973. Rates of temperature acclimation for hatchery reared striped bass fry and fingerlings. Progressive Fish-Culturist 35:214-217.

Dey, W. D. 1981. Mortality and growth of young-of-the-year striped bass in the Hudson River estuary. Transactions of the American Fisheries Society 110:151-157.

Doroshev, S. I. 1970. Biological features of the eggs, larvae, and young of the striped bass *(Roccus saxatilis* {Walbaum}) in connection with problems of its acclimatization in the U. S. S. R. Journal of Ichthyology 10:235-248.

Finger, S. E. 1985. On-site toxicity evaluations in waters from Chesapeake Bay and South Carolina. U. S. Fish and Wildlife Service, Report, Columbia, Missouri.

Finger, S. E. 1987. Striped bass contaminant studies in 1987. U. S. Fish and Wildlife Service, Report, Columbia, Missouri.

Finger, S. E. 1989. Field assessment of striped bass larval survival in the Nanticoke and Choptank Rivers. U. S. Fish and Wildlife Service, Progress Report, Columbia, Missouri.

Finger, S. E., A. C. Livingstone, and S. J. Olson. In press. Influence of contaminants on survival of striped bass in Chesapeake Bay tributaries. *in* J. E. Weaver, editor. Second US-USSR symposium on reproduction, rearing and management of anadromous fishes. U. S. Fish and Wildlife Service, Seattle.

Geiger, J. G., and N. C. Parker. 1985. Survey of striped bass hatchery management in the southeastern United States. Progressive Fish-Culturist 47:1-13.

Goodyear, C. P. 1985. Toxic materials, fishing and environmental variation: simulated effects on striped bass population trends. Transactions of the American Fisheries Society 114:107-113.

Hall, L. W., Jr. 1984. Field assessment of striped bass, *Morone saxatilis*, larval survival as related to contaminants and changes in water quality parameters. U. S. Fish and Wildlife Service, National Fisheries Center, Final Report, Leetown, West Virginia.

Hall, L. W., Jr. 1987. Acidification effects on larval striped bass, *Morone saxatilis*, in Chesapeake Bay tributaries: a review. Water, Air, and Soil Pollution 35:87-96.

Hall, L. W., Jr. 1991. A synthesis of water quality and contaminants data on early life stages of striped bass, *Morone saxatilis*. Reviews in Aquatic Sciences 4:261-288.

Hall, L. W., Jr., S. J. Bushong, M. C. Ziegenfuss, W. S. Hall, and R. L. Herman. 1988. Concurrent mobile on-site and *in situ* striped bass contaminant and water quality studies in the Choptank River and upper Chesapeake Bay. Environmental Toxicology and Chemistry 7:815-830.

Hall, L. W., Jr., W. S. Hall, S. J. Bushong, and R. L. Herman. 1987a. *In situ* striped bass *(Morone saxatilis)* contaminant and water quality studies in the Potomac River. Aquatic

Toxicology 10:73-99.

Hall, L. W., Jr., A. E. Pinkney, R. L. Herman, and S. E. Finger. 1987b. Survival of striped bass larvae and yearlings in relation to contaminants and water quality in the upper Chesapeake Bay. Archives of Environmental Contamination and Toxicology 16:391-400.

Hall, L. W. Jr., A. E. Pinkney, L. O. Horseman, and S. E. Finger. 1985. Mortality of striped bass larvae in relation to contaminants and water quality in a Chesapeake Bay tributary. Transactions of the American Fisheries Society 114:861-868.

Hall, L. W. Jr., M. C. Ziegenfuss, S. J. Bushong, J. A. Sullivan, and M. A. Unger. 1991a. Striped bass contaminant and water quality studies in the Potomac River upper Chesapeake Bay in 1989: annual contaminant and water quality evaluations in east coast striped bass habitats. U. S. Fish and Wildlife Service, National Fisheries Center, Report, Leetown, West Virginia.

Hall, L. W. Jr., M. C. Ziegenfuss, S. J. Bushong, and M. A. Unger. 1989. Studies of contaminant and water quality effects on striped bass prolarvae and yearlings in the Potomac River and upper Chesapeake Bay. Transactions of the American Fisheries Society 118:619-629.

Hall, L. W. Jr., M. C. Ziegenfuss, S. A. Fischer, R. W. Alden III, E. Deaver, J. Gooch, and N. Debert-Hastings. 1991b. A pilot study for ambient toxicity testing in Chesapeake Bay, U. S. Environmental Protection Agency, Chesapeake Bay Program, Volume 1 - Year 1 Report, Annapolis, Maryland.

Hall, L. W. Jr., M. C. Ziegenfuss, S. A. Fischer, J. A. Sullivan, and D. M. Palmer. 1992. *In situ* striped bass contaminant and water quality studies in the Potomac River and upper Chesapeake Bay in 1990. U. S. Fish and Wildlife Service, National Fisheries Center, Final Report, Leetown, West Virginia.

Kohlenstein, L. C. 1980. Aspects of the population dynamics of striped bass *(Morone saxatilis)* spawning in Maryland tributaries of the Chesapeake Bay. The State of Maryland, Power Plant Citing Program, Report JHU PPSE T-14, Annapolis, Maryland.

Lal, K., R. Lasker, and A. Juljis. 1977. Acclimation and rearing of striped bass larvae in seawater. California Fish and Game 63:210-218.

Long, E. R., and L. G. Morgan. 1990. The potential for biological effects of sediment-sorbed contaminants tested in the National Status and Trends Program. National Technical Memorandum OMA 52, Seattle.

Maryland Department of Natural Resources. 1986. Second annual status report on striped bass 1986. Tidewater Administration Fisheries Division, Annapolis, Maryland.

Mehrle, P. M. 1982. Summary of striped bass contaminant investigations at CNFRL. U. S. Fish and Wildlife Service, Columbia National Fisheries Research Laboratory, Summary Report, Columbia, Missouri.

Mehrle, P. M., D. Buckler, and S. Finger. 1986. Impacts of contaminants on striped in Chesapeake Bay: a summary of research on water quality/contaminant interactions in laboratory and field studies. U. S. Fish and Wildlife Service, National Fisheries Contaminants Research Center, Columbia, Missouri.

Mehrle, P. M., D. Buckler, S. Finger, and L. Ludke. 1984. Impact of contaminants on striped bass. U. S. Fish Wildlife Service, Columbia National Fisheries Research Laboratory, Interim Report, Columbia, Missouri.

Miller, F. P. 1976. Maryland Soils. Bulletin 212, University of Maryland, Cooperative Extension Service, College Park.

Polgar, T. T. 1977. Striped bass ichthyoplankton abundance, mortality, and production estimation for the Potomac River population. Pages 110-126 in W. Van Winkle, editor. Proceedings of the conference on assessing the effects of power plant-induced mortality on fish populations. Pergamon Press, New York.

Rago, P. 1988. Preliminary summary of fish and wildlife research program FY-1988. U. S. Fish and Wildlife Service, National Fisheries Center, Report, Leetown, West Virginia.

Rago, P. 1990. Preliminary Summary of Fish and Wildlife Research Program FY-1990. U. S. Fish and Wildlife Service, National Fisheries Center, Report, Leetown, West Virginia.

Sager, P. R., L. C. Woods III, and J. N. Kraueter. 1986. Survival of *Morone saxatilis* in low pH oligohaline waters. Journal of the Washington Academy of Sciences 76:237-243.

State Water Control Board. 1984. Water quality inventory 305(b) report. Virginia Information Bulletin 558, Richmond.

USEPA (U. S. Environmental Protection Agency). 1979. Methods for chemical analysis of water and wastes. EPA 600/4-79-020, Cincinnati, Ohio.

USEPA (U. S. Environmental Protection Agency). 1982. Methods for organic chemical analysis of municipal and industrial wastewater. EPA 660/4-82-057, Cincinnati, Ohio.

USEPA (U. S. Environmental Protection Agency). 1987. Water quality criteria summary. U. S. Environmental Protection Agency, Office of Water Regulations and Standards, Criteria and Standards Division, Washington, District of Columbia.

Ziegenfuss, M. C., L. W. Hall, Jr., S. J. Bushong, J. A. Sullivan, and M. A. Unger. 1990. A remote *in situ* apparatus for ambient toxicity testing of larval and yearling fish in river or estuarine systems. Environmental Toxicology and Chemistry 9:1311-1315.

Sublethal Effects of Methyl Parathion, Carbofuran, and Molinate on Larval Striped Bass

ALAN G. HEATH

Department of Biology, Virginia Polytechnic Institute and State University, Blacksburg, Virginia 24061, USA

JOSEPH J. CECH

Department of Wildlife and Fisheries Biology, University of California, Davis, California 95616, USA

JOSEPH G. ZINKL

Department of Clinical Pathology, School of Veterinary Medicine, University of California, Davis, California 95616, USA

BRIAN FINLAYSON

California Department of Fish and Game, Pesticides Unit, 1710 Nimbus Road, Rancho Cordova, California 95670, USA

ROBERT FUJIMURA

California Department of Fish and Game, Aquatic Toxicology Laboratory, 9300 Elk Grove-Florin Road, Elk Grove, California 95624, USA

Abstract.—In the Sacramento River, California, newly hatched striped bass *(Morone saxatilis)* larvae typically get exposed to rice pesticides (molinate, carbofuran, methyl parathion) for at least 4 d as they drift to the San Francisco Bay estuary. This was simulated in the laboratory using individual pesticides and a combination of all three. Concentrations tested were half of the 96-h LC50 values and a lower level, similar to that recorded in the Colusa Basin Drain, a major agricultural return water source. Tests on the larvae were performed immediately after the 4-d exposures and then again after they had been in non-contaminated water for 10 d. The pesticides caused no mortality during exposures but methyl parathion at the higher concentration caused delayed mortality. Sublethal tests performed on the larvae included forced swimming performance, spontaneous activity, acetylcholinesterase (AChE) activity, RNA/DNA ratio, body weight, and several morphometric measures. There was a slight decrease in swimming performance immediately after exposure to parathion at both concentrations and a decrease in AChE activity. After 10 d in clean water there was still a loss of swimming performance and decreased spontaneous activity in those larvae exposed to the higher concentration, and AChE activity failed to recover. Molinate caused decreased swimming performance at the high concentration both immediately after exposure and subsequently. Molinate also caused an inhibition of AChE activity but an apparent stimulation of growth rate. Decreases in AChE activity occurred at the high concentration of carbofuran and in fish exposed to the combined pesticides immediately after exposure, but this inhibition did not persist in the clean water. Each of the three pesticides caused a slight change in body shape resulting in a greater dorsal-ventral dimension but little or no change in length. Combining the pesticides produced a less than additive effect on the measures. Some of the sublethal effects seen might have implications for recruitment of the already declining populations of striped bass.

Introduction

Striped bass *(Morone saxatilis)* populations in the San Francisco Bay and the Sacramento - San Joaquin estuaries have experienced severe declines in the past 15 years (Stevens et al. 1985; Setzler-Hamilton et al. 1988). Possible reasons for this decline include water diversions, low fecundity, depletion of prey for larvae because of toxic pollutants, and poor larva survival due to the same or different pollutants (Stevens et al. 1985). These various reasons are not mutually exclusive and all are probably involved to at least some degree.

Spawning of striped bass occurs during May and early June in the Sacramento River between Colusa and Knights Landing, California. Several insecticides and herbicides are used extensively in rice culture in the Sacramento Valley, and runoff occurs from these fields into agricultural drains which flow into the Sacramento River concurrently with striped bass spawning (Cornacchia et al. 1984; Finlayson and Faggella 1986). The largest agricultural drain (Colusa Basin Drain) discharges into the Sacramento River at Knights Landing. The eggs and larvae drift for about 4 to 8 d, depending on river flow rates (Cornacchia et al. 1984), to the San Francisco Bay delta during which time they may experience exposure to these pesticides. Water taken directly from the Colusa Basin Drain during May and June is acutely toxic to sensitive invertebrates (Foe 1989; Finlayson et al. 1991; Norberg-King et al. 1991). Although there is little evidence to suggest direct lethal effects on larvae of striped bass, possible sublethal effects of the herbicides and insecticides have not been investigated.

Measurements of pesticide concentrations in the ambient waters (Finlayson et al. 1993) suggest that three toxic contaminants, methyl parathion, carbofuran, and molinate, may have potential to affect striped bass. Methyl parathion is an organophosphate insecticide, carbofuran (Furan®) is a carbamate insecticide, and molinate (Ordram®) is a thiocarbamate herbicide.

The concentrations of these three pesticides in the Sacramento River are well below the lethal concentrations for striped bass larvae (Fujimura et al. 1991). But it is well documented that fish may experience a variety of sublethal effects from toxic chemicals (Heath 1987; Adams 1990). These assorted biochemical and physiological changes may make the organisms more susceptible to other stressors and to predators, so the population ultimately loses members excessively but not due to a fish kill from acute toxicity. While there is a good deal of data on sublethal responses of juvenile and adult fishes to various pollutants, this is not true for larvae. Except for measures of growth, the small sizes of larvae make biochemical and physiological measurements difficult so this is a comparatively unexplored area.

The purpose of this study was to examine some of the sublethal effects of methyl parathion, carbofuran, or molinate singly and in combination, using a variety of biochemical, morphometric, and behavioral measurements. The experiments were designed to observe the effect of a 4-d pesticide exposure, such as the larvae might experience during their drift down the Sacramento River to the Sacramento-San Joaquin estuary, and then to evaluate any delayed effects remaining in the fish after 10 d in non-contaminated water.

TABLE 1.—Pesticide concentrations and water-quality data during 4-d exposures of striped bass larvae. Concentrations are means in μg/L with standard deviation in parentheses. Ranges are given for water quality.

Pesticide	Low Nominal	Low Measured	High Nominal	High Measured
Parathion	0.66	1.04 (0.59)	1,800	1,625 (150)
Carbofuran	1.3	1.1 (0.2)	110	79 (7.9)
Molinate	70.0	69.0 (2.9)	4,050	3,125 (330)
Combined:				
Parathion	0.66	0.55 (0.08)	600	470 (14)
Carbofuran	1.3	1.08 (0.18)	37	29 (4.2)
Molinate	70.0	43.0	1,350	930 (57)

	Water quality		
	Temperature (°C)	pH	Dissolved oxygen (mg/L)
Parathion	17.0-17.7	8.2-8.7	8.1-9.1
Carbofuran	16.6-17.6	8.0-8.9	7.8-8.6
Molinate	17.0-17.6	8.1-8.9	8.1-9.0
Combined	16.7-21.7	7.9-8.4	8.2-9.1

Methods

Striped bass larvae less than 1 d posthatching were obtained during May 1991 from the California Department of Fish and Game Central Valleys Hatchery at Elk Grove, California. The eggs were from artificially spawned adults taken from the Sacramento River. The larvae for any given pesticide test were from one female.

The static renewal method was used for pesticide exposures, which were carried out in the Aquatic Toxicology Laboratory of the California Department of Fish and Game (Elk Grove, California). At the beginning of an exposure period, approximately 150 larvae less than 1 d old were loaded into 2-L glass beakers containing 1,500 mL of test solution. The dilution water consisted of degassed groundwater with additions of artificial sea salt (Marine Environment®) to a salinity of 2‰ (Kane et al. 1990). Beakers were held in a water bath for temperature control. The test solutions were renewed with 1,000 mL of fresh solution daily by siphon. Before solution changes, pH, temperature, and dissolved oxygen were measured (Table 1) and any dead larvae counted and removed. Water samples were also taken and stored refrigerated for subsequent pesticide analysis by the California Department of Fish and Game Water Pollution Control Laboratory. Extractions were done by filtering water samples through SPE C-18 columns. Pesticide concentrations were determined by gas chromatography using a Varian Aerograph Model 3600 with a TSD detector. Preliminary tests indicated negligible loss of material during the 24 h between renewals. All glassware was acid washed and rinsed in distilled water.

Two concentrations of each pesticide were tested, a "high" and a "low" nominal, plus a control with no pesticide present (Table 1). Three beakers were used for each test concentration. The high nominal concentrations approximated one half the LC50 values for striped bass larvae (Fujimura et al. 1991). The low nominal concentrations were based on those observed in the Colusa Basin Drain during May-June 1990 (Finlayson et al. 1993; unpublished California Department of Fish and Game data). An additional run to test additive effects was made in which the three pesticides were combined at two concentrations. The low concentration was similar to that of the Colusa Basin Drain for all three pesticides combined. The high one was one third that used (per pesticide) for each of the other high concentration runs.

Following the 4-d exposure at the Elk Grove Laboratory, the beakers containing the now 5-d-old larvae were transported in insulated chests to the Ecology Institute Building at the University of California, Davis. Tests of swimming performance and spontaneous activity were done on the larvae the same day, or in one instance, the following morning. At the same time, samples of larvae were frozen in cryovials at -70°C for subsequent analysis of acetylcholinesterase and nucleic acids.

Swimming performance was assessed in a 9-cm diameter petri dish with an upside-down 5.5-cm diameter petri dish in the center, thus forming a circular "racetrack." Eight equally spaced radiating lines were drawn on the underside of the larger dish and it was filled with water to a depth of 0.5 cm. A

TABLE 2.—Water quality in beakers during 10-d recovery period from pesticide exposure. Values are means with ranges in parentheses.

Salinity (‰)	2.01	(1.5-2.8)
Temperature (°C)	18.3	(18.0-18.6)
pH	8.4	(8.2-8.6)
Dissolved oxygen (mg/L)	8.9	(8.6-9.1)
Ammonia (mg/L)	0.42	(0.2-0.6)

test consisted of transferring a single larva with a pipet to the petri dish and then allowing it 2 min to adjust to the dish. Then, the larva was chased with a glass rod and the number of lines crossed in 1 min was determined. The water was changed frequently to avoid evaporative cooling. A given larva would not necessarily swim consistently in one direction around the circle. There was often a good deal of back and forth movement, but we feel it gives a good indication of maximum swimming ability that would be important in evading predation. We are unaware of anyone having used this procedure before.

Spontaneous swimming activity was assessed in a pyrex pan 15 cm × 25 cm with a water depth of 1 cm. The pan was placed on a square grid; each square was 2.5 cm on a side. A test consisted of placing one larva in the middle of the pan and after 20 s, the number of lines crossed in 1 min was determined. (Crossing a corner was counted as one line.)

To maintain consistency, one experimenter was used for a given group of swimming tests. Furthermore, the experimenter did not know from which treatment the fish had come. This was also true for the subsequent biochemical and morphometric measurements so the tests were done "blind" and the results not known until the code was revealed.

Larvae used in the swimming tests were subsequently placed in the wells of plastic tissue culture plates and 5% buffered formalin added. These individually numbered larvae were used for morphometry and subsequent dry weight determination.

In order to assess the delayed effects of the pesticide exposure, after the initial tests were finished, the remaining larvae were transferred to 1-L plastic beakers. These had two Nytex® screened openings (5 cm × 2.5 cm) in the lower half of the beaker. These beakers were suspended (in a random order) at the 700-mL level in a temperature controlled 520-L water bath. This water bath received a constant flow of non-chlorinated, air-equilibrated well

TABLE 3.—Statistical analysis of those cases in which there appeared to be a treatment effect. Analysis of variance for treatment and for differences between beakers of same treatment are given. See Figures 1 and 2 and Table 2.

Measurement	Treatment	Source	P
Weight	Parathion, immediate	Treatment Beaker	0.012 0.207
	Molinate, 10 d	Treatment Beaker	0.018 0.004
RNA/DNA	Parthion, immediate	Treatment Beaker	<0.001 0.020
	Parthion, 10 d	Treatment Beaker	0.001 0.232
	Molinate, 10 d	Treatment Beaker	0.001 0.062
	Carbofuran, immediate	Treatment Beaker	0.001 0.141
	Carbofuran, 10 d	Treatment Beaker	0.023 0.743
Head length	Parathion	Treatment Beaker	0.003 0.091
Body depth (pectoral)	Parathion	Treatment Beaker	0.002 0.002
Head depth	Carbofuran	Treatment Beaker	0.001 0.044
Body depth (anus)	Molinate	Treatment Beaker	0.017 0.011
Spontaneous activity	Carbofuran, immediate	Treatment Beaker	<0.001 0.018
	Combined, immediate	Treatment Beaker	0.019 0.881
Swimming performance	Parathion, immediate	Treatment Beaker	0.054 0.752
	Parathion, 10 d	Treatment Beaker	<0.001 0.255
	Molinate, immediate	Treatment Beaker	0.001 0.331
	Molinate, 10 d	Treatment Beaker	<0.001 0.209
Acetylcholinesterase	Parathion, immediate	Treatment Beaker	0.040 0.213
	Parathion, 10 d	Treatment Beaker	<0.001 0.010
	Molinate, 10 d	Treatment Beaker	<0.001 0.051
	Combined, immediate	Treatment Beaker	0.006 0.099
	Combined, 10 d	Treatment Beaker	0.008 0.502

water at 240 mL/min from a head box kept filled with water pumped from the water bath. The bath also received a slow inflow of concentrated saline solution from a peristaltic pump to maintain the salinity at 1.5 to 2.8‰. Each beaker received a 60 mL/min flow from the head box as well as water exchange through the screens, thus keeping the water quality the same in all beakers (Table 2). Excess food and feces were drawn off with a pipet twice daily and any dead larvae were counted and removed. At the same times, the temperature and salinity were recorded and all beakers received a 10-mL syringe full of suspended brine shrimp nauplii. The sequence of feeding of the individual beakers was randomized at each feeding.

After 10 d in the non-contaminated water the same tests were performed as were done immediately after the exposure. All capturing of larvae was done with plastic pipets. Because the slowest fish might be captured first, the fish from a beaker were first dispersed into individual wells of tissue culture plates, then, using a random number table, fish were removed one by one for the tests.

Formalin-preserved larvae were morphometrically examined with a dissecting scope equipped with an ocular micrometer (Martin and Malloy 1980). The following variables were measured: notochord length, head length, head depth, body depth at the pectoral fin, and body depth at the anus. The same larvae were then dried at 40°C for several days and weighed on a Cahn microbalance.

Total RNA and total DNA were determined fluorimetrically by modifications of the enzymatic procedure proposed by Bentle et al. (1981). The modifications as developed by Sandor Kaupp (Scripps Institution of Oceanography, personal communication) and A. G. Heath permitted analysis of both RNA and DNA on single 4-d-old larva. Individual frozen larvae were measured with a dissecting microscope for notocord length and homogenized by hand in a 1-mL glass Potter-Elvehjem tissue grinder in 50 µL of cold 1 M NaCl. After transfer to a disposable glass tube suitable for use in a spectrofluorometer, the homogenizer was rinsed into the tube with 1 mL of "cocktail." The "cocktail" contained 50 mM Tris buffer (pH 7.5), 4 mM $MgCl_2$, 3.2 mM $CaCl_2$, 0.2 mg ethidium bromide/100 mL and 8 mg proteinase K/100 mL. After stirring on a vortex mixer, tubes were incubated at 37°C for 45 to 60 min. These were read in a spectrofluorometer (Perkin-Elmer Model 204) at 546 nM excitation and 590 nM emission. The 546 nM excitation wavelength yielded better sensitivity than the 390 nM wavelength proposed by Bentle et al. (1981). Tubes were marked so they could be placed in the spectrofluorometer in exactly the same position for subsequent readings.

After the initial readings, 5 µL of RNAase (5 mg/mL) was then added to all tubes and they were incubated for 30 min at 37°C. After the second reading was taken (difference between first and second readings represents RNA present), DNAase (1 mg/mL) was added followed by 30 min of incubation and the third reading then taken. The difference between second and third readings represented DNA present. Standard curves were prepared using known concentrations of DNA (0.5-3.0 µg/mL) and RNA (1.0-5 µg/mL).

Acetylcholinesterase (AChE) activity was assayed in individual whole larvae (sometimes two larvae were pooled) using the procedure of Ellman et al. (1961) adjusted for very small tissue quantities. We believe this is the first report of AChE having been assayed in larval fish (Zinkl et al. 1991).

Since there were three beakers for each of the pesticide concentrations, the statistical analysis used a nested two-way analysis of variance (ANOVA) with one factor (beakers) random, yielding P values for treatment and beaker effects (Table 3). Dunnett's multiple comparison test for statisti-

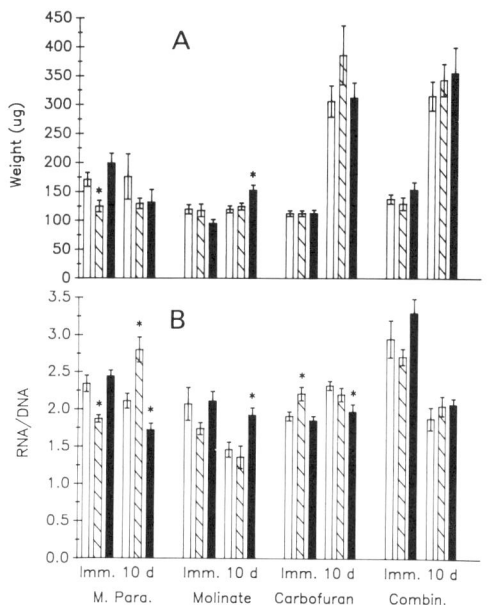

FIGURE 1.—(A) Dry weights and (B) RNA/DNA ratios of striped bass larvae immediately after (Imm.) a 4-d exposure to pesticide and after 10 d recovery in non-contaminated water (10 d). Open bars = controls, diagonal lines = low concentration, solid bars = high concentration. Those marked with an asterisk are significantly different from controls ($P \leq 0.05$). (See Table 3 for treatment and beaker statistics.) Error bars are ± 1 SE.

cal differences ($P < 0.05$) between controls and particular treatments was also used where ANOVA analysis revealed likely treatment effects. Minitab software was used for the statistical analyses.

Results

Mortalities

Mortality during the actual pesticide exposures occasionally included one or two dead in a day but usually there were none. As is usual with larval striped bass (Fujimura et al. 1991), there were high and variable (13 to 40%) mortalities in the 4- to 8-d-old fish when they began exogenous feeding. But previous pesticide exposure had no significant effect on this and control mortalities did not exceed 40%.

While there was no significant effect of methyl parathion on subsequent mortality in non-contaminated water, immediately after the exposure a majority of those from the high concentration (i.e., 50% of the 96-h LC50 value) were swimming on their sides. This behavior persisted with no discern-

TABLE 4.—Significant changes in morphometry of striped bass larvae relative to controls 10 d after exposure to pesticides.

Methyl Parathion	Carbofuran	Molinate
Increased head length (low concentration)	Increased head depth (high concentration)	Increased body depth at anus (high concentration)
Increased body depth at pectoral (low concentration)		

able change throughout the 10-d period in non-contaminated water. Had they been in a river or bay, such behavior would probably have compromised survival. Indeed, some were kept through day 11 and most of those died. So delayed mortality occurred, albeit after exposure to a fairly high pesticide concentration. Palawski et al. (1983) observed a similar delayed mortality with methyl parathion and rainbow trout *(Oncorhynchus mykiss)* larvae.

Dry Weight

There was essentially no increase in weight of larvae from the methyl parathion and molinate experiments after 10 d in non-contaminated water (Figure 1). They appeared to be eating and food was visible in their guts. However, it was learned after these experiments had been completed that the size of mesh in the beaker screens (0.3 mm square) was large enough that the brine shrimp larvae escaped through them at a rate fast enough that feeding time was reduced, so the larvae experienced essentially a maintenance diet.

Larvae from the other two experiments (i.e., carbofuran and combined pesticides) roughly tripled their mass during the 10-d period. For those fish, beaker screens with 0.2 mm square mesh were used, which greatly reduced the escape rate of the brine shrimp.

Larvae from the low concentration of methyl parathion had a reduced weight immediately after the exposure. This might be a statistical anomaly as they had not started feeding at this time and were still living on stored yolk. But the RNA/DNA (Figure 1) ratio was also low then which suggests that synthesis of new protein was depressed at that time.

An interesting phenomenon occurred with molinate in that fish from the high concentration were slightly lighter (although not statistically significant) immediately after the exposures, but then after 10 d in non-contaminated water larvae from the same group were actually heavier than controls (Figure 1).

RNA/DNA Ratio

RNA/DNA ratios for control larvae were between 1.9 and 2.4 (except for molinate at 14 d and combined pesticides immediately after exposure; Figure 1B). These control ratios are similar to those reported by Pinkney et al. (1990) for the same species. The low values for molinate controls at 14 d undoubtably reflect the restricted food availability in this group (Wright and Martin 1985). Since all three treatments in the combined group had high RNA/DNA ratios, there might have been an error in preparation of a nucleic acid standard because the same standard was always used for an entire pesticide group so that variations in the standards would not complicate comparison of treatments.

Immediately after the exposure period, larvae that had been exposed to the low concentration of methyl parathion exhibited a lower RNA/DNA ratio, but after 10 d in non-contaminated water they showed an elevated ratio. Carbofuran displayed the reverse effect in that the ratio was elevated immediately after exposure to the low concentration but depressed in larvae from the high concentration one after the 10-d recovery period (Figure 1B).

Morphometry

Only four alterations in morphometry were observed as a result of pesticide exposure. Those in which statistical differences were seen are given in Table 4. The most obvious trend is a tendency for the larvae to become deeper bodied in one or more dorsal-ventral measures.

Spontaneous Swimming Activity

Spontaneous activity was quite variable and activity in any one larva usually occurred in bursts, so 21 larvae were tested from each treatment (7 per beaker). Even so, there were significant effects seen only with the 50% LC50 concentrations (i.e., high concentrations) immediately after exposure to carbofuran and in the combined pesticide group (Figure 2).

Swimming Performance

Striped bass larvae that had been exposed to methyl parathion at both concentrations exhibited a nearly significant ($P = 0.054$) depression in swimming performance immediately after the exposure (Figure 2). Ten days later only those that had expe-

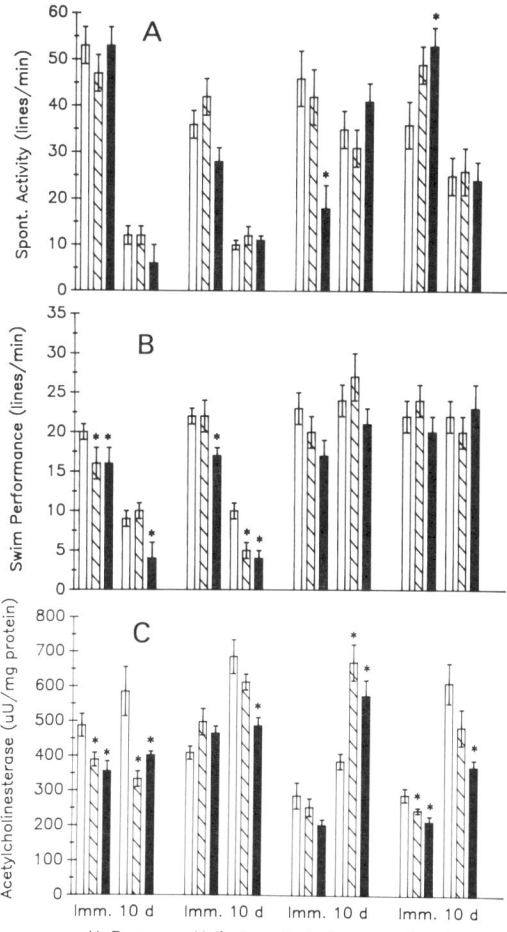

FIGURE 2.—(A) Spontaneous activity, (B) swimming performance, and (C) acetylcholinesterase activity in striped bass larvae immediately after (Imm.) 4-d exposure to pesticide and after 10 d (10 d) recovery in non-contaminated water. Open bars = controls, diagonal lines = low concentration, solid bars = high concentration. Those marked with an asterisk are significantly different from controls ($P \leq 0.05$). (See Table 3 for treatment and beaker statistics.) Error bars are ± 1 SE.

rienced the high concentration continued to exhibit a lower swimming performance compared to controls. These fish, as was mentioned above, were mostly swimming on their sides as they had done for the entire 10-d recovery period.

Molinate caused a loss of swimming performance at the high concentration and this persisted after the 10-d recovery period (Figure 2). A variety of herbicides inhibit swimming performance in older fish (Little et al. 1990) but no studies seem to have been done on larvae, nor has molinate been tested. Carbofuran caused a decline in swimming performance immediately after the exposure in the high concentration (Figure 2).

Acetylcholinesterase

Methyl parathion caused inhibition of AChE activity at both exposure concentrations and this persisted for at least 10-d in non-contaminated water (Figure 2). Inhibition of this enzyme also occurred in those exposed to the high concentrations of carbofuran and combined pesticides immediately after the exposures. Inhibition persisted in the combined pesticide group after 10-d and there appeared to be an inhibition at that time in the high molinate group.

Discussion

Sublethal effects on fish of contaminants are usually subtle, otherwise they would probably cause mortality. None of the changes we saw from the pesticide exposures were large and dose-dependent relationships, as usually seen with more typical toxicity tests where mortality is the endpoint, were not generally evident.

The only pesticide-induced mortalities seen in this study occurred after the termination of the high-concentration methyl parathion group. This was 11 d after the actual exposures ended and was probably not due to persistence of the pesticide in the fish tissues. Bruijn and Hermens (1991) found that methyl parathion was eliminated from the guppy (*Poecilia reticulata*) within 24 h of transfer to non-contaminated water. A more likely explanation is the continued depression of AChE activity in these fish, which we observed. There could conceivably also be osmoregulatory disturbances since the abdomens of the larvae appeared bloated. Our data suggest that larvae do not recover from methyl parathion toxicity as well as older fish. Babu et al. (1986) reported that juvenile common carp (*Cyprinus carpio*) quickly recovered from exposure to lethal doses of this pesticide.

The body weight data are interesting but not in the expected way. These data indicated a failure of the fish in the parathion and molinate groups (including controls) to grow while in the non-contaminated water. As mentioned earlier, we later learned this was due to reduced food availability in the beakers. While this is not an especially desirable experimental situation, in some ways it may more accurately reflect the situation in the river and bay. There is evidence that zooplankton availability in the larva nursery areas of the Sacramento and San Joaquin estuaries has been declining, which may cause a severe shortage of food for striped bass larvae (Stevens et al. 1985).

The apparent stimulation of growth following exposure to molinate at the higher concentration is interesting. Furthermore, the RNA/DNA ratio was elevated at this time (Figure 1), which can be interpreted as an indication of growth stimulation (Bulow 1987). In a parallel study in this same laboratory using larvae of the Japanese medaka *(Oryzias latipes)* and molinate at the same concentrations (Heath et al. in press), stimulation of growth at the high concentration was also observed.

Such a stimulation of growth by an environmental challenge is not unheard of. Crab zoeae grew faster after exposure for 5 d to petroleum hydrocarbons (Laughlin et al. 1981). McLeay and Brown (1975) observed growth stimulation of juvenile coho salmon *(Oncorhynchus kisutch)* by sublethal levels of bleached kraft pulp mill waste, and Bengtsson (1979) observed the same thing in minnows *(Phoxinus phoxinus)* with PCB exposure. This could represent a form of hormesis wherein an organism overcompensates for some exogenous stressor (Stebbing 1987).

As mentioned above, the RNA/DNA ratio has been used as a relative measure of growth rate, especially in rapidly growing organisms such as fish larvae (Buckley 1984; Bulow 1987; Miglavs and Jobling 1989; Pinkney et al. 1990). It is based on the theory that tissues that are actively synthesizing protein will have more messenger RNA and ribosome numbers compared with the amount of DNA. The elevated RNA/DNA ratio in the parathion-treated animals after 10 d in non-contaminated water may have been another example of hormesis. At the high parathion concentration, however, it was suppressed. Overall there was little evidence of suppression of growth by any of the pesticides as reflected in RNA/DNA ratios; this fits with the data on larva weights.

Morphometry changes produced by the pesticides were slight in our study. This can be contrasted with the findings of Pinkney et al. (1990) who, using the same morphometric measurements, reported that dry weight, body depth, head length, and notochord length of striped bass larvae were the most sensitive parameters to tributyltin exposure (i.e., no changes occurred in the others). In their study, decreases occurred in all these parameters, but since their exposures were chronic for 6 or 7 d, and ours were 4-d exposures followed by a 10-d "recovery," it is perhaps not surprising that there are large differences.

The pesticides tested in this study include two known neurotoxins, methyl parathion and carbofuran (Hassal 1990). Since molinate is a thiolcarbamate, it may also be neurotoxic, although it is not generally characterized as such. In any case, neurotoxins are likely to affect muscular activity in fishes (Drummond et al. 1986; Heath 1987). Alterations in spontaneous activity may interfere with normal foraging behavior, and by that, decrease energy intake or make the individuals more susceptible to predation (Farr 1977).

In studies where spontaneous activity has been measured, most report decreases in activity with chemical exposure (Rand 1977; Drummond et al. 1986; Finger and Bulak 1988; Little et al. 1990), although Ellgaard et al. (1977) found an increased level of activity with DDT exposure and Farr (1977) observed the same in grass shrimp *(Palaemonetes pugio)* with exposure to a concentration of methyl parathion only 3% of the LC50 value. Interestingly, an increase in activity was seen in our study when the pesticides were combined, whereas when exposures were to individual pesticides, there was a slowing (if any effect at all) which suggests a complex interaction between the pesticides when combined (Figure 2). We found significant changes in spontaneous activity only at the higher concentrations, which suggests this parameter (at least measured as was done here) is not especially sensitive to stress from these pesticides.

Spontaneous activity appears to be especially sensitive to food availability. Larvae that had received a maintenance diet (i.e., methyl parathion and molinate groups) showed a much lower rate of activity than those fed ad libitum (Figure 2), even though there was no food in the activity chamber at the time of the measurements. Thus, this dynamic measurement might be diagnostic of larvae that have experienced reduced food availability.

There have been many studies of various aspects of swimming performance in juvenile and adult fishes (Beamish 1978; Randall and Brauner 1991) using a variety of swimming chambers, usually some sort of water tunnel. The various devices that have been developed for assessing swimming performance of larvae are either quite tedious to use or they do not lend themselves to work with 4-d-old striped bass larvae (McLeod 1967; Wieser and Forstner 1986; Kaufmann 1990). Our system of chasing a larva around a petri dish simulates being chased by a predator. It thus probably integrates both neural and metabolic aspects of the animal since nervous coordination and capacity to rapidly expend energy are both involved.

Our data suggest that methyl parathion and molinate both cause reductions in swimming performance and the effect persists even 10-d after the exposures have ceased. Little et al. (1990) found no effect of methyl parathion on swimming capacity of small (0.5 to 1.0 g) rainbow trout but they used a concentration that was only 3% of the LC50 value. In addition, they assessed swimming capacity by measuring maximum sustained speed for 3 min in a water tunnel, so the findings are probably not comparable to ours. Cripe et al. (1984) reported a reduction in critical swimming speed in sheepshead minnow *(Cyprinodon variegatus)* after prolonged exposures to the organophosphate insecticides Guthion and EPN at concentrations of 0.5 to 4.0 µg/L.

There are multiple mechanisms by which a pesticide might affect swimming performance. Since swimming involves adjustments in nervous function, muscle cell energy expenditure, respiratory gas exchange and circulation, there are several points where a chemical might have impact. Attempts by others to correlate organophosphate insecticide suppression of brain AChE activity with swimming performance have been mixed (reviewed in Heath 1987). The same might be said for our larva AChE data in which there are times that the two factors go together (e.g., methyl parathion) but so many exceptions exist that these may be merely coincidental (Figure 2).

As was found in the spontaneous activity experiments, those fish that received the reduced food abundance (methyl parathion and molinate groups) had reduced swimming performance compared to the other two groups (Figure 2). The fish were not starving, which clearly will reduce swimming performance (Laurence 1972), but rather they were merely receiving a maintenance ration, which caused a big reduction in swimming performance. This serendipitous finding might provide insight into an important mechanism contributing to reduced recruitment of striped bass. If, as has been suggested by others (Stevens et al. 1985), larval striped bass in the Sacramento-San Joaquin estuary are experiencing reductions in food availability, the resulting impaired swimming ability could have a marked impact on their ability to avoid predation (Webb 1986; Little et al. 1990).

AChE activity (expressed as units per gram of protein) in controls generally exhibited an increase between day 4 (immediately after pesticide exposure) and day 14 (Figure 2). This probably reflects development of the brain and muscular system since this enzyme is located at synapses and myoneural junctions. The low control values for the carbofuran group in 14-d-old fish were probably experimental artifacts.

Methyl parathion is an organophosphate insecticide while carbofuran is a carbamate insecticide, both of which are well known AChE inhibitors (see Zinkl et al. 1990 for review). The extent of inhibition we observed is not especially large compared to that reported by others for juvenile and adult fish brain, but ours is whole-body activity, which might be more refractory to pesticide effect. While herbicides are not generally considered to be in this group, the apparent inhibition by molinate suggests that they should not be ignored. Since it is a thiocarbamate, perhaps its ability to suppress AChE is not too surprising.

Recovery of AChE activity did not generally occur with our larvae during the 10 d in non-contaminated water, and this was especially noteworthy in the methyl parathion groups. Compared to rodents, fish recover AChE activity very slowly (Wallace and Herzberg 1988). Recovery rates in adult fish almost always require days or weeks depending on degree of inhibition, fish species, and pesticide (Morgan et al. 1990). Our larvae would certainly appear to be no exception.

General Conclusions

Exposure of newly hatched striped bass larvae to individual pesticides for 4 d resulted in some sublethal effects, even at pesticide concentrations several orders of magnitude below the LC50 value. Some of these effects persisted in larvae kept for 10 d in non-contaminated water. Swimming performance and AChE inhibition were the two most sensitive parameters examined; swimming performance was especially sensitive to food shortage. Combining the three pesticides produced a less than additive effect on most parameters measured.

Acknowledgments

This project was supported by contract FGO479 with the California Department of Fish and Game. We thank Scott Cech, Paul DeVries, Jon Hamm, Laurie Martin, Kevin Reimer, and Mike Steele for technical assistance. The Cahn microbalance was borrowed from Barbara Horowitz and the spectrofluorometer from Jerry Hedrick, both of University of California, Davis. Dr. George Terrill and Patty Kitchen of the Virginia Tech statistics department provided statistical advice.

References

Adams, S. M., editor. 1990. Biological indicators of stress in fish. American Fisheries Society Symposium 8, Bethesda, Maryland.

Babu, R., B. Jayasundaramma, and B. Ramaurthi. 1986. Behavioral abnormalities of juveniles of the fish *Cyprinus carpio* exposed to methyl parathion. Environmental Ecology 4:95-97.

Beamish, F. W. H. 1978. Swimming capacity. Pages 101-188 *in* W. S. Hoar and D. J. Randall, editors. Fish physiology, Vol. VII. Academic Press, New York.

Bengtsson, D. E. 1979. Increased growth in minnows exposed to PCBs. Ambio 8:169-170.

Bentle, L. A., S. Dutta, and J. Metcoff. 1981. The sequential enzymatic determination of DNA and RNA. Analytical Biochemistry 116:5-16.

Bruijn, J., and J. Hermens. 1991. Uptake and elimination kinetics of organophosphorous pesticides in the guppy *(Poecilia reticulata)*: correlations with the octanol/water partition coefficient. Environmental Toxicology and Chemistry 10:791-804.

Buckley, L. J. 1984. RNA-DNA ratio: an index of larval fish growth in the sea. Marine Biology 80:291-298.

Bulow, F. J. 1987. RNA-DNA ratios as indicators of growth in fish: a review. Pages 45-64 *in* R. C. Summerfelt and G. E. Hall, editors. Age and growth of fish. Iowa State Univ. Press, Ames, Iowa.

Cornacchia, J. W., D. B. Cohen, G. W. Bowes, R. J. Schanagl, and B. L. Montoya. 1984. Rice herbicides: Molinate (Ordram) and Thiobencarb (Bolero). California State Water Resources Control Board, Special Projects Report 84-4SP.

Cripe, G. M., L. R. Goodman, and D. J. Hansen. 1984. Effect of chronic exposure to EPN and to guthion on the critical swimming speed and brain AChE activity of *Cyprinodon variegatus*. Aquatic Toxicology 5:255-266.

Drummond, R. A., C. L. Russom, D. L. Geiger, and D. L. DeFoe. 1986. Behavioral and morphological changes in fathead minnow *(Pimephales promelas)* as diagnostic endpoints for screening chemicals according to mode of action. Pages 415-435 *in* T. M. Poston and R. Purdy, editors. Aquatic toxicology and environmental fate: ninth volume. American Society for Testing and Materials, STP 921, Philadelphia.

Ellgaard, E. G., J. C. Ochsner, and J. K. Cox. 1977. Locomotor hyeractivity induced in the bluegill sunfish, *Lepomis macrochirus*, by sublethal concentrations of DDT. Canadian Journal of Zoology 55:1077-1081.

Ellman, G. L., K. D. Courtney, O. Andres, and R. Featherstone. 1961. A new and rapid colorimetric determination of AChE activity. Biochemical Pharmacology 7:88-95.

Farr, J. A. 1977. Impairment of antipredator behavior in *Palaeomonetes pugio* by exposure to sublethal concentration of parathion. Transactions of the American Fisheries Society 106:287-290.

Finger, S. E., and J. S. Bulak. 1988. Toxicity of water from three South Carolina rivers to larval striped bass. Transactions of the American Fisheries Society 117:521-528.

Finlayson, B. J., and G. A. Faggella. 1986. Comparison of laboratory and field observations of fish exposed to the herbicides molinate and thiobencarb. Transactions of the American Fisheries Society 115:882-890.

Finlayson, B., J. Harrington, R. Fujimura, and G. Issac. 1993. Identification of methyl parathion toxicity in Colusa Basin Drain water. Environmental Toxicology and Chemistry 12:291-303.

Foe, C. 1989. 1989 rice season toxicity monitoring results. Internal memorandum to J. Bruns and R. Schnagle. California Regional Water Quality Control Board., Central Valley Region, Sacramento.

Fujimura, R., B. Finlayson, and G. Chapman. 1991. Evaluation of acute and chronic toxicity tests with larval striped bass. Pages 193-211 *in* M. A. Mayes and M. G. Barron, editors. Aquatic toxicology and risk assessment. 14th volume. American Society for Testing and Materials, STP 1124, Philadelphia.

Hassal, K. A. 1990. The Biochemistry and Uses of Pesticides, Second Edition. VCH Verlagsgesellschaft, New York.

Heath, A. G. 1987. Water Pollution and Fish Physiology. CRC Press, Boca Raton, Florida.

Heath, A. G., J. J. Cech, J. G. Zinkl, and M. D. Steele. In press. Sublethal effects of three pesticides on Japanese medaka. Archives of Environmental Contamination and Toxicology.

Kane, A. S., R. O. Bennett, and E. B. May. 1990. Effect of hardness and salinity on survival of

striped bass larvae. North American Journal of Fisheries Management 10:67-71.
Kaufmann, R. 1990. Respiratory cost of swimming in larval and juvenile cyprinids. Journal of Experimental Biology 150:343-366.
Laughlin, R. B., J. Ng, and H. E. Guard. 1981. Hormesis: a response to low environmental concentrations of petroleum hydrocarbons. Science 211:705-706.
Laurence, G. C. 1972. Comparative swimming abilities of fed and starved larval largemouth bass *(Micropterus salmoides)*. Journal of Fish Biology 4:73-78.
Little, E. E., R. D. Archeski, B. Flerov, and V. I. Kozlovskaya. 1990. Behavioral indicators of sublethal toxicity in rainbow trout. Archives of Environmental Contamination and Toxicology 19:380-385.
Martin, F. D., and R. Malloy. 1980. Histologic and morphometric criteria for assessing nutritional state in larval striped bass, *Morone saxatilis*. Pages 157-161 *in* L.A. Fuiman, editor. Proceedings of the fourth annual larval fish conference, U. S. Fish and Wildlife Service, FWS/OBS-80/43, Washington, District of Columbia.
McLeay, D. J., and D. A. Brown. 1975. Growth stimulation and biochemical changes in juvenile coho salmon *(Oncorhynchus kisutch)* exposed to sublethal levels of neutralized bleached Kraft pulpmill effluent for 200 days. Journal of the Fisheries Research Board of Canada 31:1043-1049.
McLeod, J. C. 1967. A new apparatus for measuring maximum swimming speeds of small fishes. Journal of the Fisheries Research Board of Canada 24:1241-1252.
Miglavs, I., and M. Jobling. 1989. Effects of feeding regime on food consumption, growth rates and tissue nucleic acids in juvenile Arctic charr, *Salvelinus alpinus*, with particular respect to compensatory growth. Journal of Fish Biology 34:947-957.
Morgan, M. J., L. L. Fancey, and J. W. Kiceniuk. 1990. Response and recovery of brain AChE activity in Atlantic salmon *(Salmo salar)* exposed to fenitrothion. Canadian Journal of Fisheries Aquatic Sciences 47:1652-1654.
Norberg-King, T. J., E. J. Durhan, and G. T. Ankley. 1991. Application of toxicity identification evaluation procedures to the ambient waters of the Colusa Basin Drain, California. Environmental Toxicology and Chemistry 10:891-900.
Palawski, D., D. R. Buckler, and F. L. Mayer. 1983. Survival and condition of rainbow trout *(Salmo gairdneri)* after acute exposures to methyl parathion, triphenyl phosphate, and DEF. Bulletin of Environmental Contamination and Toxicology 30:614-620.
Pinkney, A. E., L. L. Matteson, and D. A. Wright. 1990. Effects of tributyltin on survival, growth, morphometry, and RNA-DNA ratio of larval striped bass, *Morone saxatilis*. Archives of Environmental Contamination and Toxicology 19:235-240.
Rand, G. M. 1977. The effect of subacute parathion exposure on the locomotor behavior of the bluegill sunfish and largemouth bass. Pages 253-268 *in* F. L. Mayer and J. L. Hamelink, Editors. Aquatic toxicology and hazard evaluation. American Society for Testing and Materials, STP 634, Philadelphia.
Randall, D., and C. Brauner. 1991. Effects of environmental factors on exercise in fish. Journal of Experimental Biology 160:113-126.
Setzler-Hamilton, E. M., J. A. Whipple, and B. MacFarlane. 1988. Striped bass populations in Chesapeake and San Francisco Bays: two environmentally impacted estuaries. Marine Pollution Bulletin 19:466-477.
Stebbing, A. R. D. 1987. Growth hormesis: a by-product of control. Health Physics 52:543-547.
Stevens, D. E., D. W. Kohlhorst, and L. W. Miller. 1985. The decline of striped bass in the Sacramento-San Joaquin estuary, California. Transactions of the American Fisheries Society 114:12-30.
Wallace, K. B., and U. Herzberg. 1988. Reactivation and aging of phosphorylated brain AChE from fish and rodents. Toxicology and Applied Pharmacology 92:307-314.
Webb, P. W. 1986. Locomotion and predator-prey relationships. Chap. 3 *in* M. E. Feder and G. V. Lauder, Editors. Predator prey relationships, perspectives and approaches from study of lower vertebrates. University of Chicago Press.
Wieser, W., and H. Forstner. 1986. Effects of temperature and size on the routine rate of oxygen consumption and on the relative scope for activity in larval cyprinids. Journal of Comparative Physiology 156B:791-796.
Wright, D. A., and F. D. Martin. 1985. The effect of starvation on RNA:DNA ratios and growth of larval striped bass, *Morone saxatilis*. Journal of Fish Biology 27:479-485.

Zinkl, J., W. Lockhart, S. Kenny, and F. Ward. 1991. Effects of cholinesterase-inhibiting insecticides on fish. Pages 233-254 *in* P. Mineau, editor. Cholinesterase-inhibiting insecticides — impacts on wildlife and the environment, Elsevier Science Publications, Amsterdam, The Netherlands.

Production, Mortality, and Transport of Striped Bass Eggs in Congaree and Wateree Rivers, South Carolina

JAMES S. BULAK

South Carolina Wildlife and Marine Resources Department
1921 Van Boklen Road, Eastover, South Carolina 29044, USA

NOEL M. HURLEY, JR.

U. S. Geological Survey
720 Gracern Road, Columbia, South Carolina 29210, USA

JOHN S. CRANE

South Carolina Wildlife and Marine Resources Department
1921 Van Boklen Road, Eastover, South Carolina 29044, USA

Abstract.—From 1988 to 1990, annual production and mortality of striped bass *(Morone saxatilis)* eggs were investigated in the two freshwater spawning tributaries of the Santee-Cooper system, South Carolina. Eggs were collected every 8 or 12 h at two sites on both the Congaree and Wateree rivers. A striped bass egg transport model was developed to link biological events with physical processes. Total combined egg production for the Congaree and Wateree rivers ranged from 10.99×10^9 to 22.93×10^9 during the 3 years. Egg mortality averaged 80%/d on the Congaree River and 94%/d on the Wateree River. Mean specific gravity of eggs was 1.00109. The importance of sampling interval and location in obtaining precise and accurate estimates of total egg production was demonstrated. The study suggests that the magnitude of egg production and egg mortality rates are meaningful contributors to eventual recruitment.

The Santee-Cooper river system in South Carolina provided the first documentation that a striped bass *(Morone saxatilis)* population was successfully completing its entire life cycle in fresh water (Scruggs 1957). An expanding population, created by impounding the Santee River in the 1940s, fostered a sport fishery of national significance (Stevens 1958). Artificial propagation techniques for striped bass were later developed in the Santee-Cooper system by Stevens (1967). A substantial number of existing inland populations of striped bass were parented by Santee-Cooper stock (Harrell et al. 1990).

The primary spawning tributaries for Santee-Cooper striped bass are the Congaree and Wateree rivers (May and Fuller 1965). The Congaree (82 km long) and Wateree (120 km long) rivers merge to form the Santee River which then flows 26 km to Lake Marion, a 45,000 hectare freshwater impoundment (Figure 1). Annual discharge averages 260 and 176 m³/s on the Congaree and Wateree rivers, respectively, and is regulated by upstream hydroelectric facilities. Each spring, striped bass migrate from downstream impoundments to spawn in the Congaree and Wateree rivers. Eggs are broadcast into the main channel and develop as they drift downstream.

McCoy (1959) and Cheek (1961), working on the Roanoke River, North Carolina, developed a method of estimating the standing stock of striped bass eggs that drifted by a set sampling location. They determined that estimated variance was minimized by sampling at 3-h intervals during the spawning season. Hassler et al. (1981) employed this technique on the Roanoke River and reported that egg standing stock ranged from 870×10^6 to $4,932 \times 10^6$ during a 20-year period. May and Fuller (1965) also used this method to estimate standing stock of eggs near the confluence of the Congaree and Wateree rivers. Estimates obtained in 1961 and 1962 were 542×10^6 and 307×10^6 on the Congaree River and 22×10^6 and 17×10^6 on the Wateree River.

Measurement of total production and subsequent mortality of eggs is an essential step in understanding life-history processes that control population

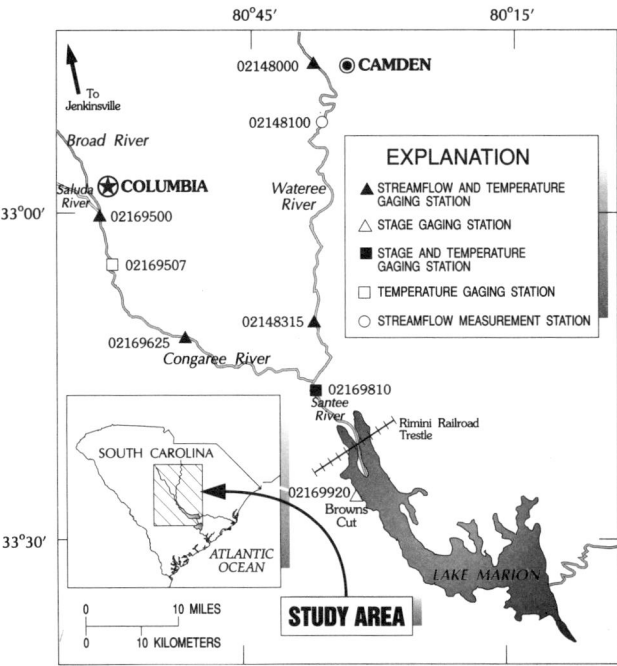

FIGURE 1.—Map of upper impoundment and striped bass spawning tributaries of the Santee-Cooper system. U. S. Geological Survey physical and chemical monitoring stations are identified with 8-digit code.

size. Egg production and mortality of pelagic marine spawners are frequently estimated and exhibit considerable temporal and spatial variation (Smith et al. 1989; Coombs et al. 1990). Dahlberg (1979) reviewed the survival of fish eggs and noted that data extracted from Polgar (1977) revealed a very high mortality (90 to 94%/d) of striped bass eggs in the Potomac River, Maryland. Olney et al. (1991) used dye to follow a cohort of striped bass eggs in the Pamunkey River, Virginia, and obtained mortality estimates of 10 to 91%/d. Mean mortality was 68%/d.

Beach seining surveys indicated that Santee-Cooper striped bass experienced a decline in juvenile recruitment in the 1970s (M. G. White III, South Carolina Wildlife and Marine Resources Department, personal communication). This decline concerned fishery managers and investigations into factors affecting reproductive success were initiated in the 1980s. Finger and Bulak (1988) implicated water quality as a factor affecting survival of striped bass prolarvae in the Wateree River. Rulifson and Manooch (1990) implicated water management strategies at upstream dams as a factor affecting striped bass recruitment on the Roanoke River. As an additional investigative step on the Santee-Cooper system, the egg studies of May and Fuller (1965) were repeated in 1983 and 1984.

Standing stock estimates of 984×10^6 and $2,220 \times 10^6$ eggs in the Congaree River and 60×10^6 and 9×10^6 eggs in the Wateree River were obtained (Bulak et al. 1985). However, these estimates represented standing stock measured at a fixed location and not total production, which was needed to estimate adult spawning biomass.

In the late 1980s efforts were initiated to construct an individual based model of striped bass in order to test the sensitivity of various factors that had potential to affect recruitment. To construct the model, estimates of total egg production and mortality were needed. The objectives of the present study were (1) to develop a striped bass egg transport model so that (2) estimates of total egg production and mortality could be obtained for each of the spawning rivers. Study results and other relevant data are reviewed to determine what management actions could halt the observed recruitment decline.

Methods

The sampling scheme used in this study established two fixed collection sites for eggs on each spawning tributary. Egg aging data and egg transport equations (described below) were used to calculate total production and mortality of striped bass eggs.

Development of Egg Transport Model

Streamflow was modeled using the BRANCH model (Schaffranek et al. 1981), a one-dimensional numerical model for simulation of flow in singular and interconnected channels. Transport of the eggs was modeled using a branched Lagrangian transport model, or BLTM, developed by Jobson and Schoellhamer (1987).

The BRANCH model uses stage or discharge at the study boundaries to compute stage and discharge throughout the study area. Historic U. S. Geological Survey stage-discharge information acquired near the origin of the Congaree and Wateree rivers was used to drive the model. An automated stage recorder was placed at the downstream limit of the study, the confluence of Santee River with Lake Marion.

The study area was divided into 13 branches with 100 cross sections. Channel morphometry was measured at each cross section with a graphic fathometer. The width of the main channel and the cross sectional geometry for approximately 50 m on each side of the channel were determined with conventional surveying equipment. Additionally, U. S. Geological Survey 7.5-minute topographic maps were used to extrapolate the floodplain beyond the surveyed section. The BRANCH model was calibrated with a time step of 1 h for the 1988 striped bass spawning season. The computed streamflow hydrographs were then compared to observed streamflows at four U. S. Geological Survey stations with automated stage recorders.

The BRANCH model computed instantaneous discharge, area, and top width at specific cross-section locations and supplied these conditions to the BLTM to simulate egg transport. Dye studies were conducted on five river segments on August 10 to 25, 1987, to calibrate the BLTM. Procedures generally followed those of Hubbard et al. (1982). For calibration, the BLTM was operated at a 30-min time step. Observed dye concentrations were entered into the BLTM model at the uppermost cross section. Adjustments to the dispersion coefficient allowed calibration of peak dye concentrations and travel time. The BLTM was developed as a transport model for dissolved water-quality constituents and, therefore, assumes that striped bass eggs are transported at the same rate as a dissolved substance.

Following calibration of the egg transport model, stepwise regression equations that predicted hatching and spawning location of an egg were computed for each sample site. Discharge, egg age, remaining egg development time to hatching, and water temperature were explanatory variables. All variables were transformed into logarithms and statistical testing was done prior to analysis to obtain a linear regression model and achieve equal variance about the regression line. A 95% confidence limit was specified to select the significant independent variables. Technical and specific details of the transport model development process are provided by Hurley (1991).

Egg Collections

Two locations were sampled at regular intervals throughout the striped bass spawning season on both the Congaree and Wateree rivers. The Congaree sites were at river km 8 and 44, while Wateree sites were at river km 18 and 88. Each location was sampled at 8-h intervals in 1988 and 1989 and at 12-h intervals in 1990.

A 0.5-m plankton net with a mesh of 505 μm was lowered to mid-depth from an anchored boat to collect eggs. Samples were collected from the center of the channel. Sample duration was 5 min and a flowmeter mounted in the net frame was used to calculate the volume of water filtered. Surface water temperature and river stage were recorded for each sample. A stage-discharge relationship was calculated for each sampling site using standard hydrologic techniques.

The field sample was preserved in 7% unbuffered formalin (Gates et al. 1987) and brought back to the lab where eggs were sorted and enumerated. A 10% subsample of viable eggs from each collection was aged under a dissecting microscope using the nomograph of Hassler et al. (1981). Age structure of the subsample was directly expanded to the entire sample.

Egg viability was assessed daily for at least one sampling location per river using eggs collected in a separate 1 min sample. Fresh eggs were inspected in the field with a dissecting microscope. The percentage of viable eggs was determined based on criteria provided by Hassler et al. (1981). Daily measures of viability were used to adjust daily egg collection data for the estimation of standing crop, production, and mortality. If an estimate of viability was not available for a day, weighted annual average egg viability for each year was used as the daily estimate.

The specific gravity of striped bass eggs was determined by inserting viable eggs into solutions of known concentrations, made by dissolving sodium chloride in distilled water. Pre-mixed solutions

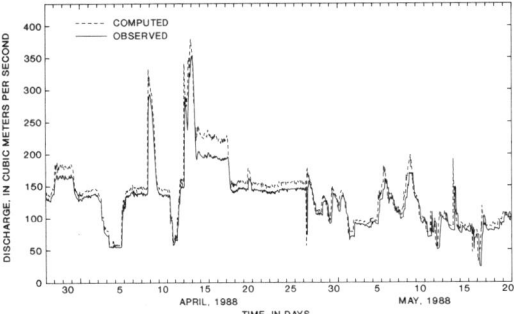

FIGURE 2.—Observed and computed discharge at U.S. Geological Survey gaging station 02169500, Congaree River at Columbia. The BRANCH model was used to compute discharge.

varied in specific gravity from 1.00025 to 1.00140 and were mixed in increments of 0.00005 units. Specific gravity was determined by defining the specific gravity of the solution in which the egg was neutrally buoyant. Congaree River eggs were evaluated on April 7, 1988, while Wateree River eggs were assessed on May 10, 1988.

The minimum velocity required to keep striped bass eggs in suspension was determined by visual inspection in a sand channel flume (10 m × 0.3 m × 0.3 m). Eggs were inserted in the middle of the water column and observed as they traveled downstream. Flume velocity was measured with an electromagnetic flowmeter set at 0.80 of full flume depth. Tests were conducted on April 29, 1988, and May 1, 1989.

The total number of eggs that passed each sampling site during a spawning season (i.e., standing crop) was determined by multiplying egg density by the total discharge that occurred during a sampling interval. Interval estimates were summed for a spawning season to obtain an annual estimate. Estimates were calculated for all eggs and for viable eggs only. Standard errors associated with these estimates were determined by the mean square successive difference procedure (Cochran 1977).

To calculate mortality estimates of viable eggs, those eggs that were spawned above the upstream site and would have hatched below the downstream site were determined with transport equations for both sampling sites on a river. Restricting analysis to this group of eggs, the standing crop estimates and their associated standard errors were calculated for the upstream and downstream site. The reduction in standing crop estimates from the upstream to the downstream site was defined as mortality. To convert this annualized estimate of mortality to an instantaneous rate (Z), the weighted (by egg density per interval) mean travel time (t, in days) of an egg between sampling stations was calculated. The instantaneous rate of mortality was calculated with the equation (Ricker 1975)

$$Z = [\log_e (N_0) - \log_e (N_t)] / t$$

where, N_t is the standing crop estimate at downstream site and N_0 is the standing crop estimate at upstream site.

The actual total mortality rate (A) per day was then derived through the equation (Ricker 1975)

$$A = 1 - e^{-Z}$$

Using a two-step process, standing crop egg estimates were adjusted for transport and pre-sampling mortality to estimate total egg production in each river. First, transport equations were used to determine an estimate of the standing crop of eggs at the downstream station that was attributable to spawning below the upstream station. This estimate was added to the standing crop of eggs measured at the upstream station to yield an estimate of the total standing crop of eggs for each river. Calculation of spawning location used temperature data from each river to adjust development rates of eggs. Estimates were derived for all eggs and for viable eggs only. Second, instantaneous mortality rates were derived for each river and year and were used to estimate the initial egg quantity (N_0) prior to mortality above the collecting sites in the equation

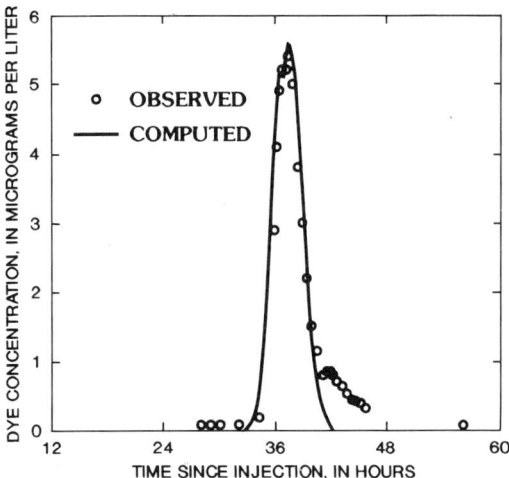

FIGURE 3.—Observed and computed dye concentrations in Wateree River at river km 0.40. A branched Lagrangian transport model was used to compute dye concentrations.

TABLE 1.—Viability determinations of striped bass eggs collected from Congaree and Wateree rivers.

River	Year	N	% Viable
Congaree	1988	5,395	69
Congaree	1989	3,069	74
Congaree	1990	2,216	78
Wateree	1988	1,618	61
Wateree	1989	1,095	60
Wateree	1990	473	70

$$\log_e (N_0) = Z \cdot T + \log_e (N_t)$$

where, N_0 is the estimated number of spawned eggs, Z is the instantaneous mortality rate, T is egg age, in days, and N_t is the number of sampled eggs. The equation was solved for each sampling interval and age class to properly weight the contribution of each age class of eggs. Each year, the quotient

$$\Sigma N_0 / \Sigma N_t$$

was calculated for each site's data. This quotient was multiplied by corrected standing stock estimates for each sampling site to derive total egg production.

Results

Egg transport model

Computed streamflow hydrographs produced by the BRANCH model were in close agreement with observed data (Figure 2). Computed discharges were adjusted by modifying cross-sectional areas, varying flow resistance (Manning's n) with stage, and by accounting for storage in the stream reach.

Computed and observed dye concentrations and travel times agreed closely (Figure 3). Discharge during the dye studies ranged from 79.6 to 175 m³/s on the Congaree River and from 33.1 to 93.2 m³/s on the Wateree River. A complete description of modelling results is given in Hurley (1991).

Regression equations that predicted hatching and spawning location of eggs sampled at each of four sampling locations were developed. Hatching distance from the sample site was a function of discharge and time remaining to hatch. Spawning distance from the sample site was a function of discharge and the age of the egg. The R^2 values for the eight equations ranged from 0.93 to 0.99. Plots of residuals and the explanatory variables indicated that the regression models were not biased. The specific equations are reported in Hurley (1991).

Biological Estimates

The mean specific gravity of striped bass eggs was 1.00115 (SD = 0.00008) on the Congaree River ($N = 13$) and 1.00101 (SD = 0.00007) on the Wateree River ($N = 10$). Flume tests conducted on two dates indicated a mean channel velocity of 0.06 m/s was sufficient to keep the eggs in suspension. Two of the 10 eggs examined at 0.06 m/s in 1989 did contact the bottom sand but bounced away. Of 13,866 eggs assessed for viability during the 3 study years, 69.9% were viable (Table 1).

Site-specific estimates of egg standing crop showed that the Congaree River was the primary spawning tributary for Santee-Cooper striped bass in 1988 through 1990 (Table 2). In all years, Congaree River egg abundance was greater at the upstream station than at the downstream station. Standard error as a proportion of a viable egg estimate ranged from 0.26 to 0.41 for the Congaree River and from 0.18 to 0.79 for the Wateree River.

Egg mortality averaged 80%/d on the Congaree River and 94%/d on the Wateree River (Table 3). The mean weighted transport time between sampling sites during the 3 study years was 19.2 h (SD = 1.0) on the Congaree River and 27.9 h (SD = 7.5) on the Wateree River.

Total estimated production of striped bass eggs in the Congaree and Wateree rivers ranged from 10.99×10^9 in 1990 to 22.93×10^9 in 1988 (Table 4). Average total production for the 3 study years was 18.43×10^9 eggs. In the 3 study years, Congaree River accounted for an average of 82% of the total production from the two rivers. On the Congaree River, an average of 0.80 (SD = 0.09) of total production occurred above the upstream station.

Discussion

The sampling design provided accurate and precise estimates of the standing stock of eggs in the two Santee-Cooper spawning tributaries. Pelagic spawning in a riverine environment promotes effective sampling as eggs occur within a confined chan-

TABLE 2.—Standing crop estimates ($\times 10^9$) of all striped bass eggs and viable eggs at two sampling locations on both the Congaree and Wateree rivers. Standard errors of estimates are in parentheses.

		Congaree River		Wateree River	
Year	Egg type	km 44	km 8	km 88	km 18
1988	All	7.3 (2.43)	4.34 (1.20)	0.78 (0.26)	0.33 (0.06)
1989	All	8.16 (2.47)	3.03 (1.13)	0.71 (0.18)	0.38 (0.26)
1990	All	4.32 (1.03)	1.74 (0.49)	0.11 (0.02)	0.48 (0.17)
1988	Viable	4.36 (1.45)	3.18 (1.07)	0.50 (0.15)	0.20 (0.04)
1989	Viable	5.90 (1.87)	2.31 (0.94)	0.39 (0.12)	0.32 (0.26)
1990	Viable	3.54 (0.93)	1.23 (0.37)	0.08 (0.02)	0.31 (0.13)

TABLE 3.—Annual mortality estimates for viable striped bass eggs on the Congaree and Wateree rivers obtained by identifying the cohort of eggs that had equal probability of caprture at an upstream and a downstream collection site.

Year	River	Estimated number of eggs ($\times 10^9$)		Time[a]	Z[b]	A[c]
		Upstream	Downstream			
1988	Congaree	4.256	1.299	0.84	1.41	0.76
1989	Congaree	4.655	1.207	0.79	1.71	0.82
1990	Congaree	3.429	0.921	0.76	1.73	0.82
1988	Wateree	0.227	0.002	1.53	3.09	0.95
1989	Wateree	0.082	0.003	0.97	3.41	0.97
1990	Wateree	0.053	0.006	1.00	2.18	0.85

[a] Mean time of travel (in d) between stations.
[b] Instantaneous rate of mortality.
[c] Daily percent mortality.

nel and are transported past fixed sites. Brander and Thompson (1989) pointed out that an advantage to sampling eggs is that they have no escape response to avoid sampling gear. Dudley and Black (1978) and Hassler et al. (1981) have also successfully employed a riverine sampling strategy for striped bass eggs on the Savannah and Roanoke rivers.

The time interval between samples is of major importance in obtaining precise estimates of striped bass spawning in riverine environments. It is well known that striped bass egg production occurs in sharp spawning peaks of somewhat short duration (Johnson and Koo 1975). Cheek (1961) showed that sampling every 3 h minimized the variance of estimates. Analysis of data from a study using a 3-h sampling interval on the Congaree River in 1983 showed that sampling intervals of up to 12 h produced standard errors that were less than 25% of the estimate (M. Udevitz, Institute of Statistics, North Carolina State University, unpublished data). The interval between samples must be sufficiently small to ensure that peak events are sampled.

The location of sampling sites is also important in obtaining reasonable estimates of production. The downstream site used in this study for the Congaree River was close to the sampling location used by May and Fuller (1965). However, this study showed that hatching and mortality of eggs may occur prior to reaching this site. In this study, egg mortality upstream from the sampling sites was accounted for and a mortality-adjusted estimate of total egg production was produced. Mortality from time of spawning and hatching location must be considered carefully if a sound estimate of total production is desired. Future studies should determine if the rate of egg mortality varies within a season; for this study, we assumed egg mortality was constant.

Accurate estimates of total egg production of a population allow calculation of the abundance and size composition of the adult spawning stock if fecundity-size relationships and relative size composition of the stock are known. Determining composition of the adult spawning stock is essential for

TABLE 4.—Estimates of total striped bass egg production which compensate for egg mortality at four sampling sites on the Congaree and Wateree rivers, South Carolina. Standing stock estimates at each river's downstream site only consider eggs spawned downstream from the upstream site. Mortality rates are given in Table 3.

Year	River	Standing Stock of eggs ($\times 10^9$)	Average age (h)	Total production of eggs ($\times 10^9$)
1988	Congaree (44[a])	7.30	8.5	12.33
1988	Congaree (8)	2.71	9.7	4.98
1988	Wateree (88)	0.78	6.0	1.79
1988	Wateree (18)	0.33	13.2	3.83
1989	Congaree (44)	8.16	8.2	15.10
1989	Congaree (8)	1.38	12.5	3.48
1989	Wateree (88)	0.71	6.9	2.05
1989	Wateree (18)	0.36	3.4	0.80
1990	Congaree (44)	4.32	8.8	8.43
1990	Comgaree (8)	0.45	9.8	0.95
1990	Wateree (88)	0.11	6.4	0.20
1990	Wateree (18)	0.47	10.5	1.42

[a] Kilometers upstream from mouth of river.

formulating management strategies for exploited populations. However, high egg mortality rates, as observed in this study, cause calculation of adult stock estimates that are below their actual values unless egg mortality rate is taken into account. For example, if eggs are not vulnerable to sampling for 8 h and have an instantaneous mortality rate of -1.61 (i.e., 80%/d), then an estimate of egg standing stock is only 59% of the actual production. Efforts to account for egg mortality within the sampling and analysis strategy are needed to obtain non-biased estimates of total production.

The development of the transport model was a key element that linked biological, chemical, and physical processes. Finger and Bulak (1988) identified water quality as a factor influencing prolarva survival in Congaree and Wateree rivers. Future management efforts should focus on developing a real-time monitoring strategy that will ensure acceptable water quality for striped bass eggs and larvae throughout the spawning season. Future research efforts should focus on further defining the interaction between the temporal variability in water quality and the survival of eggs and larvae. Other beneficial applications of the transport equations include delineation of hatching locations and their importance to survivorship (Methot 1983), identification of spawning sites and their relation to streamflow, and delineation of possible changes in egg mortality during the spawning season.

The assumption that transport of eggs occurred at the same rate as dissolved water-quality constituents may demand further inquiry. Specific gravity and settling velocity of eggs defined in this study suggest that in streamflow conditions generally encountered during the spawning season, egg transport is very close to dye transport. Yet Cobb (1989) reported that striped bass eggs were transported at 87% of observed streamflow in Santee River, South Carolina. Testing the sensitivity of transport equations to this level of error appears justified.

Crance (1984) stated that the settling velocity of striped bass eggs was 0.30 m/s based on the studies of Albrecht (1964) on the Sacramento-San Joaquin population. Settling velocity is important in determining river length and streamflow required to keep eggs suspended and allow hatching. Flume observations during our study suggest that a settling velocity closer to 0.06 m/s may be more realistic for the Santee-Cooper system. Stock differences in egg buoyancy may account for differences in settling velocities.

Results from this study support other studies that derived a high egg mortality rate for striped bass eggs. Evidence from other pelagic spawners suggests that the magnitude of egg production and egg mortality rates set bounds on eventual recruitment but do not determine recruitment (Clark and Marr 1955; Zijlstra and Witte 1985; Smith et al. 1989; Thompson 1989). Ulanowicz and Polgar (1980) examined the relation between egg production and recruitment in striped bass and concluded that recruitment is primarily driven by density-independent events. They also observed a mismatch between spawning site selection by adults and survival potential of spawned eggs and hypothesized a life-history strategy adapted to persistence. Future efforts in Santee-Cooper will focus on integrating life-history information into an individual based model (Cowan et al. 1993) to identify those factors that provide the greatest influence on striped bass recruitment potential.

With available egg production information, what can managers do to improve the average recruitment potential over years of the Santee-Cooper stock? Goodyear (1985) showed that increased egg production will increase long-term average recruitment where density-independent factors dominate. Winemiller and Rose (1992) point out that maintenance of some critical density of the adult stock and protection of spawners and spawning habitat during a relatively short reproductive period is crucial in the management of long-lived, periodic strategists such as striped bass. Management agencies should ensure adequate habitat and design and implement strategies that maximize the number of eggs produced and the temporal variability in egg production to maximize the potential for matching eggs with an optimal survival window (Cushing 1975).

Acknowledgments

This work is part of a doctoral dissertation by JSB at the University of South Carolina; a fellowship was provided by the Electric Power Research Institute. Biological investigations were funded through the Sport Fish Restoration Act of the U. S. Fish and Wildlife Service. Funds for transport investigations were provided by U. S. Geological Survey, South Carolina Water Resources Commission, South Carolina Department of Health and Environmental Control, South Carolina Public Service Authority, South Carolina Electric and Gas Co., South Carolina Wildlife and Marine Resources Department and Duke Power Company. Glenn

Patterson with the U. S. Geological Survey was instrumental in securing funding for transport investigations. Mark Udevitz and Gerry Gray, Institute of Statistics, North Carolina State University, wrote the programs used to quantify production and mortality. W. Van Winkle provided substantial editorial assistance. Jean Kinard, Drew Robb, Donald Gates and Gerrit Jobsis collected, sorted, and aged many striped bass eggs. Access to sampling locations was provided by Congaree National Monument, Kingsville Hunt Club, South Carolina Electric and Gas Co., and Mr. and Mrs. Ben Shirley.

References

Albrecht, A. B. 1964. Some observations of factors associated with survival of striped bass eggs and larvae. California Fish and Game 50:100-113.

Brander, K. M., and A. B. Thompson. 1989. Diel differences in avoidance of three vertical profile sampling gears by herring larvae. Journal of Plankton Research 11:775-784.

Bulak, J. S., J. S. Crane, and D. W. Gates. 1985. Natural reproduction of striped bass. Study completion report WC-1. South Carolina Wildlife and Marine Resources Department, Columbia.

Cheek, R. P. 1961. Quantitative aspects of striped bass spawning in the Roanoke River, North Carolina. Master's thesis. North Carolina State University, Raleigh.

Clark, F. N., and J. C. Marr. 1955. Population dynamics of the Pacific sardine. Progress report, California Cooperative Oceanic Fisheries Investigations, July 1, 1953 to March 31, 1955.

Cobb, C. S. 1989. Factors influencing the spawning of striped bass in the Santee River. Master's thesis. Clemson University, Clemson, South Carolina.

Cochran, W. G. 1977. Sampling techniques, 3rd edition. John Wiley and Sons, New York.

Coombs, S. H., J. H. Nichols, and C. A. Fosh. 1990. Plaice eggs in the southern North Sea: abundance, spawning area, vertical distribution and buoyancy. Journal du Conseil, Conseil International pour l'Exploration de la Mer 47: 133-139.

Cowan, J. H., Jr., K. A. Rose, E. S. Rutherford, and E. D. Houde. 1993. Individual-based model of young of the year striped bass population dynamics. II. Factors affecting recruitment in the Potomac River, Maryland. Transactions of the American Fisheries Society 122:439-458.

Crance, J. H. 1984. Habitat suitability index models and instream flow suitability curves: inland stocks of striped bass. U. S. Fish and Wildlife Service, FWS/OBS-82/10.85.

Cushing, D. H. 1975. Marine ecology and fisheries. Cambridge University Press, New York.

Dahlberg, M. D. 1979. A review of survival rates of fish eggs and larvae in relation to impact assessments. Marine Fisheries Review 41(3):1-12.

Dudley, R. G., and K. N. Black. 1978. Distribution of striped bass eggs and larvae in the Savannah River estuary. Proceedings of the Annual Conference Southeastern Association of Fish and Wildlife Agencies 32:561-570.

Finger, S. E., and J. S. Bulak. 1988. Toxicity of water from three South Carolina rivers to larval striped bass. Transactions of the American Fisheries Society 117:521-528.

Gates, D. W., J. S. Bulak, and J. S. Crane. 1987. Preservation of striped bass eggs collected from a low hardness freshwater system in South Carolina. The Progressive Fish-Culturist 49:230-232.

Goodyear, C. P. 1985. Toxic materials, fishing and environmental variation: simulated effects on striped bass population trends. Transactions of the American Fisheries Society 114:107-113.

Harrell, R. M., J. H. Kerby, and R. V. Minton. 1990. Culture and propagation of striped bass and its hybrids. Striped Bass Committee, Southern Division, American Fisheries Society, Bethesda, Maryland.

Hassler, W. W., N. L. Hill, and J. T. Brown. 1981. The status and abundance of striped bass in the Roanoke River and Albemarle Sound, North Carolina, 1966-1980. North Carolina Division of Marine Fisheries, Special Scientific Report 38, Morehead City.

Hubbard, E. F., F. A. Kilpatrick, C. A. Martens, and J. F. Wilson, Jr. 1982. Measurement of time of travel and dispersion in streams by dye tracing. U. S. Geological Survey Techniques of Water-Resources Investigations, Book 3, Chapter A9.

Hurley, N. M. Jr. 1991. Transport simulation of striped bass eggs in the Congaree, Wateree, and Santee rivers, South Carolina. U. S. Geological Survey Water-Resources Investigations Report 91-4088.

Jobson, H. E., and D. H. Schoellhamer. 1987. Users

manual for a branched lagrangian transport model. U. S. Geological Survey, Water-Resources Investigations Report 87-4163.

Johnson, R. K., and T. S. Koo. 1975. Production and distribution of striped bass eggs in the Chesapeake and Delaware Canal. Chesapeake Science 16:39-55.

May, O. D., Jr., and J. C. Fuller, Jr. 1965. A study on striped bass egg production in the Congaree and Wateree rivers. Proceedings of the Annual Conference Southeastern Association of Game and Fish Commissioners 16(1962):285-301.

McCoy, E. G. 1959. Quantitative sampling of striped bass eggs in the Roanoke River, North Carolina. Master's thesis. North Carolina State University, Raleigh.

Methot, R. D. 1983. Seasonal variation in survival of larval *Engraulis mordax* estimated from the age distribution of juveniles. Fishery Bulletin, U. S. 81:741-750.

Olney, J. E., J. D. Field, and J. C. McGovern. 1991. Striped bass egg mortality, production, and female biomass in Virginia rivers, 1980-1989. Transactions of the American Fisheries Society 120:354-367.

Polgar, T. T. 1977. Striped bass ichthyoplankton abundance, mortality, and production estimation for the Potomac River population. Pages 110-126 *in* W. Van Winkle, editor. Assessing the effects of power plant induced mortality on dish populations. Pergamon Press, New York.

Ricker, W. E. 1975. Computation and interpretation of biological statistics of fish populations. Fisheries Research Board of Canada Bulletin 191.

Rulifson, R. A., and C. S. Manooch III. 1990. Recruitment of juvenile striped bass in the Roanoke River, North Carolina, as related to reservoir discharge. North American Journal of Fisheries Management 10:397-407.

Schaffranek, R. W., R. A. Baltzer, and D. C. Goldberg. 1981. A model for simulation of flow in singular and interconnected channels. U. S. Geological Survey Techniques of Water-Resources Investigations, Book 7, Chapter C3.

Scruggs, G. D., Jr. 1957. Reproduction of resident striped bass in Santee-Cooper reservoir, South Carolina. Transactions of the American Fisheries Society 85:144-159.

Smith, P. E., H. Santander, and J. Alheit. 1989. Comparison of the mortality rates of Pacific sardine and Peruvian anchovy eggs off Peru. Fishery Bulletin, U. S. 87:497-508.

Stevens, R. E. 1958. The striped bass of the Santee-Cooper reservoir. Proceedings of the Annual Conference Southeastern Association of Game and Fish Commissioners 11(1957):253-264.

Stevens, R. E. 1967. A final report on the use of hormones to ovulate striped bass. Proceedings of the Annual Conference Southeastern Association of Game and Fish Commissioners 18(1964):525-538.

Thompson, A. B. 1989. Mackerel egg mortality: the Western mackerel stock in Biscay and the western approaches in 1977, 1980, 1983, and 1986. Journal of Plankton Research 11:1297-1306.

Ulanowicz, R. E., and T. T. Polgar. 1980. Influences of anadromous spawning behavior and optimal environmental conditions upon striped bass *(Morone saxatilis)* year-class success. Canadian Journal of Fisheries and Aquatic Sciences 37:143-154.

Winemiller, K. O., and K. A. Rose. 1992. Patterns of life-history diversification in North American fisheries: implications for population regulation. Canadian Journal of Fisheries and Aquatic Sciences 49:2196-2218.

Zijlstra, J. J., and J. I. Witte. 1985. On the recruitment of 0-group plaice in the North Sea. Netherlands Journal of Zoology 31:360-376.

Effects of Changes in Age-0 Survival and Fishing Mortality on Egg Production of Winter Flounder in Cape Cod Bay

JOHN BOREMAN

*UMass/NOAA CMER Program, Blaisdell House
University of Massachusetts, Amherst, Massachusetts 01003-0820, USA*

STEVEN J. CORREIA AND DAVID B. WITHERELL[1]

*Massachusetts Division of Marine Fisheries
18 Route 6A, Sandwich, Massachusetts 02563, USA*

Abstract.—Eggs-per-recruit methodology was used to examine potential effectiveness of improving habitat quality, versus reducing fishing mortality, for increasing abundance of winter flounder *(Pleuronectes americanus)* in Cape Cod Bay, Massachusetts. The analysis compared the change in potential lifetime fecundity of an age-1 female due to changes in fishing mortality rates versus changes in age-0 survival, presumably due to alteration of the habitat for age-0 life stages. The relative change in the baseline rate of age-0 survival, both positive (due to habitat restoration) and negative (due to habitat loss), was found to approximate the relative change in fishing mortality rate necessary for an age-1 female to attain a specified level of potential egg production during her lifetime. Regulating fisheries to increase stock abundance and yield may represent less risk to managers than undertaking habitat restoration programs, in terms of predicting the outcome. However, undertaking habitat restoration programs should not be dismissed outright, since they may result in longer term benefits to reproductive success and provide managers with a firmer basis to increase fishery yield in the future.

Winter flounder *(Pleuronectes americanus)* is one of the more commercially and recreationally valuable fish species to the northeastern United States (Howe et al. 1976). It has been the focus of numerous environmental impact studies in coastal waters of New England because of its economic importance and apparent susceptibility to poor water quality (Murchelano and Wolke 1985). Early life stages (egg, larva, and juvenile) of inshore populations of winter flounder are susceptible to mortality induced by water withdrawal for power plants and domestic consumption (Hess et al. 1975; Marine Research 1989; Crecco and Howell 1990), by discharge of toxic substances contaminating water and sediments (Nelson et al. 1991), and by physical loss or degradation of habitat (Briggs and O'Connor 1971). Later life stages (age 1 and older) are subjected to high levels of recreational and commercial fishing mortality (NOAA 1991).

In Massachusetts, commercial landings of winter flounder from inshore waters declined 84% between 1978 and 1989 (Witherell et al. 1990). The decline in landings and abundance of inshore stocks of winter flounder in Massachusetts and other coastal states prompted the Atlantic States Marine Fisheries Commission (ASMFC) to develop a coastwide management plan for the species (ASMFC 1992). The goal of the management plan is to reduce fishing- and habitat-induced mortality in order to increase abundance of inshore stocks to sustainable levels in the presence of moderate exploitation. Fishery managers have two options for achieving this goal. Firstly, they can undertake a restoration program by adjusting fishing regulations to reduce fishing-induced mortality. Secondly, they can increase survival by improving habitat quality.

Although data for fish stocks are often inadequate for estimating fishing mortality rates with high precision, devising and implementing fishery regulations to lower fishing mortality are straightforward when political and economic aspects are not considered. Fishing regulations can be adjusted as soon as new information warrants. In contrast, determining how to restore habitat in order to improve survival rates for fish populations is not as

---

[1]Present address: North Pacific Fishery Management Council, Post Office Box 10316, Anchorage, Alaska 99510, USA

easy. Habitat changes which intuitively seem positive, such as reducing organic loading in spawning areas, may decrease larva growth and survival rates. Additionally, separating natural environmental effects on recruitment from effects due to habitat alterations can be technically difficult because of the poorly understood factors that determine year-class size. Effects of altering water and habitat quality are difficult to quantify because estimates of effects may have to be extrapolated from the early life stages, which are already experiencing high rates of natural mortality, to adult age groups. Even if increased mortality of early life stages is the true reason for a stock decline, managers need to determine if, and to what degree, fishing mortality exacerbates this decline.

In his analysis of the implications of power-plant-induced mortality on the striped bass *(Morone saxatilis)* fishery in the Hudson River, Goodyear (1988) described a method for equating effects on early life stages of a fish species to subsequent loss in fishing opportunity. In the method, potential lifetime egg production from a female recruit, or eggs per recruit (EPR), is calculated under a variety of fishing and early life stage mortalities. A closely related method uses spawning stock biomass per recruit as a biological reference point to analyze effects of changes in fishing regulations on subsequent recruitment (Gabriel et al. 1989). In addition to evaluating power plant impacts, the EPR method has also been used to estimate the effectiveness of various options for restoring striped bass populations in Chesapeake Bay (Boreman and Goodyear 1984) and for regulating the striped bass fishery in Rhode Island (Prager et al. 1987).

The EPR method is also useful for evaluating two options for restoring abundance of winter flounder — reducing fishing mortality or reducing habitat-induced mortality. In this paper, we use the EPR method to examine the relative effectiveness of improving habitat quality for early life stages versus reducing fishing mortality to increase abundance of winter flounder inhabiting Cape Cod Bay, Massachusetts.

The Eggs-per-Recruit Method

The total number of eggs that an age-1 female will produce in her lifetime (E) is the sum of the number of eggs she is likely to produce at each spawning age times the probability she will survive to that age. The summation is applied across all ages:

$$E = \sum_{i=2}^{n} \lambda_i \phi_i \prod_{j=1}^{i-1} S_j;$$

$$S_j = e^{-(F_j + M_j + P_j)}$$

λ_i is the proportion of females mature at age i, ϕ_i is the average fecundity of an age-i female, S_j is the survival rate for period j, F_j is the instantaneous rate of fishing mortality during period j, M_j is the instantaneous rate of natural mortality during period j, and P_j is the instantaneous rate of mortality from "other" sources during period j, such as pollution and habitat loss. We assume that $P_j = 0$ for age 1 and older. The maximum E-value (E_{max}) is defined as the lifetime egg production of an age-1 female when $F = 0$ and other sources of mortality are unchanged.

To maintain a constant abundance of females, the product of the lifetime fecundity of an age-1 female (E) and the survival rate from egg to age 1 (S_0) must be equal to 1:

$$E \cdot S_0 = R$$

R = number of age-1 female progeny produced

TABLE 1.—Life history data for female winter flounder in Cape Cod Bay (see text for data sources).

Age	Average length (cm)	Average weight (g)	Fecundity ($\times 10^{-3}$)	Maturity (%)	Vulnerability to F (%)
1	14.2	35	52	0	0
2	19.4	91	144	3	10
3	26.2	230	385	32	26
4	32.6	450	790	86	80
5	36.5	638	1,146	99	100
6	39.5	813	1,485	100	100
7	40.8	899	1,652	100	100
8	42.0	983	1,816	100	100
9	42.9	1,052	1,954	100	100
10	43.6	1,105	2,059	100	100
11	44.1	1,145	2,138	100	100
12	44.5	1,175	2,197	100	100

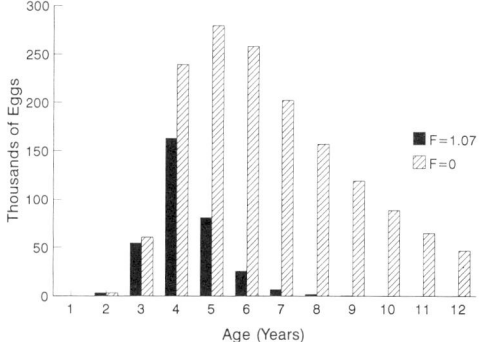

FIGURE 1.—Egg production by age for winter flounder in Cape Cod Bay with and without fishing mortality.

from the lifetime egg production of an age-1 female parent. If $R > 1$, female abundance is increasing, and if $R < 1$ it is decreasing. With this premise, changes in fishing mortality, which affect estimates of E, were tested against changes in survival during age 0, which affect the number of age-1 recruits, to determine what combinations result in equivalent levels of egg production from a cohort, expressed as a percentage of E_{max} (i.e., maintaining $R = 1$).

Data Sources

The winter flounder stock used in our analysis is from Cape Cod Bay, Massachusetts. This stock represents populations from the northernmost of three management units defined in the fishery management plan (ASMFC 1992). In the plan, the instantaneous rate of natural mortality (M) for age-1 and older females in the northernmost management unit is assumed to be 0.35. The average instantaneous rate of fishing mortality (F) for fully recruited individuals in the unit is estimated to be 1.07, based on data collected between 1984 and 1990. Fish begin to recruit to the fishery at age 2 and are fully recruited by age 5 (Table 1).

Age-specific vulnerability to exploitation was derived using a model developed by Correia (1991), which incorporates growth, discard mortality, mesh selectivity, minimum legal size, and age-specific vulnerability to exploitation by recreational hook-and-line and commercial otter-trawl fisheries. The mesh size for the trawl fisheries is assumed to be 14 cm (5.5 inches), which is the minimum size allowed under current regulations.

Female winter flounder in the stock begin to mature at age 2 and are fully mature by age 6 (Witherell, unpublished data). Estimates for age-specific fecundity were derived from length data through a series of conversions. Length-at-age (L_i) data for the population (Table 1), published by Howe and Coates (1975), were converted to weight-at-age i (W_i) using a relationship between length (in mm) and weight (in g) developed for females in the stock (Howe et al. 1990):

$$W_i = (8.13 \times 10^{-6}) L_i^{3.081}$$

More recent data for the stock (Witherell et al. 1990) suggest that the growth rate has not changed significantly from that reported by Howe and Coates (1975).

Weight at age i (in g) was converted to fecundity at age i (Φ_i) using a corrected relationship for the stock published by Topp (1968):

$$\log_{10} \phi_i = 3.0697 + 1.0659 \log_{10} W_i$$

We assume 100% egg viability and hatchability for all spawning ages.

Results and Discussion

With no fishing mortality imposed on winter flounder in Cape Cod Bay ($F = 0$), seven ages account for 90% of the potential lifetime egg production of an age 1 female (Figure 1). Only three age groups contribute to 90% of the lifetime egg production when fishing mortality is set to the current level of $F = 1.07$. The low number of age classes contributing to egg production at the current rate of fishing suggests that the stock is more vulnerable to collapse from a succession of poor year classes than it would be at lower rates of fishing mortality.

The E-value at the currently estimated F level for fully exploited ages of 1.07 is approximately 22% of E_{max} (Figure 2). The management plan for winter flounder (ASMFC 1992) lists a level of 25% of the

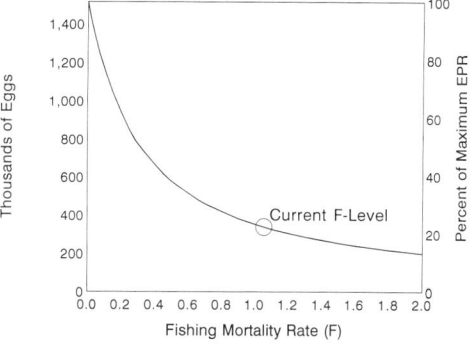

FIGURE 2.—Potential lifetime egg production versus fishing mortality for an age-1 female winter flounder in Cape Cod Bay.

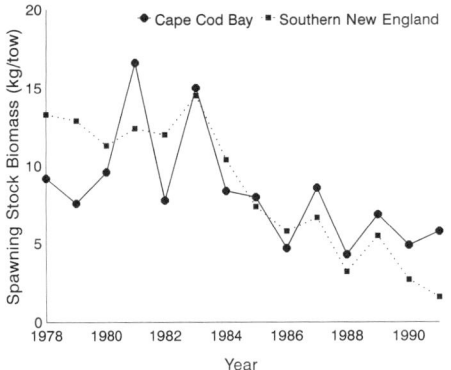

FIGURE 3.—Indices of spawning stock biomass for winter flounder in Cape Cod Bay and southern New England waters, expressed as stratified mean kg/tow, 1978-1991 (from Howe et al. 1992).

maximum spawning potential as a threshold below which year classes of winter flounder do not, on average, achieve replacement ($R < 1$) and a long-term decline in stock abundance is probable. Maximum spawning potential, as defined in the plan, is approximately equivalent to E_{max}, assuming a linear relationship between female body weight and fecundity. In the management plan, the threshold value for spawning potential is based upon analogy with the relationship between percent maximum spawning potential and recruitment for other finfish species of commercial importance. Although spawning stock biomass has declined for winter flounder in Cape Cod Bay (Figure 3), declines in recruitment have yet to be observed; however, declines in recruitment concomitant with declines in spawning stock biomass have been observed for winter flounder in adjacent waters off southern New England (Howe et al. 1992).

The significance of the current level of lifetime eggs-per-recruit cannot be evaluated without supplemental information on stock status and the relationship between stock abundance and recruitment. This limitation does not necessarily prevent managers from formulating various strategies for achieving target E-values based on combinations of reduced fishing mortality and habitat improvements. The strategies then can be evaluated using other criteria, such as the probability of achieving the desired E-level and the economic benefits from the fishery versus anticipated costs. Additionally, general benefits to society which cannot be quantified, such as improving habitat aesthetics or maintaining societal cultures, may be included in the decision process.

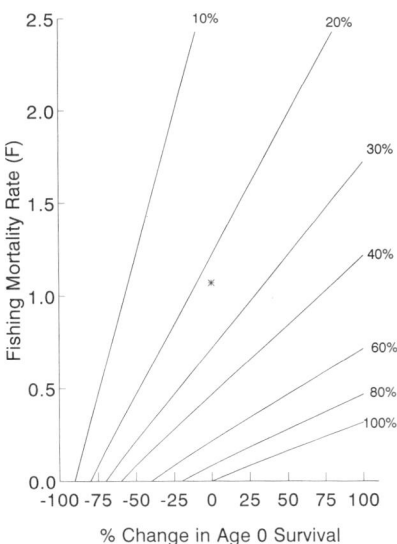

FIGURE 4.—Isopleths representing combinations of changes in fishing mortality and baseline rates of age-0 survival necessary to maintain constant lifetime egg production for female winter flounder in Cape Cod Bay, expressed as percentages of the potential lifetime egg production given the baseline age-0 survival rate and the absence of fishing mortality (asterisk indicates current levels).

Several studies have described pollution-induced reduction in survival of age-0 winter flounder. A study of winter flounder in Long Island Sound indicated that 60% of the egg production may be lost due to contaminant-related mortality (Nelson et al. 1991). Rusanowsky et al. (in press) found that the viable hatch of winter flounder eggs declined almost 80%, from an average 29% to 6%, upon exposure to suspended dredged material from Milford Harbor, Connecticut. As such, halving or doubling age-0 survival to reflect possible conditions induced by habitat degradation or restoration is within the range of possibility. A reduction of 50% in age-0 survival of winter flounder from Cape Cod Bay, perhaps caused by habitat degradation, is equivalent to holding age-0 survival constant and increasing the current fishing mortality rate 2.3-fold, from $F = 1.07$ to $F = 2.44$, in terms of changes to egg production (Figure 4). Conversely, doubling age-0 survival, perhaps through a habitat restoration program, would yield the same egg production as holding age-0 survival constant and reducing fishing mortality 63%, from $F = 1.07$ to $F = 0.40$.

The apparent conclusion that doubling survival of age-0 winter flounder through restoring habitat would be more effective than halving fishing mortality rates for winter flounder in Cape Cod Bay is potentially misleading for several reasons. Effects of habitat improvement on survival of winter flounder is undocumented, as is the type of habitat improvement that would be necessary to increase survival by a desired amount. Effects of changing fishing mortality on stock abundance are better understood. Habitat improvement is a gradual process that can extend from a period of years to decades; fishing mortality rates can be adjusted as soon as new regulations can be imposed, resulting in almost immediate impacts on spawner abundance. Improving habitat to increase survival of age-0 winter flounder introduces the necessity to understand the little known, but potentially overriding influence of annual variation in environmental conditions (temperature, salinity, turbidity, abundance of predators and prey, availability of habitat) on population regulation. Directly assessing fishery-level effects that would be obtained by improving a specific estuarine habitat remains difficult because many inshore flounder populations may contribute to a single fishery (Saila 1961; Howe and Coates 1975).

The EPR method compares changes in potential egg production due to losses of age-0 fish to equivalent losses to age-1 and older individuals. As such, it gives managers a means to compare potential effects of habitat loss or gain with increases or decreases in fishing mortality. The management plan for winter flounder (ASMFC 1991) contains evidence suggesting that inshore stocks are declining in abundance due to both excessive fishing mortality and habitat degradation. Instantaneous rates of fishing mortality for winter flounder stocks covered in the plan (Gulf of Maine to Delaware) vary from $F = 0.9$ to $F = 1.1$, which are more than double the plan's target rates ($F = 0.4$-0.5). The average loss from dredging or filling of shallow water acreage along the coast from Massachusetts to New Jersey, considered prime nursery habitat for winter flounder, was 6 to 15% per state from 1954 to 1968 (Gusey 1978).

The EPR method, as used in our analysis, assumes that mortality mechanisms only act on the stock in a density-independent fashion, and does not consider the effects of natural variation in the physical and biological environments. Other assumptions which may or may not be realistic include constant relationships between age and growth rate, fecundity, and maturity of females. Density-dependent effects have been observed in other winter flounder populations. Kennedy and Steele (1971) found evidence of cannibalism in a Newfoundland population, and Tyler and Dunn (1976) found that winter flounder sacrificed egg production to maintain body weight when food was scarce. Growth and developmental rates for larval winter flounder have been linked to temperature (Laurence 1975).

As stock size increases and effects of density possibly become significant, the gain in recruits per unit of habitat improvement will probably decrease. However, this assumption also applies to the methodologies used in developing fishery reference points and stock rebuilding schedules in the management plan (yield per recruit and spawning stock biomass per recruit; ASMFC 1992). The relationship between fishing mortality and egg production will change when density-dependent mechanisms become active at higher levels of abundance; recruit production per spawner will likely decline as spawning stock increases. Both the eggs-per-recruit methodology and the methodologies used in the management plan may have to be modified as information concerning spawner and recruitment relationships, interspecific and intraspecific interactions, and the role of environmental factors in determining year class size becomes available.

Incorporating spawner and recruitment relationships into the EPR methodology may be difficult for several reasons. The relationship between spawning-stock size and recruitment for the population in Cape Cod Bay remains unclear. Inability of stock and recruitment data to fit any of the common stock-recruitment models can be attributed to: random environmental effects, unknown mechanisms of population regulation, lack of knowledge concerning community dynamics, and the inappropriateness of the theoretical framework of the models (Fletcher and Deriso 1988). Even if the relationship between stock and recruitment were known, habitat improvement may alter it. Habitat improvement could affect population dynamics in at least two ways: by creating new suitable habitat or by increasing the carrying capacity of existing habitat. Both ways may increase recruitment but differentially affect the stock-recruitment relationship. In creating new suitable habitat, the range of stock densities over which density-dependent mechanisms have distinguishable influence remains the same, but these mechanisms would occur at larger stock abundance levels. In contrast, increasing the carrying capacity per unit of habitat may increase survival rates, and may fundamentally alter the stock-recruitment relationship by delaying the onset of density-dependent effects.

Most habitat areas needing improvement may now serve as nursery habitat for pre-adult winter flounder and spawning areas for adults. We assume that restoration of nursery habitat will increase the number of recruits to age 1 and older because growth in abundance is limited by a carrying-capacity bottleneck occurring sometime during the pre-adult life stages. This assumption is reasonable for a long-lived, highly fecund species like winter flounder that exhibits no parental care (Pitcher and Hart 1982). An increase in carrying capacity for age-0 fish, combined with reduced fishing mortality on age-1 and older individuals, may shift the carrying-capacity bottleneck to the adult ages and simultaneously reduce recruitment gains through habitat improvement. Conversely, reductions in fishing mortality may increase yield per recruit and egg production per recruit, but not increase the number of recruits, if the bottleneck to growth in a population of winter flounder remains during the age-0 stages.

Acknowledgments

Penny Howell, Connecticut Department of Environmental Protection, Wendy Gabriel and Frank Almeida, Northeast Fisheries Science Center, National Marine Fisheries Service, and Arnold Howe, Massachusetts Division of Marine Fisheries, provided many useful comments on earlier versions of the paper. We also wish to thank two anonymous reviewers for their helpful comments and suggestions.

References

ASMFC (Atlantic States Marine Fisheries Commission). 1992. Interstate management plan for inshore stocks of winter flounder *Pseudopleuronectes americanus*. Washington, District of Columbia.

Boreman, J., and C. P. Goodyear. 1984. Effects of fishing on the reproductive capacity of striped bass in Chesapeake Bay, Maryland. National Marine Fisheries Service, Northeast Fisheries Center, Woods Hole Laboratory Document 84-29.

Briggs, P. T., and J. S. O'Connor. 1971. Comparison of shore-zone fishes over naturally vegetated and sand-filled bottoms in Great South Bay. New York Fish and Game Journal 18:15-41.

Correia, S. J. 1991. BIOREF: a model to estimate the effects of discard mortality on biological reference points. Papers of the Northeast Regional Stock Assessment Workshop, Research Document SAW13/2. US Department of Commerce, National Marine Fisheries Service, Northeast Fisheries Science Center, Woods Hole, Massachusetts.

Crecco, V., and P. Howell. 1990. Potential effects of current larval entrainment mortality from the Millstone nuclear power station on the winter flounder, *Pseudopleuronectes americanus*, spawning population in the Niantic River. Connecticut Department of Environmental Protection, Bureau of Fish and Wildlife, Marine Fisheries Division, Waterford.

Fletcher, R. I., and R. B. Deriso. 1988. Fishing in dangerous waters: remarks on a controversial appeal to spawner-recruit theory for long-term impact assessment. American Fisheries Society Monograph 4:232-244.

Gabriel, W. L., M. P. Sissenwine, and W. J. Overholtz. 1989. Analysis of spawning stock biomass per recruit: an example for Georges Bank haddock. North American Journal of Fisheries Management 9:383-391.

Goodyear, C. P. 1988. Implications of power plant mortality for management of the Hudson River striped bass fishery. American Fisheries Society Monograph 4:245-254.

Gusey, W. 1978. The fish and wildlife resources of the Middle Atlantic Bight. Shell Oil Company, Houston, Texas, 582 pp.

Hess, K. W., M. P. Sissenwine, and S. B. Saila. 1975. Simulating the impact of the entrainment of winter flounder larvae. Pages 1-29 *in* S. B. Saila, editor. Fisheries and energy production: a symposium. Lexington Books, D. C. Heath and Company, Lexington, Massachusetts.

Howe, A. B., and P. G. Coates. 1975. Winter flounder movements, growth, and mortality off Massachusetts. Transactions of the American Fisheries Society 104:13-29.

Howe, A. B., P. Coates, and D. Pierce. 1976. Winter flounder estuarine year class abundance, mortality, and recruitment. Transactions of the American Fisheries Society 105:647-657.

Howe, A. B., T. P. Currier, S. J. Correia, and D. B. Witherell. 1990. Massachusetts fishery resource assessment. Annual Report 1990, Project Number 3-IJ-3-2. Massachusetts Division of Marine Fisheries, Sandwich,

Massachusetts.

Howe, A. B., T. P. Currier, S. J. Correia, and D. B. Witherell. 1992. Massachusetts fishery resource assessment. Job Performance Report, Project Number F-56-R-1, Massachusetts Division of Marine Fisheries, Sandwich, Massachusetts.

Kennedy, V. S., and D. H. Steele. 1971. The winter flounder *(Pseudopleuronectes americanus)* in Long Pond, Conception Bay, Newfoundland. Journal of the Fisheries Research Board of Canada 28:1153-1165.

Laurence, G. C. 1975. Laboratory growth and metabolism of winter flounder *(Pseudopleuronectes americanus)* from hatching through metamorphosis at three temperatures. Marine Biology 32:223-229.

Marine Research, Inc. 1989. Ichthyoplankton entrainment monitoring at Pilgrim Nuclear Power Station. Report to Boston Edison Company, Inc., Boston, Massachusetts.

Murchelano, R. A., and R. E. Wolke. 1985. Epizootic carcinoma in the winter flounder *Pseudopleuronectes americanus*. Science 228:587-589.

Nelson, D. A., and nine co-authors. 1991. Comparative reproductive success of winter flounder in Long Island Sound: a three-year study (biology, biochemistry, and chemistry). Estuaries 14:318-331.

NOAA (National Oceanic and Atmospheric Administration). 1990. A three-year assessment of reproductive success in winter flounder in Long Island Sound, with comparisons to Boston Harbor, 1986-1988. National Marine Fisheries Service, Northeast Fisheries Center, Milford Laboratory Report prepared for the U. S. Environmental Protection Agency, Long Island Sound Project, Boston, Massachusetts.

NOAA (National Oceanic and Atmospheric Administration). 1991. Status of the fishery resources off the northeastern United States. Conservation and Utilization Division, Northeast Fisheries Center. NOAA Technical Memorandum NMFS-F/NEC-86.

Pitcher, T. J., and P. J. B. Hart. 1982. Fisheries ecology. The AVI Publishing Company, Westport, Connecticut.

Prager, M. H., J. F. O'Brien, and S. B. Saila. 1987. Using lifetime fecundity to compare management strategies: a case history for striped bass. North American Journal of Fisheries Management 7:403-409.

Rusanowsky, D., D. Nelson, and M. Ludwig. In press. Impacts of dredging: a winter flounder case study. NOAA Technical Memorandum.

Saila, S. B. 1961. A study of winter flounder movements. Limnology and Oceanography 6:292-298.

Topp, R. W. 1968. An estimate of fecundity of the winter flounder. Journal of the Fisheries Research Board of Canada 25:1299-1302.

Tyler, A., and R. Dunn. 1976. Ration, growth, and measures of somatic and organ condition in relation to meal frequency in winter flounder, *Pseudopleuronectes americanus*, with hypotheses regarding population homeostasis. Journal of the Fisheries Research Board of Canada 33:63-75.

Witherell, D. B., S. J. Correia, A. B. Howe, and T. P. Currier. 1990. Stock assessment of winter flounder in Massachusetts waters. Massachusetts Division of Marine Fisheries, Sandwich, Massachusetts.

Pathological Conditions of Narragansett Bay Young-of-the-Year Winter Flounder

SHARON A. MACLEAN

National Oceanic and Atmospheric Administration, National Marine Fisheries Service
28 Tarzwell Drive, Narragansett, Rhode Island 02882-1199, USA

Abstract.—A total of 107 young-of-the-year winter flounder *(Pleuronectes americanus)* was collected in June and August from three sites representing a pollution gradient in Narragansett Bay. Three- to 5-month-old fish were afflicted with a variety of parasites and pathologic conditions. A few conditions, including gastritis, inflammatory heart conditions, and the only occurrence of gill bifurcation, appeared only at Gaspee Point, the most contaminated site sampled in the bay. The microsporidan *Glugea* was found in only one fish from the least contaminated site. Some conditions appeared more frequently in the older fish than in the younger ones; trematode metacercaria *(Cryptocotyle)* infections increased at two stations from 20% in June to 100% in August, while infection intensities increased from 1 to 2 cysts of metacercaria per fish in June to 188 cysts per fish in August (it is notable that metacercariae were not found on fish from the most polluted site); epitheliocystis-like lesions in gills increased from 2% in June to 59% in August; gill hyperplasia and fusion increased from 24% in June to 41% in August. This study reports on melanized fibrosis in the liver and parasites (coccidians and myxosporidans) not previously recorded in winter flounder. These conditions are discussed relative to year-class success.

Despite the interest in winter flounder biology, most reports of parasites and pathologic conditions of this species have been concerned with adults or late juveniles (see Klein-MacPhee 1978 for review). Some reports on parasites include examinations of young-of-the-year (YOY or age-0) fish; for example, studies of *Glugea stephani*, a microsporidan of flatfishes, in winter flounder of the New York-New Jersey area and of Martha's Vineyard included data on age-0 flounder (Takvorian and Cali 1984; Stunkard and Lux 1965). Frequently, data on grossly obvious lesions are reported as an aside to the focus of the research. In a study of growth and biochemical composition of age-0 winter flounder, Buckley and Caldarone (1988) found nearly 30% *Glugea* infection in Narragansett Bay flounder by October. Similarly, data collected during a comparative ecology study of winter and smooth flounders *(Pleuronectes putnami)* in Great Bay, New Hampshire, showed 65% of age-0 winter flounder infected with *Glugea* (M. P. Armstrong, University of New Hampshire, personal communication).

Although studies of specific parasites and other studies of fishes may include age-0 fishes, emphasis rarely is placed on the pathologic conditions specifically of that age class. The present study was undertaken to examine pathologic conditions occurring in age-0 winter flounder collected at sites considered to represent a pollution gradient in Narragansett Bay. Information is presented on previously unreported conditions occurring in young winter flounder, some of which may be important in year-class success.

Methods

Winter flounder YOY were collected during June and August of 1990 from three sites (Figure 1) representative of the pollution gradient in Narragansett Bay (Santschi et al. 1984; Pruell and Quinn 1985). The sites were Fogland on the eastern side of the Bay (least contaminated), Prudence-Patience Islands located mid-bay in the West Passage (moderately contaminated), and Gaspee Point at the mouth of the Providence River (most contaminated). Based on typical spawning times in Narragansett Bay and size of YOYs collected, these fish were considered to be 3 (June collection) and 5 (August collection) months old (T. Halavik, National Marine Fisheries Service, personal communication).

Fish were collected either by beach seine or by small dipnet while snorkeling. Temperature and salinity data were recorded during each collection except salinity for August collections at Fogland

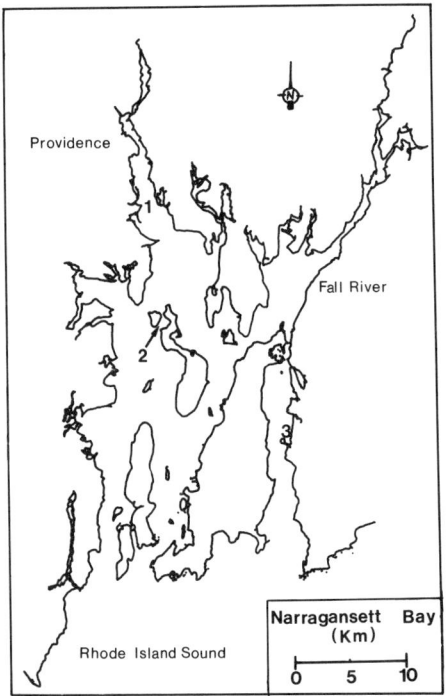

FIGURE 1.—Young-of-the-year winter flounder sampling sites within Narragansett Bay, Rhode Island. 1, Gaspee Point; 2, Prudence-Patience Islands; 3, Fogland.

and Prudence-Patience. Attempts to collect 30 fish during each sampling were unsuccessful. Table 1 presents the collection data.

Specimens were preserved in 10% neutral-buffered formalin and examined grossly and dissected in the laboratory under a stereo microscope. Excised tissues were processed by standard histologic procedures, embedded in paraffin, sectioned at 6 µm and stained with hematoxylin and eosin (Humason 1967). Two slides of each block of tissues were examined, representing a section of the outer portion of tissues and an advancement of 150 µm into the tissue. Several tissues were placed in one block depending on the size of tissue pieces. Various tissues, therefore, were not all on the same plane and consequently, some tissues, particularly kidney, were sectioned through before reaching the appropriate level for taking sections of other tissues. As a result, the number of tissues examined is not always the same as the number of animals sampled.

Results

External examination of the YOYs revealed a trematode *(Cryptocotyle)* metacercarial infection as the only gross abnormality. Metacercariae, commonly referred to as "pepperspots," were found in 21% of Fogland and 13% of Prudence-Patience winter flounder collected in June (Table 1). This prevalence increased to 97% and 100%, respectively, by August. Similarly, intensity of infection increased dramatically from 1 to 2 metacercariae per fish in June to 4 to 188 per fish (mean: 34) in August. Metacercariae were not observed in fish

TABLE 1.—Collection data and prevalence of trematode metacercariae in young-of-the-year winter flounder taken from Narragansett Bay in 1990 (TL = total length of fish).

Site/Date	N	Mean TL (mm) (range)	Bottom temperature (°C)	Salinity (‰)	Fish with metacercariae number	%
Fogland						
15 Jun	29	38 (29-47)	19.1	29	6	21
24 Aug	32	56 (41-78)	21.3	-	31	97
Prudence-Patience						
15 Jun	8	40 (32-49)	18.5	25	1	13
22 Aug	10	70 (53-84)	22.1	-	10	100
Gaspee Point						
15 Jun	19	42 (33-51)	17.5	16	0	0
28 Aug	9	78 (62-85)	23.5	23	0	0

TABLE 2.—Histopathologic conditions observed in 107 young-of-the-year winter flounder collected in Narragansett Bay, June and August 1990.

Tissue	Conditions
Liver	Vasculitis Hepatitis Melanized periductular/perivascular fibrosis Myxosporidiosis *(Kudoa)*
Kidney	Myxosporidiosis (bivalvulate) Cytopenia of hematopoietic tissue Atrophic glomeruli
Intestine	Trematodiasis (adult) Enteritis Coccidiosis (eimeriid) Microsporidiosis *(Glugea)* Myxosporidiosis *(Kudoa)* Nematodiasis
Stomach	Gastritis Myxosporidiosis *(Kudoa)*
Gill	Epitheliocystis-like Trematodiasis (larval) Hyperplasia Fusion Bifurcation Myxosporidiosis *(Kudoa)*
Heart	Epicarditis Myocarditis

TABLE 3.—Pathologic conditions observed in the intestine and liver of young-of-the-year winter flounder collected in Narragansett Bay, 1990. Values are frequencies with percentages in parentheses.

Date/Site	N	Intestine			Liver		
		Adult trematodes	Enteritis	Coccidiosis	Vasculitis	Hepatitis	Melanized fibrosis
15 Jun							
Fogland	29	16 (55)	2 (7)	1 (3)	5 (17)	0	0
Prudence - Patience	8	7 (88)	3 (38)	0	0	0	0
Gaspee Point	19	15 (79)	6 (32)	3 (16)	4 (21)	0	1 (5)
Total	56	38 (68)	11 (20)	4 (7)	9 (16)	0	1 (2)
22-28 Aug							
Fogland	32	14 (44)	6 (19)	5 (16)	2 (6)	2 (6)	0
Prudence - Patience	10	5 (50)	2 (20)	0	1 (10)	1 (10)	3 (30)
Gaspee Point	9	5 (56)	2 (22)	0	0	2 (22)	7 (78)
Total	51	24 (47)	10 (20)	5 (10)	3 (6)	5 (10)	10 (20)

collected at Gaspee Point either in June or in August. A single case of *Glugea* infection occurred in a fish from Fogland.

Microscopic examination of various tissues revealed numerous parasites and histopathologic conditions (Table 2). Not all conditions occurred at all sites although some did, such as hepatitis, adult trematodiasis, enteritis, epitheliocystis-like lesions, gill hyperplasia and fusion, and myxosporidiosis *(Kudoa)* in muscle. Some lesions occurred at both the least and most contaminated sites (for example, hepatic vasculitis and intestinal coccidiosis) and some occurred at the least and moderately contaminated sites (larval trematodiasis [metacercariae] and cytopenia of kidney hematopoietic tissue). Gill bifurcation, gastritis, and the cardiac lesions were found in fish from Gaspee Point only.

Numerically important lesions observed in various tissues sampled from each site and sampling date are presented in Tables 3 and 4. Histopathologic conditions noted in the intestine did not appear related to environmental quality of the sites or to the age of the fish (Table 3). Adult trematodes in the gut lumen were common at all sites ($\geq 44\%$) and at both sampling times. In June, enteritis, inflammation of the gut, appeared to be more prevalent among Prudence-Patience and Gaspee Point fish than YOYs from Fogland, but by August the occurrence of enteritis was comparable at all three sites. Sporulated coccidian oocysts (Figure 2), resembling *Goussia*, were found in the mucosa of the gut of fish from Fogland and Gaspee Point. The intestinal mucosa of infected fish appeared disrupted, particularly in areas of heavy infection. No oocysts were observed in the 18 fish examined from Prudence-Patience islands.

The most commonly observed liver lesions included vasculitis, hepatitis, and melanized periductular and perivascular fibrosis (Table 3). Vasculitis, inflammation of blood vessels, appeared more prevalent in June than in August at Fogland and Gaspee Point, whereas hepatitis was observed only in the 5-month-old fish at all three sites. In general, inflammation of the blood vessels and among hepatocytes was not extensive. Periductular and perivascular fibrosis and melanization (Figure 3) occurred in fishes from Prudence-Patience and Gaspee Point but not from Fogland, and was particularly prevalent among the older juveniles at Gaspee Point (78%).

Among the gill lesions (Table 4), only infection with metacercariae shows any relationship to site of collection and reflects the geographic distribution of grossly observed metacercariae. Therefore, no metacercariae were found in gills of fish from Gaspee Point and prevalences in fish from Fogland and Prudence-Patience increased from June to August. Metacercariae elicited a marked host response characterized by melanization of encapsulating host tissue. When the site of infection was at the base of a gill lamella or in the filament near the base of the lamellae, the physical presence of the metacercaria and the host encapsulation response frequently resulted in fusion of the lamellae. Large epitheliocystis-like lesions, caused by chlamydial or rickettsial infection, also resulted in fusion of lamellae (Figure 4). The lesions were not extensive and ranged in number from 1 to 6 per gill section. Hyperplasia of gill epithelium occurred commonly in fish at all sites. Seventy five percent of hyperpla-

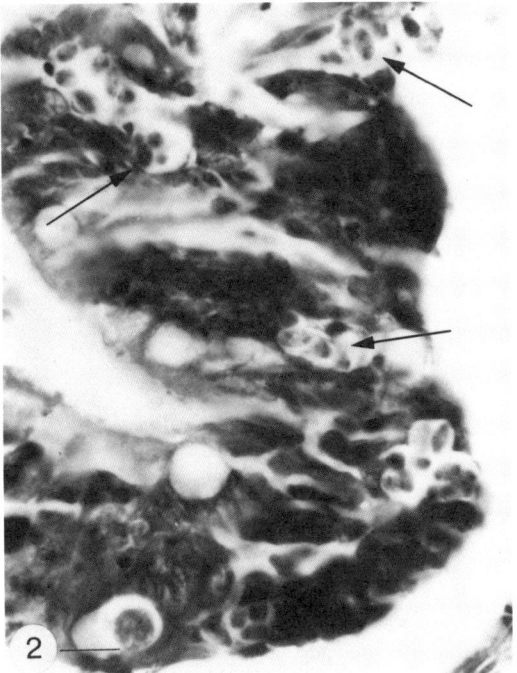

FIGURE 2.—Sporulated oocysts (arrows) of coccidian parasites in the intestinal mucosa of age-0 winter flounder. Scale = 10 μm.

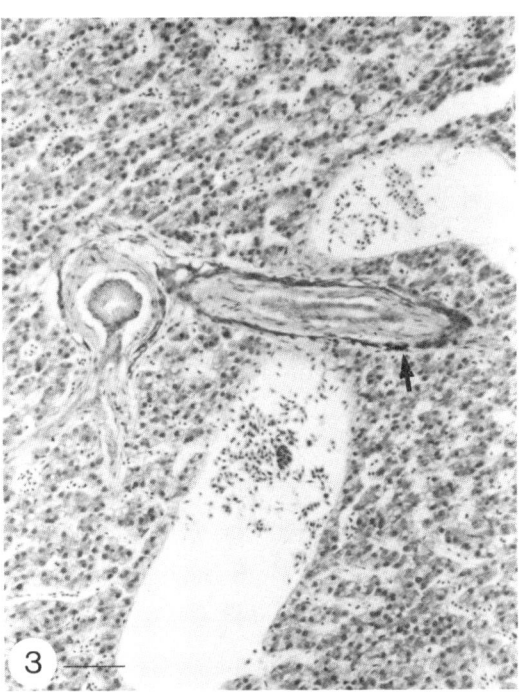

FIGURE 3.—Dark area around bile ducts (arrow) is melanized fibrotic tissue in the liver of age-0 winter flounder. Scale = 50 μm.

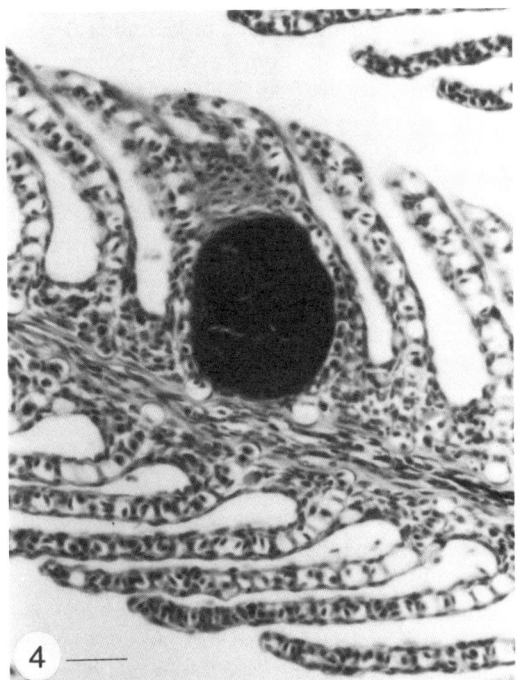

FIGURE 4.—A large, densely staining epitheliocystis-like lesion has caused fusion of adjacent gill lamellae. Normal lamellae are below the lesion. Scale = 25 μm.

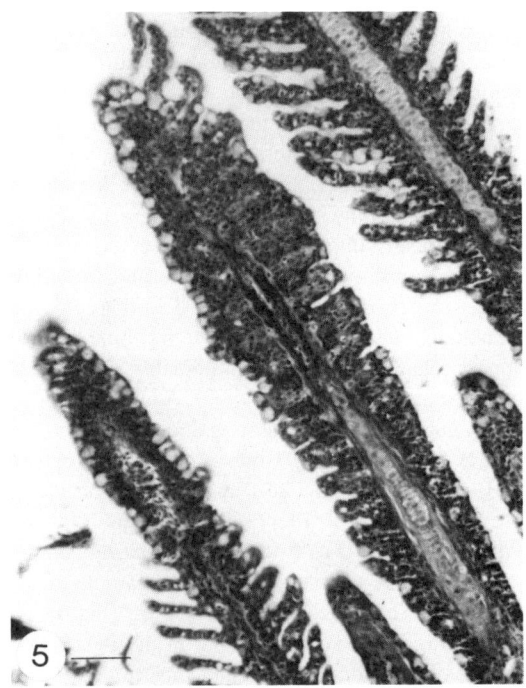

FIGURE 5.—Hyperplasia of the gill epithelium has resulted in the fusion of many lamellae. Normal lamellae are to the lower left. Scale = 50 μm.

TABLE 4.—Pathologic conditions observed in the gill of young-of-the-year winter flounder collected in Narragansett Bay, 1990. Values are frequencies with percentages in parentheses.

Date/Site	N	Epitheliocystis-like lesions	Trematode metacercariae	Epithelial hyperplasia and fusion
15 Jun				
Fogland	28	0	2 (7)	2 (7)
Prudence - Patience	8	0	1 (13)	6 (75)
Gaspee Point	19	1 (5)	0	5 (26)
Total	55	1 (2)	3 (5)	13 (24)
22-28 Aug				
Fogland	32	16 (50)	7 (22)	13 (40)
Prudence - Patience	10	8 (80)	1 (10)	2 (20)
Gaspee Point	9	6 (67)	0	6 (67)
Total	51	30 (59)	8 (16)	21 (41)

sia events were so extensive as to result in lamellar fusion (Figure 5), usually encompassing large areas of the filament. As did metacercarial infection, the latter two lesions increased in prevalence from June to August; from 2% to 59% for epitheliocystis-like lesions and from 24% to 41% for lamellar hyperplasia. Bifurcation of the gill filament was observed in one fish collected from Gaspee Point.

Lesions of the heart were found only in fish collected from Gaspee Point. In June, four of 19 (21%) winter flounder had epicarditis, inflammation of the outer epicardial layer of the heart, and one had myocarditis, inflammation of the heart muscle, in the absence of apparent bacterial or parasitic infection. These lesions were not seen in fish from other sites or in Gaspee Point fish collected in August.

Two types of myxosporidan parasites were observed in this study (Table 5). Presporogonic, sporogonic, and spore stages of a *Kudoa* infected smooth muscle of the alimentary tract, particularly of the esophagus and stomach, and, rarely, blood vessels in the liver. In some cases extensive areas of esophageal and gastric muscle were replaced by parasites (Figure 6) and no host response was evident. Large areas of hypobranchial muscle at the attachment site of the gill arch also were infected with this parasite and were often seen in sections of gill tissue. This organism occurred at all three sites, but only at Prudence-Patience during both sampling times. The second was a bivalvulate myxosporidan that was observed in sporogonic and spore stages in the lumina of kidney tubules. This was common in June at both Fogland (83%) and Gaspee Point (57%). The lumina of tubules were partially occluded and in most cases the tubule epithelium was affected such that mitotic figures, evidence of repair, were frequently seen in the epithelial cells (Figure 7).

Discussion

Most pathologic conditions found in this study occurred at all sites or at the least (Fogland) and most (Gaspee Point) contaminated sites, indicating little association between the occurrence of the lesions and contaminant exposure. A few conditions of low prevalences occurred only at Gaspee Point, for example, gastritis, gill bifurcation, and inflammatory heart conditions. Although gastritis is not uncommon in adult fish, based on this study it appears unusual in age-0 winter flounder. Several parasites and bacterial infections cause gastritis but none of these were observed in this study. Gill bifurcations in late juvenile winter flounder have been correlated with residence in contaminated environments (Pereira et al. 1992). Although the observation of one bifurcated gill in this study does not substantiate the findings of Pereira et al., it is notable that the bifurcation was observed in a fish sampled from the most contaminated site. Twenty one percent of the Gaspee Point YOYs in June had epicarditis, a lesion frequently due to parasitic or bacterial infection in fishes (Ferguson 1989). No infectious agents were observed in these preparations, therefore the cause of the lesion remains uncertain. Fibrosis, as seen in the hepatic periductular and perivascular lesions of Gaspee Point and Prudence-Patience fishes, is a non-specific response to injury or insult to the tissues (Cawson et al.

TABLE 5.—Myxosporidans in Narragansett Bay young-of-the-year winter flounder, 1990. (N = sample size).

Date/Site	*Kudoa* in muscle			Myxosporidan in kidney tubules		
	N	Frequency	%	N	Frequency	%
Fogland						
15 Jun	29	0	0	12	10	83
24 Aug	32	3	9	13	2	15
Prudence - Patience						
15 Jun	8	2	25	3	0	0
22 Aug	10	1	10	4	0	0
Gaspee Point						
15 Jun	19	2	10	7	4	57
28 Aug	9	0	0	4	0	0

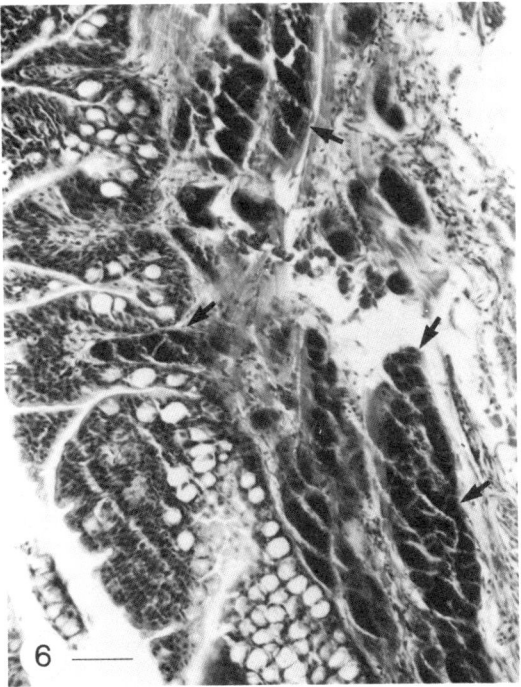
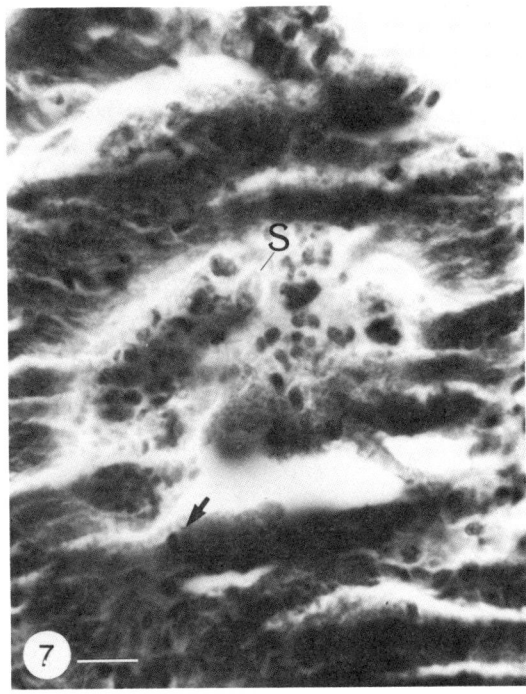

FIGURE 6.—Presporogonic and sporogonic stages of the myxosporidan *Kudoa* (arrows) infecting large areas of smooth muscle in the esophagus. Scale = 50 μm.

FIGURE 7.—Sporogonic and spore (S) stages of a bivalvulate myxosporidan in the lumen of a kidney tubule. Mitotic figures (arrow) in the epithelium were common in many affected tubules. Scale = 10 μm.

1989) which may be caused by pathogens or toxicants. Deposition of melanin is not unusual in fishes (all trematode metacercariae were surrounded by melanin), however, the site of deposition in these fish is rather unusual (personal observation). The location of these lesions, around the blood vessels and bile ducts, in the absence of pathogens suggests that toxicants may be causal. The fact that these lesions, gastritis, gill bifurcation, melanized periductular/perivascular fibrosis, and cardiac lesions, were found in Gaspee Point fish ($N = 28$) and not in the more heavily sampled yet least contaminant-exposed Fogland fish ($N = 61$), suggests that a more detailed study of YOYs from contaminated areas may warranted. Recent work by Saucerman and Deegan (1991) demonstrated little movement during the summer of age-0 winter flounder from the habitat in which they settle. Therefore, although the data collected in this study are not statistically valid, the lesions observed in YOYs, particularly those that occurred at Gaspee Point, should be considered unique to the site.

The effect of parasites on larval and early juvenile fishes may be quite serious as suggested by Rosenthal (1967) who attributed 10% of larval her- ring mortalities to infection by parasites acquired by feeding on wild plankton. Burreson and Zwerner (1984) showed 100% mortality among age-1 summer flounder *(Paralichthys dentatus)* experimentally infected with the blood parasite *Trypanoplasma bullocki*, particularly when held in cold water, and concluded that mortalities observed concurrently in infected field-caught juvenile flounder were due to the parasite and low ambient water temperature. Several studies on *Glugea stephani* in flatfish have shown age-0 and age-1 fish to be particularly susceptible to infection and that they experience high mortalities (Stunkard and Lux 1965; McVicar 1975; Cali et al. 1986).

Previously unrecorded parasites (myxoporidans and coccidians) for winter flounder were found in this study of age-0 fish and some of these may have the potential to affect year-class success. Both the kidney-inhabiting and muscle myxosporidans have obvious effects on their hosts, such as kidney tubule occlusion, disruption of tubule epithelium, and destruction and replacement of muscle tissue. Compensatory mechanisms may occur, such as repair of tubule epithelium by increased mitosis, however, regeneration of large areas of destroyed muscle tissue was not evident and is unlikely. Since no tissue response to control the spread of the infection was apparent, it is possible the YOYs infected with the *Kudoa* may succumb to the disease. Myxosporidans of the genera *Myxidium* and *Ceratomyxa* have been reported by Mavor (1915) in the gall bladder of adult winter flounder. Their distinctive spore morphologies exclude the possibility that the kidney myxosporidan of this study is one of them. The kidney myxosporidan and the *Kudoa* species observed in the musculature of age-0 winter flounder are believed to be previously unreported parasites. Although coccidians are very common in coastal marine fishes, intestinal coccidiosis has not been reported from winter flounder. Coccidiosis in adult fishes appears to be benign (Lom 1984), but nothing is known of the effect on age-0 fish. It is conceivable that the effect may be quite deleterious, particularly if the intensity of infection is severe and large areas of intestinal mucosa are rendered non-functional due to tissue disruption.

In regards to *Glugea* infection, discrepancies in the results of this study and those of Buckley and Caldarone (1988) probably are based on the fact that the observations of Buckley and Caldarone were made on age-0 winter flounder collected in October (estimated to be 7 months old), whereas this study examined fish collected in June (estimated to be 3 months old) and in August (estimated to be 5 months old). Water temperature is critical in establishing the initial *Glugea* infection and in the growth rate of the lesions (McVicar 1975; Olson 1976; Takvorian and Cali 1984). Longer term exposure to warm water temperatures and longer time for parasite growth in 7 month old fish may account for the higher prevalence of *Glugea* observed by Buckley and Caldarone versus the very low prevalence found in this study. Recent work on the comparative ecology of smooth and winter flounders in Great Bay, New Hampshire revealed a *Glugea* infection rate of 65% in age-0 winter flounder, as compared to 3% in age-0 smooth flounder (M. P. Armstrong, University of New Hampshire, personal communication). Stomach content analysis showed the winter flounder fed primarily on amphipods, whereas smooth flounder fed on polychaetes, and occupied a less saline habitat than the winter flounder (M. P. Armstrong, personal communication). This suggests that prey items and habitat may play roles in the infection dynamics of *Glugea* in flatfish.

Although Sindermann and Rosenfield (1954) demonstrated 100% mortality within 15 days of age-0 Atlantic herring *(Clupea harengus)* being massively infected with *Cryptocotyle* metacercariae, massive infections in the wild seem unlikely. Nearly 100% of the age-0 winter flounder in this study were infected with metacercariae by June of the year, however, the mean number of parasites per fish (34) was unlikely to cause mortality. The absence of trematode metacercariae in age-0 winter flounder collected at Gaspee Point (Buckley and Caldarone 1988, and this study) may be due to the lack of intermediate hosts (periwinkle and gull) required to complete the life cycle of the trematode *Cryptocotyle*. It has been suggested that water temperature above 21°C in summer restricts habitation of an area by *Littorina* (Wells 1965), one of the intermediate hosts of *Cryptocotyle*, and that absence in the area of gulls, the final host of the trematode, restricts infection of snails by the parasite (Murray 1980). Both of these conditions may be met at Gaspee Point. Therefore, ecologic factors, rather than pollution effects, may have substantial roles in the distribution of these parasites in Narragansett Bay. Khan et al. (1992) found the highest prevalence of metacercarial infection in winter flounder collected at a polluted pulp and paper mill site as compared to the control site. Pollution, then, at least by pulp mill effluent, seems to have little effect on metacercarial infections. All this suggests that caution is necessary in interpreting results of pollution studies which include parasites with other than a direct life cycle.

Acknowledgments

My appreciation goes to Tom Halavik, Peter Long, and Jerome Prezioso for help in field collections, Dorothy Howard for histologic slide preparation, and Sheila Polofsky for help with graphics.

References

Buckley, L. J., and E. M. Caldarone. 1988. Recent growth and biochemical composition of juvenile, young-of-year winter flounder from different areas of Narragansett Bay. Final Report to the Narragansett Bay Project, 291 Promenade Street, Providence, Rhode Island, 02908.

Burreson, E. M., and D. E. Zwerner. 1984. Juvenile summer flounder, *Paralichthys dentatus*, mortalities in the western Atlantic Ocean caused by the hemoflagellate *Trypanoplasma bullocki*:

evidence from field and experimental studies. Helgoländer Meeresuntersuchungen 37:343-352.

Cali, A., P. M. Takvorian, J. J. Ziskowski, and T. K. Sawyer. 1986. Experimental infection of American winter flounder *(Pseudopleuronectes americanus)* with *Glugea stephani* (Microsporida). Journal of Fish Biology 28:199-206.

Cawson, R. A., A. W. McCracken, P. B. Marcus, and G. S. Zaatari. 1989. Pathology: the mechanisms of disease. The C.V. Mosby Company, St. Louis, Missouri.

Ferguson, H. W. 1989. Systemic Pathology of Fish. Iowa State University Press, Ames, Iowa.

Humason, G. L. 1967. Animal tissue techniques. 2nd Edition. W. H. Freeman, San Francisco.

Khan, R. A., D. Barker, R. Hooper, and E. M. Lee. 1992. Effect of pulp and paper effluent on a marine fish, *Pseudopleuronectes americanus*. Bulletin of Environmental Contamination and Toxicology 48:449-456.

Klein-MacPhee, G. 1978. Synopsis of biological data for the winter flounder, *Pseudopleuronectes americanus* (Walbaum). NOAA Technical Report NMFS Circular 414, FAO Fisheries Synopsis 117., Washington, District of Columbia.

Lom, J. 1984. Diseases caused by protistans. Page 129 *in* O. Kinne, editor. Diseases of marine animals, Volume 1, Part 1. Biologische Anstalt Helgoland, Hamburg.

Mavor, J. W. 1915. Studies on the Sporozoa of the fishes of the St. Andrews region. Contributions to Canadian Biology 1911-1914:25-38.

McVicar, A. H. 1975. Infection of plaice *Pleuronectes platessa* L. with *Glugea (Nosema) stephani* (Hagenmuller 1899) (Protozoa: Microsporidia) in a fish farm and under experimental conditions. Journal of Fish Biology 7:611-619.

Murray, T. E. 1980. Biological versus environmental sources of morphologic variation in estuarine populations of *Littorina littorea* in New England. Ph.D. Thesis. University of Rhode Island, Kingston.

Olson, R. E. 1976. Laboratory and field studies on *Glugea stephani* (Hagenmuller), a microsporidan parasite of pleuronectid flatfishes. Journal of Protozoology 23:158-164.

Pereira, J. J., E. J. Lewis, R. L. Spallone, and C. Sword. 1992. Gill bifurcations in winter flounder *(Pleuronectes americanus)* from Long Island Sound. Journal of Fish Biology 41:327-338.

Pruell, R. J., and J. G. Quinn. 1985. Geochemistry of organic contaminants in Narragansett Bay sediments. East Coast Shelf Science 21:195-312.

Rosenthal, H. 1967. Parasites in larvae of herring *(Clupea harengus* L*.)* fed with wild plankton. Marine Biology 1:10-15.

Santschi, P. H., S. Nixon, M. Pilson, and C. Hunt. 1984. Accumulation of sediments, trace metals (Pb, Cu) and total hydrocarbons in Narragansett Bay, Rhode Island. East Coast Shelf Science 19:427-449.

Saucerman, S. E., and L. A. Deegan. 1991. Lateral and cross-channel movement of young-of-the-year winter flounder *(Pseudopleuronectes americanus)* in Waquoit Bay, Massachusetts. Estuaries 14:440-446.

Sindermann, C. J., and A. Rosenfield. 1954. Diseases of fishes of the western North Atlantic. III. Mortalities of sea herring *(Clupea harengus)* caused by larval trematode invasion. Maine Department of Sea and Shore Fisheries Research Bulletin 21.

Stunkard, H. W., and F. E. Lux. 1965. A microsporidian infection of the digestive tract of the winter flounder, *Pseudopleuronectes americanus*. Biological Bulletin 129:371-387.

Takvorian, P. M., and A. Cali. 1984. Seasonal prevalence of the microsporidan, *Glugea stephani* (Hagenmuller), in winter flounder, *Pseudopleuronectes americanus* (Walbaum), from the New York-New Jersey lower bay complex. Journal of Fish Biology 24:655-663.

Wells, H. W. 1965. Maryland records of the gastropod, *Littorina littorea*, with a discussion of factors controlling its southern distribution. Chesapeake Science 6:38-42.

Effects of Chemical Stresses on Behavior of Larval and Juvenile Fishes and Amphibians

W. J. Birge

*School of Biological Sciences, and Graduate Center for Toxicology,
University of Kentucky, Lexington, Kentucky 40506-0225, USA*

R. D. Hoyt

*Department of Biology, Western Kentucky University,
Bowling Green, Kentucky 42101, USA*

J. A. Black

*EA Engineering Science and Technology Inc.,
Sparks, Maryland 21152, USA*

M. D. Kercher

*School of Biological Sciences, and Graduate Center for Toxicology,
University of Kentucky, Lexington, Kentucky 40506-0225, USA*

W. A. Robison

*U. S. Fish and Wildlife Service,
Ecological Services, Cookeville, Tennessee 38501, USA
and Graduate Center for Toxicology,
University of Kentucky, Lexington, Kentucky 40506-0054, USA*

Abstract.—A series of investigations was conducted to determine the effects of chemical stresses on avoidance/attraction and feeding behavior of larval and juvenile fishes and amphibians. Avoidance/attraction experiments were conducted with cadmium (Cd), copper (Cu), chloroform, dioctyl phthalate (DOP), trisodium nitrilotriacetic acid (NTA), mercury (Hg), phenol, and zinc (Zn). Tests were performed with a dual channel fluviarium. Juvenile largemouth bass *(Micropterus salmoides)*, bluegill *(Lepomis macrochirus)*, and rainbow trout *(Oncorhynchus mykiss)*, and American toad *(Bufo americanus)* tadpoles proved to be suitable for evaluating avoidance/attraction responses. With one or more species, significant avoidance ($P < 0.01$) occurred in tests with Cd, Zn, and phenol. Significant attraction resulted from exposures to chloroform, DOP, and Hg. Animals generally avoided lower concentrations of Cu, but were attracted to higher exposure levels. NTA produced more variable responses. The trout was the most sensitive species tested. Aluminum (Al) exposure and pH affected feeding behavior in larval fathead minnows *(Pimephales promelas)*. Larvae in low pH (< 6.5) and moderately hard water exhibited reduced feeding. Aluminum at 50 µg/L further reduced feeding under acidic conditions, and 100 µg Al/L affected feeding under acidic and alkaline pH conditions. Additional feeding trials were conducted on 18-d-old fathead minnows at pH levels of 3.5 to 11.5. Total mortality was observed at a pH of 3.5 and 11.5. A pH of 4.0 to 5.0 did not impair appetitive behavior but reduced feeding success, especially when live shrimp were fed in the dark. The major conclusions were: (1) sublethal pH did not affect appetitive behavior; (2) fathead minnow larvae were predominantly daylight, visual feeders; (3) a pH of 5.0 or lower significantly affected chemoreception and reduced feeding; and (4) chemoreception and mechanoreception were strongly correlated with vision in feeding.

Aquatic risk assessment has been based largely on determining toxicant concentrations that produce lethality or impair the health of aquatic species (NAS-NAE Committee 1973; Cairns et al. 1978; Weber et al. 1989). However, it is becoming increasingly apparent that trace levels of aquatic pollutants that do not produce mortality may limit distribution, migration, feeding, or reproduction of fish and other species (Sprague et al. 1965; Sprague 1968; Weir and Hine 1970; Birge et al. 1981). Geckler et al. (1976) conducted a correlated field and laboratory study involving the effects of copper (Cu) on the green sunfish *(Lepomis cyanellus)*. They found that trace levels of Cu which were not toxic in laboratory tests affected the in-stream distribution of sunfish. The sunfish populations were restricted to stream areas that were most free of Cu. This reduced Cu exposure, but limited feeding and spawning habitats. Zinc (Zn) and other metals also have been shown to disrupt distribution and migration of fish species due to avoidance or attraction responses (Beitinger 1990; Sprague and Drury 1969).

Avoidance and attraction phenomena in fish have been reviewed by Sprague (1968), Cherry et al. (1976), Geckler et al. (1976), Kleerekoper (1976), McCauley (1977), Beitinger (1990), and others. These behavioral patterns are thought to be initiated by olfactory or gustatory reception, though avoidance may be induced by less specific irritability responses mediated through general exteroception. Avoidance behavior in salmon and trout is first observed at 5 weeks of age (Bishai 1962). As noted by Hasler (1957), olfactory sacs open at about this stage (21 mm), and removal of olfactory tissue is known to eliminate avoidance responses to certain toxicants (Hoglund 1961). In trout, acuity of the avoidance response increases rapidly with growth, and high sensitivity to Zn is exhibited by juveniles of 30 to 60 mm in length (Bishai 1962, Sprague 1968). Lemly and Smith (1987) reported changes in patterns of behavior to be sensitive indicators of sublethal toxicity.

Although acid precipitation and depressed pH conditions have been studied at length (Zischke et al. 1983; Leino et al. 1987; Mills et al. 1987; Jansen and Gee 1988, among others), little attention has been given to the effects of acid stress on fish behavior (Jones et al. 1985). Little (1990) reported that the effects of contaminants on feeding behavior could be readily documented in the laboratory. He suggested that the development of behavioral endpoints would produce results more sensitive and versatile than toxicological studies based on lethality.

The principal objectives of our studies were: (1) to provide further documentation on the nature and development of stress-related behavioral patterns in juvenile fishes and amphibian tadpoles; (2) to define further avoidance and attraction phenomena and determine the extent to which behavior potentiates or minimizes adverse effects of chemical stresses; (3) to describe the effects of acid stress and Al on the feeding behavior of larval and juvenile fishes; and (4) to examine the feasibility and design of behavioral test systems for evaluating environmental impact from metals, organic compounds, and complex (NPDES) effluents.

Methods

Rainbow trout *(Oncorhynchus mykiss)* were provided by the Erwin National Fish Hatchery, Erwin, Tennessee. Largemouth bass *(Micropterus salmoides)*, fathead minnows *(Pimephales promelas)*, channel catfish *(Ictalurus punctatus)*, bluegill *(Lepomis macrochirus)*, and American toads *(Bufo americanus)* were obtained from the Frankfort National Fish Hatchery near Frankfort, Kentucky, or the University of Kentucky Biomonitoring Laboratory.

Experiments were conducted with aluminum chloride ($AlCl_3$), cadmium chloride ($CdCl_2$), copper sulfate ($CuSO_4$), chloroform (CCl_4), dioctyl phthalate (DOP), mercuric chloride ($HgCl_2$), trisodium nitrilotriacetic acid (NTA), phenol, and zinc chloride ($ZnCl_2$). Exposure concentrations (mg/L) of the inorganic compounds were based on metal content.

Avoidance responses were evaluated using a flow-through, dual channel fluviarium modified from those described by Hoglund (1961) and Wilson (1973). The chamber was constructed of one-quarter-inch plexiglass, with dimensions of 60 cm × 30 cm × 6.5 cm and an overall capacity of 10 L (Figure 1). Carbon-filtered water was provided to a constant head box, using an in-line pressure regulator. Adjustable standpipes were used to regulate flow rates for control and experimental channels of the test system, and flow rates were monitored with high-resolution flowmeters (Gilmont, Inc.). For all test compounds except DOP, a variable speed peristaltic pump (Brinkmann Model 131900) was used to inject toxicant into a mixing chamber placed immediately ahead of the test channel (Figure 1). Dilution water and toxicant were thoroughly blend-

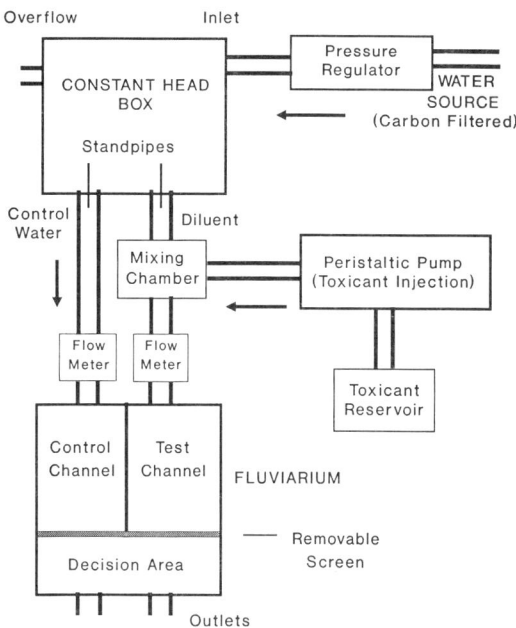

FIGURE 1.—Schematic of the flow-through, dual channeled fluviarium used in avoidance-attraction studies. Arrows indicate direction of flow. Capacity of the test chambers, including control and test channels and decision area was 10 L. Flow rate was 250 mL/min per channel, and detention time was 20 min.

ed in the mixing chamber (i.e., 125-mL side-arm flask) using a magnetic stirrer. The supply of toxicant to the test chamber was regulated by adjusting the concentration in the toxicant reservoir. For DOP, the only insoluble compound tested, toxicant was injected with a syringe pump (Sage Model 355) into a 400-mL mixing chamber of a high-speed blender (Osterizer Pulsematic). Intermittent blending (1 min/0.5 h) was sufficient to maintain DOP in suspension (Birge et al. 1979a). This precluded the need for a carrier solvent.

Juvenile organisms used in avoidance/attraction studies were maintained for a minimum of 4 d in temperature-regulated fiberglass tanks (200 L) which received a continuous flow (15 L/h) of the same carbon-filtered water that was supplied to the fluviarium. Water was monitored at regular intervals for temperature, dissolved oxygen, water hardness, pH, and specific conductivity using a YSI tele-thermometer with thermocouple (Model 42SC), YSI oxygen meter (Model 51A), Orion divalent cation electrode (Model 93-92), Corning digital pH meter (Model 110), and a Radiometer conductivity meter (Model DCM 2e). Mean water temperatures (± SE) were 17.8 ± 0.2°C, 19.1 ± 0.2°C, 23.0 ± 0.1°C, and 23.2 ± 0.2°C during experiments with the trout, toad, bluegill, and bass, respectively. Dissolved oxygen was maintained at 85 to 90% of saturation by moderate, continuous aeration to the constant head box. Dilution water values (mean ± SD) for pH, alkalinity, hardness and conductivity were 7.6 ± 0.3, 63.0 ± 5.0 mg $CaCO_3$/L, 112.4 ± 30.8 mg $CaCO_3$/L, and 136.0 ± 20.2 µmhos/cm, respectively.

The standard length for the fish species ranged as follows: 35 to 60 mm for bluegill; 35 to 40 mm for bass; 75 to 100 mm for catfish; and 30 to 50 mm for trout. American toad tadpoles varied in length from 12 to 17 mm. Avoidance tests were performed using a sample size of 10 organisms per exposure concentration, as follows: (1) Withholding toxicant, water flow was initiated in control and test channels. A flow rate of 250 mL/min was used, giving a detention time of 20 min for each of the two 5-L channels. (2) Test organisms were introduced into the decision area and maintained there for 20 min. (3) The screen enclosing the decision area was removed, and the locations of test organisms on the control and test sides of the fluviarium were recorded at 30-s intervals for 10 min to ascertain any significant non-random channel preferences. (4) Test organisms were relocated in the decision area and toxicant was administered to the test channel for 20 min (one detention interval). (5) The screen was again removed from the decision area, and the locations of organisms were recorded at 30-s intervals for 20 min. Water samples were collected from each channel at the beginning and end of each 20-min test period and analyzed for toxicant concentration. (6) The control and test channels were reversed and steps 1 through 5 were repeated using the same organisms selected for each toxicant concentration.

Using mean scores for control (C) and test (T) sides of the fluviarium, percent avoidance (A) was determined for each toxicant concentration as $A = 100 \times (C - T)/(C + T)$. Negative avoidance was defined as attraction. Significance for percent avoidance or attraction ($P < 0.01$) was determined with the Student's t-test. When it was possible to establish no observed effect (NOEC) and lowest observed effect (LOEC) concentrations, geometric means were used to approximate threshold concentrations.

In feeding experiments with pH and aluminum, fathead minnows were cultured according to standard laboratory methods (Klemm 1985). Minnows used for pH exposure were 18 d old and averaged

11.6 mm in length (range 10.1 mm to 13.0 mm). Experiments were conducted with ASTM 100 water (Weber et al. 1989), and dilute nitric acid and sodium hydroxide were used to adjust pH. Acidic (4.0 to 5.0), basic (10.0 to 11.5), and neutral (7.0) pH ranges were used for each of four feeding regimes: living brine shrimp fed in the light (light/live); living brine shrimp fed in the dark (dark/live); dead brine shrimp fed in the light (light/dead); and dead brine shrimp fed in the dark (dark/dead). There were three replicates for each feeding regime, and this experiment was repeated three times with new test organisms. Minnows were selected at approximately 1600 hours the day before feeding trials, separated into 12 groups of 15, and each group was placed in a 500-mL beaker containing water of a specific pH. The fish were maintained at 25.0°C with an 8-h dark period (2200 to 0600 hours) for acclimation until approximately 1300 hours the following day. Five fish from each pH range were then placed in each of three 150-mL finger bowls containing fresh mixtures of the desired pH's for replicate trials. The fish were allowed to acclimate in the finger bowls for 10 min in either light or dark before food was added. Fifty brine shrimp were then added by syringe to each group of five fish, and the feeding time for all tests was 10 min. Brine shrimp for dead feeding trials were killed in an ultrasonic bath.

Immediately following the feeding trials, the number of fish having consumed at least one brine shrimp was recorded with the aid of microscopic examination, and the number of unconsumed brine shrimp remaining in each test chamber was determined. The feeding study with Al was performed with the same procedures, except that 8-d-old fathead minnow larvae were used. All feeding studies were analyzed statistically using a one-way analysis of variance (significance, $P \leq 0.05$).

Exposure concentrations for toxicants were confirmed by analysis of test water. Aqueous metal concentrations were determined by atomic absorption spectrophotometry (AAS) using a Perkin-Elmer AAS (Model 503) equipped with a Hg analysis system and a graphite furnace (Model HGA 2100). The cold vapor technique of Hatch and Ott (1968), with a detection limit of 0.1 µg/L, was used for Hg determinations. Cadmium, Zn, and Cu were analyzed with an air-acetylene flame, providing detection limits of 1.0, 10, and 50 µg/L, respectively. Lower concentrations of Cu were quantified with the graphite furnace, and the detection limit was 1.0 µg/L (ASTM 1989).

Chloroform was analyzed from 1- to 15-mL aliquots of test water with a Hewlett Packard gas chromatograph (Model 5838A) equipped with a Purge and Trap system (Model 7675A). Each sample was purged with dry, pre-purified nitrogen at 10 mL/min for 10 min. Chloroform was adsorbed on a Tenex GC trap at ambient temperature, desorbed at 200°C, and analyzed at programmed temperatures of 70 to 105°C on a 2-m × 2-mm I.D. glass column. The stationary phase was 10% Carbowax 20 M on 80/100 Anakrom U, and the detector temperature was 250°C. Nitrogen was used as the carrier gas and the detection limit was 1 µg/L (USEPA 1979).

Dioctyl phthalate (DOP) was extracted from 0.5-L aliquots of test water using reagent grade chloroform. Chloroform extracts were dried with anhydrous sodium sulfate and concentrated with an air stream. DOP was then quantitatively reconstituted in ethyl acetate and analyzed on a Packard gas chromatograph (Model 7400) equipped with a flame ionization detector (Model 881) and a glass column (46 cm × 2 mm I.D.). The stationary phase was 1.5% OV-17/1.95% QF-1 on 80/100 Chromosorb W HP (Supelco, Inc.). The oven, inlet, and detector temperatures were 235, 250, and 260°C, respectively. The detection limit for DOP was 25 µg/L (Longbottom and Leichtenberg 1982).

Trisodium nitrilotriacetic acid (NTA) was analyzed by the zinc-zincon method (USEPA 1979), with a detection limit of 0.5 mg/L. Absorbance at 620 nm was measured using a Model 635 Varian-Techtron spectrophotometer. To preclude interference from Ca and Mg ions, NTA test water samples were batch-treated with an ion exchange resin (Dowex 50W-X8, 50-100 mesh).

Phenol concentrations were determined using the chloroform extraction, 4-aminoantipyrine procedure (APHA 1989). Prepared samples were quantified spectrophotometrically at a wavelength of 460 nm, with a detection limit of 2.0 µg/L. Additional details on methods and monitoring data observed in the avoidance/attraction studies were presented earlier (Birge and Black 1980).

Results

The most sensitive avoidance responses to metals were obtained with the trout (Tables 1 and 2). Juvenile trout exhibited consistent avoidance to cadmium, with NOEC and LOEC concentrations of 0.01 and 0.05 mg Cd/L The avoidance threshold (geometric mean) was 0.03 mg Cd/L and this was equivalent to the LC1 concentration established in

TABLE 1.—Avoidance/attraction responses of juvenile rainbow trout to cadmium, copper, and zinc. Trout ranged in length from 35 to 50 mm. An asterisk denotes significance using the Student's t-test ($P < 0.01$).

Metal	Metal concentration (mg/L) mean ± SE	Percent avoidance (+) or attraction (-)
Cadmium	0.01 ± 0.00	+37.3
	0.05 ± 0.00	+51.7*
	0.10 ± 0.02	+55.3*
	1.03 ± 0.04	+72.3*
Copper	0.07 ± 0.02	+13.2*
	0.37 ± 0.03	-12.8
	0.77 ± 0.06	+11.3
	4.56 ± 0.22	-37.0*
	7.56 ± 0.63	-26.8*
Zinc	0.01 ± 0.00	+11.0
	0.05 ± 0.01	+94.5*
	0.10 ± 0.01	+95.0*
	1.13 ± 0.12	+99.8*

28-d tests with trout embryo-larva stages (Birge et al. 1979b). Test concentrations of 0.09, 0.90, and 8.8 mg Cd/L were administered to the bass, and bluegill were exposed to 0.8, 8.3, and 41.1 mg Cd/L, but no consistent trends or significant responses were observed for either species.

Tests with Cu were conducted on juvenile bluegill, trout, and tadpoles of the American toad. Rainbow trout was the species most sensitive to Cu treatment, exhibiting significant avoidance at 0.07 mg/L, the lowest concentration tested (LOEC). However, as concentrations were elevated to 4.6 and 7.6 mg Cu/L, trout showed significant attraction. Similarly, American toad tadpoles avoided a concentration of 0.10 mg Cu/L, but were attracted to 0.93 mg Cu/L ($P < 0.01$). Cu was the only toxicant tested which produced this pattern of response with at least two animal species. In tests with the bluegill, levels of 8.5 and 43.2 mg Cu/L resulted in significant attraction.

Zinc was administered to bluegill, largemouth bass, and rainbow trout. Trout was the species most sensitive to Zn exposure (Table 1). At concentrations of 0.01, 0.05, 0.10, and 1.1 mg Zn/L, juvenile trout avoided Zn at net frequencies of 11.0%, 94.5%, 95.0%, and 99.8%, respectively. The last three values were statistically significant and the avoidance threshold was calculated to be 0.02 mg Zn/L. Juvenile largemouth bass also were affected by Zn exposure and avoided concentrations of 7.0 to 39.2 mg/L at net frequencies of 15.5% to 56.8% ($P < 0.01$). The most tolerant species tested was the bluegill, which exhibited no significant response to exposures as high as 43.7 mg Zn/L.

Mercury was administered at 0.0002 and 0.0074 mg/L and elicited attraction at frequencies of 24.5% and 18.0%, respectively. Therefore, the threshold for Hg was at or below 0.2 µg/L, and this was the lowest exposure concentration at which any of the selected toxicants produced significant responses ($P < 0.01$).

More limited avoidance experiments were conducted with four organic compounds. Bluegill and rainbow trout were exposed to chloroform, and both species showed significant attraction to this highly volatile organic compound. Bluegill responded at a frequency of 93.5% to a concentration of 33.2 mg CCl_4/L. By comparison, trout juveniles were attracted at a frequency of 21.0% to chloroform at 11.9 mg CCl_4/L (Table 2). Calculated attraction thresholds (geometric means) for chloroform were 7.05 and 10.0 mg CCl_4/L for the trout and bluegill, respectively.

Dioctyl phthalate (DOP) was administered only to juvenile bluegill at a concentration of 112.4 mg/L, and significant attraction occurred at a frequency of 41.1%. Because only one concentration was tested, an accurate threshold for DOP could not be estimated. In addition, the homogenized DOP suspension produced some turbidity in the test channel that may have affected responses.

Trisodium nitrilotriacetic acid (NTA) was administered to juvenile bluegill and rainbow trout. No significant attraction or avoidance was observed with the bluegill at exposure concentrations of 0.23, 6.65, and 39.8 mg/L. In tests with the trout, NTA at 57 mg/L produced 23.8% avoidance. However, when the concentration was increased to 101 mg/L,

TABLE 2.—Comparison of avoidance/attraction thresholds with toxicity values for rainbow trout. Trout ranged in length from 35 to 50 mm. An asterisk denotes significance using the Student's t-test ($P < 0.01$). Geometric means of NOEC and LOEC values are given in parentheses. Toxicity data taken from Birge et al. (1979a, 1979b, 1979c).

Toxicant	Threshold concentration (mg/L)		Lethal concentration (mg/L)		
	avoidance	attraction	LC1	LC10	LC50
Cd	0.01-0.05* (0.03)	0.008	0.03	0.14	
Cu	0.07*	4.56* (1.9)	0.003	0.02	0.11
Hg		0.0002*	<0.0001	–	–
Zn	0.01-0.05* (0.02)	0.216	0.45	1.12	
CCl_4		4.18-11.9* (7.05)	0.005	0.09	1.24
NTA	57*	101*	20.2	43.9	114

TABLE 3.—Effects of acidity on the feeding behavior of juvenile fathead minnows. Means represent three replicates.

Experiment number	Feeding conditions	Mean percentage of brine shrimp remaining after feeding trials		
		pH 5.0	pH 7.0	pH 10.0
I	Light-Live	0 ± 0	0 ± 0	0 ± 0
	Light-Dead	0 ± 0	0 ± 0	0 ± 0
	Dark-Live	1.3 ± 2.3	0 ± 0	4.3 ± 1.1[a,b]
	Dark-Dead	10.3 ± 3.2[b]	3.7 ± 4.0	0.7 ± 1.1
		pH 4.5	pH 7.0	pH 10.5
II	Light-Live	1.0 ± 1.0	0 ± 0	0 ± 0
	Light-Dead	0 ± 0	0 ± 0	0 ± 0
	Dark-Live	20.7 ± 3.2[a,b]	1.7 ± 1.1	0 ± 0
	Dark-Dead	13.7 ± 8.1[b]	3.0 ± 5.2	0.7 ± 1.1
		pH 4.0	pH 7.0	pH 11.0
III	Light-Live	0 ± 0	0 ± 0	0 ± 0
	Light-Dead	0 ± 0	0 ± 0	0 ± 0
	Dark-Live	24.7 ± 14.7[a,b]	0 ± 0	0.7 ± 1.1
	Dark-Dead	3.7 ± 3.1	0.7 ± 0.6	2.3 ± 0.6[a,b]

[a]Significantly different ($P \leq 0.05$, one-way ANOVA) from corresponding treatments at pH 7.0.
[b]Significantly different ($P \leq 0.05$, one-way ANOVA) from corresponding feeding conditions in light at the same pH.

test animals were attracted to this compound at a frequency of 24.5% (Table 2). These results with the trout were considered to be inconclusive.

Avoidance tests with phenol were conducted on juvenile bluegill and concentrations of 0.76, 6.76, and 39.0 mg/L produced responses of -2.8%, +2.9%, and +75.3%, respectively. Phenol avoidance was significant only at the highest concentration and the calculated threshold was 39.0 mg/L. In summary, avoidance was significant in tests with Cd, Zn, and phenol, and significant attraction resulted from exposure to Hg, chloroform, and DOP for at least one species. In tests with Cu, animals generally avoided lower concentrations but were attracted to higher exposure levels. NTA produced variable responses.

Effects of pH on feeding in juvenile fathead minnows are summarized in Table 3. Complete mortality was observed at pH 3.5 and 11.5. The number of fish ingesting at least one brine shrimp during the feeding trials ranged from 93 to 100 percent for all other pH levels. Feeding success was not impaired significantly in light trials, as 99.9% of all live and dead brine shrimp fed were consumed. Consequently, vision did not appear to be affected. In the dark, however, live or dead brine shrimp remained after feeding in 15 of 18 trials, and numbers of residual brine shrimp per individual replicate ranged from 1.3% to 49.3%. The greatest reduction in feeding success occurred with live brine shrimp fed under dark conditions at pH levels of 4.0 and 4.5 and, to a lesser extent, dead brine shrimp fed under dark conditions at pH levels of 4.5 and 5.0 (Table 3).

The effect of pH on feeding behavior in fathead minnows was more pronounced when combined with Al treatments. Feeding suppression in 8-d-old larvae at acidic pH levels without Al was generally similar to results reported for 18-d-old juveniles (Table 3). Addition of 50 μg Al/L reduced feeding as acidity increased (Figure 2). Results obtained at pH 4 for all feeding regimes differed significantly from those at pH 7 ($P < 0.01$). Dark live feeding was most affected, based on the greater number of unconsumed shrimp, and results observed for this feeding treatment at pH levels of 4.0, 4.5, and 5.0 were significantly different from responses at pH 7. When 100 μg Al/L was added, feeding efficiency was reduced at both acidic and alkaline pH levels. At pH 4.0 to 4.5, 40% of the brine shrimp were not consumed, while 20 to 25% remained at pH 10.0 to 11.0.

Discussion

Carlander (1969) reported the fathead minnow to have a broad tolerance of pH. In the present study, mortality was infrequent at pH 4.0 to 10.5 for 21 h in moderately hard water. Complete mortality was observed, however, at pH 3.5 and 11.5, indicating sharp thresholds for pH lethality. Mount (1973) observed that acid-induced mortality in fathead minnows occurred below pH 4.0.

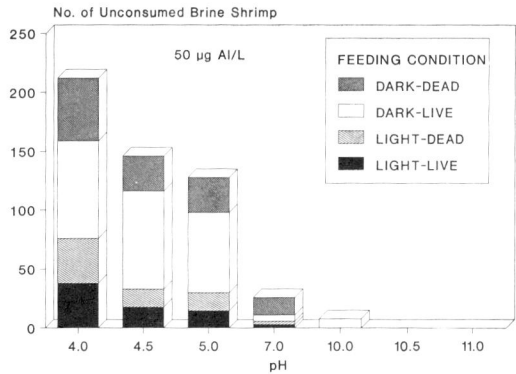

FIGURE 2.—Effects of combined pH and Al exposures on brine shrimp consumption of 8-d-old fathead minnow larvae. Feeding success was scored as the number of shrimp remaining after 10-min feeding trials conducted with live and dead shrimp under light and dark conditions with three replicates per treatment. Fifty shrimp were fed to five fish per test chamber.

Based upon the high frequencies of feeding observed at all survivable pH levels, appetitive behavior did not appear to be impaired in the range of 4.0 to 11.0. Hill (1989), in a chronic study of low-pH effects on feeding behavior of smallmouth bass (*Micropterus dolomieu*), observed no loss of desire to feed at pH 4.2. Since the desire to feed was present, reductions in feeding activity were considered to result from the impairment of sensory perception by pH and/or specific test conditions (i.e., darkness). Jones et al. (1985), in studying the effects of sublethal pH levels on the behavior of Arctic char (*Salvelinus alpinus*), reported that acid stress suppressed chemoreception. He attributed this to the reduced stimulatory nature of amino acids and the damage of olfactory epithelial tissues by acidic conditions. Lemly and Smith (1985, 1987) found acidification to affect the ability of fathead minnows to detect or respond to chemical stimuli. They suggested that increased thickness of olfactory mucous in response to lowered pH prevented normal stimulus-receptor interaction and/or that chemical interaction at the subcellular level was impaired because of stearic/charge changes at the receptor cells.

That the fathead minnow is predominantly a visual, daylight feeder was supported by findings in this study. Klemm (1985) reported the fathead minnow to be primarily omnivorous and to possess functional, pigmented eyes at the time of hatching. The efficient feeding of 14-d-old larvae on brine shrimp, live or dead, in the light at all pH levels supports this premise. However, the limited reaction distance in the test chambers (150 mL) and the number of brine shrimp used per trial (50) could have compensated for some degree of eye stress. Hill (1989) reported that low pH impaired visual acuity, coordination, and agility of the smallmouth bass, and subsequently resulted in lower growth in chronic trials.

Considering the number of uneaten live and dead brine shrimp remaining after all dark feeding trials, the fathead minnow was not observed to be an adept nocturnal feeder. Feeding on dead food in the dark was considered to be primarily a function of chemoreception, as stimuli for visual and mechanoreceptive senses were not present. Chemoreception may have been impaired at pH levels of 5.0 and 4.5 when 20.7% and 27.3% of the dead brine shrimp remained, respectively. Live prey in the dark represented the maximum sensory challenge. Although live brine shrimp were successfully preyed upon in the dark at pH 7.0 and higher, only about 50% were captured at the lower pH levels, indicating probable effects on chemoreception and/or mechanoreception.

Jones et al. (1985) reported that pH 5.0 suppressed chemo-orientation in Arctic char, and Lemly and Smith (1985, 1987) found that pH levels of 6.0 and lower eliminated fathead minnow responses to chemical stimuli. However, Yoshii and Kurihara (1983) reported that bluegill without functional lateral lines did not produce successful feeding strikes in the dark based on chemoreception alone. Hara (1986) summarized the extensive literature on the role of chemoreception in feeding behavior, and identified the first step in feeding as arousal to the presence of food mediated primarily by olfaction. Atema (1980) excepted anosmic sticklebacks, the most visual fish species, from the above behavior and suggested that accompanying senses such as mechanoreception and vision may also be involved in initial prey recognition. Enger et al. (1989) and Montgomery (1989) substantiated the role of mechanoreception in detecting moving prey. Montgomery (1989) further described mechanoreception as operating synergistically with vision in daylight planktivory and singly in total darkness. Whether or not mechanoreception in juvenile fathead minnows was affected by pH could not be determined with certainty in the present study.

Aluminum at 50 µg/L further reduced food consumption rates under acidic conditions (pH 4.0 to

5.0) as shown in Figure 2, and 100 µg Al/L reduced feeding efficiency under both acidic and alkaline conditions. This interaction between Al and pH was synergistic. The most pronounced effects in this 24-h feeding test occurred when live brine shrimp were fed under dark conditions. These findings indicate that adverse effects on behavior may result at aluminum concentrations well below EPA (1988) acute (750 µg/L) and chronic (87 µg/L) criteria for aquatic life. By comparison, Leino et al. (1990) found that 30 µg Al/L did not significantly elevate reproductive and histopathological disorders in fathead minnows exposed for 109 d to pH 5.2 and pH 5.5 in soft water (20-24 mg $CaCO_3$/L).

Juvenile bluegill, largemouth bass, rainbow trout, and American toad tadpoles were found to be suitable test animals for use in avoidance response bioassays. Each of these species exhibited the capacity to discriminate between control and toxicant exposures in tests with one or more compounds during the course of this study. However, due to its predilection for schooling behavior, the channel catfish could not be used in the test system designed for these experiments (Figure 1). The most sensitive species was the rainbow trout, which responded significantly at LOECs of 0.0002 mg Hg/L, 0.05 mg Zn/L, 0.05 mg Cd/L, 0.07 mg Cu/L, and 11.9 mg chloroform/L. The high relative sensitivity of the rainbow trout was consistent with results from earlier embryo-larva toxicity tests, in which trout developmental stages generally suffered higher rates of lethality and teratogenesis than did other fish species (Birge et al. 1979a, 1979c). Chronic studies also indicate the rainbow trout to be among the aquatic species most sensitive to toxicological stress, as reviewed by Birge et al. (1981). Conversely, those species which most often are tolerant of toxicological stress (i.e., bluegill, largemouth bass) exhibited less defined avoidance/attraction behavior or higher thresholds when such behavior was observed.

In Table 2, results of avoidance/attraction studies with the trout are compared with toxicity values determined in 28-d embryo-larva exposures (Birge et al. 1979b). The LC1 values determined with trout in flow-through systems for metals and some organic compounds were shown earlier to approximate MATCs and chronic values (Birge et al. 1981). The only clear case in which the behavioral response was more sensitive than toxicological endpoints was with Zn. The calculated Zn threshold for avoidance was an order of magnitude below the toxicological threshold (LC1). Behavioral responses with Cd and Hg also were measurable at concentrations close to LC1 or LC10 values. However, for the other chemicals, toxicity values were better indicators of stress to trout than behavioral thresholds, and this generally held true for results obtained with the other test species. It has been proposed that behavioral responses are often detectable at chemical concentrations below acute LC50 levels. Although this trend was observed within the present study, it is suggested that such comparisons have more relevance to ecological risk assessment when chronic values are used as the toxicological reference. Based on observations presented above, behavioral responses to chemicals are estimated to occur up to one-third of the time at or below concentrations that impair fish early life stages or produce chronic effects on growth and reproduction. This premise is generally consistent with the view presented by Beitinger (1990).

In agreement with Beitinger (1990), Little (1990), and others, results of this investigation support the use of behavioral tests for evaluating effects of chemical stresses on aquatic biota. Avoidance/attraction, feeding responses, and other behavioral endpoints may be particularly useful in characterizing the effects of complex NPDES effluents and receiving waters, for which time and other constraints limit alternatives for toxicological testing (Weber et al. 1989). Fish avoidance of kraft and pulpmill effluents shown by Sprague and Drury (1969), Lewis and Livingston (1977), and Wildish et al. (1977) support this premise.

Beitinger (1990) recently presented an extensive review of the literature on fish behavioral responses to chemical stimuli. Published data on avoidance/attraction responses to chemicals were grouped into five categories, including: (1) consistent avoidance; (2) attraction to low concentrations but avoidance of high concentrations; (3) attraction to, or no avoidance at levels above the avoidance threshold; (4) no avoidance; and (5) inconsistent responses. A surprising finding was the near absence of cases of consistent avoidance to metals (category 1). This is in contrast to findings presented in Table 1 that show clear and consistent avoidance to Cd and Zn by juvenile trout. Consistent avoidance of phenol also was observed with the bluegill. DeGraeve (1982) reported avoidance by rainbow trout of phenol at 3.2 to 6.5 mg/L. Results presented above for copper fit category 3, and are in general agreement with studies of Sprague (1964) and Pedder and Maly (1985). Similar results were obtained for NTA in tests with juvenile trout. The attraction of bluegill and trout to chloroform and trout to Hg does not appear to fit within Beitinger's classification. Attraction of juvenile fishes to such toxic chemicals potentiates the prospects for deleterious environmental effects. Alternatively, in agreement with Beitinger, avoidance affords some protection from chemical stresses but may limit available habitat (Geckler et al. 1976).

Better comprehension and predictability of behavioral responses to chemical stimuli are fundamental to risk assessment, resource management, and stress ecology. Procedures described above and reviewed elsewhere for conducting behavioral studies with aquatic organisms (Beitinger 1990; Little 1990; Little and Finger 1990) should be useful in standardizing test methodology, and present knowledge supports the integration of behavioral considerations into environmental risk assessment strategies.

Acknowledgments

We are grateful to D. J. Price, J. E. Hudson, A. G. Westerman, and H. Abdul-Rahim for assistance in conducting this investigation, and to Shale Johnson for preparing this manuscript. Funding was provided by the United States Department of the Interior, the National Science Foundation Kentucky EPSCoR Program, and the NIH program for Summer Research Experience for Minorities.

References

APHA (American Public Health Association), American Water Works Association, and Water Pollution Control Federation. 1989. Standard methods for the examination of water and wastewater, 17th Edition. APHA, Washington, District of Columbia.

ASTM (American Society for Testing and Materials). 1989. Annual Book of ASTM Standards. Section 11. Water and Environmental Technology. Volume 11.01. Water (1). American Society for Testing and Materials, Philadelphia.

Atema, J. 1980. Chemical senses, chemical signals, and feeding behavior in fishes. Pages 57-94 in J. E. Bardach et al., editors. Fish behavior and its use in the capture and culture of fishes. International Center for Living Aquatic Resources Management, Manila.

Beitinger, T. L. 1990. Behavioral reactions for the assessment of stress in fishes. Journal of Great Lakes Research 16:495-528.

Birge, W. J., and J. A. Black. 1980. An avoidance response bioassay for aquatic pollutants. University of Kentucky, Water Research Institute. Research Report Number 123. Lexington.

Birge, W. J., J. A. Black, and D. M. Bruser. 1979a. Toxicity of organic chemicals to embryo-larval stages of fish. EPA, Report 560/11-79-007, Washington, District of Columbia.

Birge, W. J., J. A. Black, and B. A. Ramey. 1981. The reproductive toxicology of aquatic contaminants. Pages 59-115 in J. Saxena, editor. Hazard assessment of chemicals: current developments. Volume 1. Academic Press, New York.

Birge, W. J., J. A. Black, A. G. Westerman, and J. E. Hudson. 1979b. Aquatic toxicity tests on inorganic elements occurring in oil shale. Pages 519-534 in C. Gale, editor. Oil shale symposium: sampling, analysis, and quality assurance. U. S. Environmental Protection Agency, Cincinnati, Ohio.

Birge, W. J., J. A. Black, A. G. Westerman, and J. E. Hudson. 1979c. The effects of mercury on reproduction of fish and amphibians. Pages 629-655 in J. O. Nriagu, editor. The biogeochemistry of mercury in the environment. Elsevier/North Holland Biomedical Press, Amsterdam, The Netherlands.

Bishai, H. M. 1962. Reactions of larval and young salmonids to different hydrogen ion concentrations. Journal du Conseil 27:181-191.

Cairns, J., Jr., K. L. Dickson, and A. W. Maki, editors. 1978. Estimating the hazard of chemical substances to aquatic life. American Society for Testing and Materials, Special Technical Publication 657. Philadelphia.

Carlander, K. D. 1969. Handbook of freshwater fishery biology, volume 1. Iowa State University Press, Ames.

Cherry, D. S., R. K. Guthrie, J. H. Rodgers, J. Cairns, Jr., and K. L. Dickson. 1976. Responses of mosquitofish *(Gambusia affinis)* to ash effluent and thermal stress. Transactions of the American Fisheries Society 105:686-694.

DeGraeve, G. M. 1982. Avoidance response of rainbow trout to phenol. The Progressive Fish-Culturist 44:82-87.

Enger, P. S., A. J. Kalmijn, and O. Sand. 1989. Behavioral investigations on the functions of the lateral line and inner ear in predation. Pages 575-587 in S. Coombs et al., editors. The mechanosensory lateral line. Springer-Verlag, New York.

Geckler, J. R., W. B. Horning, T. M. Neiheisel, Q. H. Pickering, E. L. Robinson, and C. E. Stephan. 1976. Validity of laboratory tests for predicting copper toxicity in streams. EPA, Report 600/3-76-116, Environmental Research Laboratory, Duluth, Minnesota.

Hara, T. J. 1986. Role of olfaction in fish behavior. Pages 152-176 in T. J. Pitcher, editor. The behavior of teleost fishes. Croom Helm, London.

Hasler, A. D. 1957. The sense organs: olfactory and gustatory senses of fishes. Pages 187-209 in M. E. Brown, editor. Physiology of fishes, Academic Press, New York.

Hatch, R., and W. L. Ott. 1968. Determination of sub-microgram quantities of mercury by atomic absorption spectrophotometry. Analytical Chemistry 40:2085-2087.

Hill, J. 1989. Analysis of six foraging behaviors as toxicity indicators, using juvenile smallmouth bass exposed to low environmental pH. Archives of Environmental Contamination and Toxicology 18:895-899.

Hoglund, L. B. 1961. The reactions of fish in concentration gradients. Report of the Institute of Freshwater Research, Drottningholm 43:1-147.

Jansen, W. A., and J. H. Gee. 1988. Effects of water acidity on swimbladder function and swimming in the fathead minnow, *Pimephales promelas*. Canadian Journal of Fisheries and Aquatic Sciences 45:65-77.

Jones, K. A., T. J. Hara, and E. Scherer. 1985. Behavioral modifications in arctic char *(Salvelinus alpinus)* chronically exposed to sublethal pH. Physiological Zoology 58:400-412.

Kleerekoper, H. 1976. Effects of sublethal concentrations of pollutants on the behavior of fish. Journal of the Fisheries Research Board of Canada 33:2036-2039.

Klemm, D. J. 1985. Distribution, life cycle, taxonomy, and culture methods. 3. Fathead minnow *(Pimephales promelas)*. Pages 112-125 in U. S. Environmental Protection Agency. Methods for measuring the acute toxicity of effluents to freshwater and marine organisms. EPA, Report 600/4-85/013, Cincinnati, Ohio.

Leino, R. L., J. H. McCormick, and K. M. Jensen. 1990. Multiple effects of acid and aluminum on brood stock and progeny of fathead minnows, with emphasis on histopathology. Canadian Journal of Zoology 68:234-244.

Leino, R. L., P. Wilkinson, and J. G. Anderson. 1987. Histopathological changes in the gills of pearl dace, *Semotilus margarita*, and fathead minnows, *Pimephales promelas*, from experimentally acidified Canadian lakes. Canadian Journal of Fisheries and Aquatic Sciences 44 (Supplement 1):126-134.

Lemly, A. D., and R. J. F. Smith. 1985. Effects of acute exposure to acidified water on the behavioral response of fathead minnows, *Pimephales promelas*, to chemical feeding stimuli. Aquatic Toxicology 6:25-36.

Lemly, A. D., and R. J. F. Smith. 1987. Effects of chronic exposure to acidified water on chemoreception of feeding stimuli in fathead minnows *(Pimephales promelas)*: mechanisms and ecological implications. Environmental Toxicology and Chemistry 6:225-238.

Lewis, F. G., III, and R. J. Livingston. 1977. Avoidance of bleached kraft pulpmill effluent by pinfish *(Lagodon rhomboides)* and gulf killifish *(Fundulus grandis)*. Journal of the Fisheries Research Board of Canada 34:568-570.

Little, E. E. 1990. Behavioral toxicology: stimulating challenges for a growing discipline. Environmental Toxicology and Chemistry 9:1-2.

Little, E. E., and S. E. Finger. 1990. Swimming behavior as an indicator of sublethal toxicity in fish. Environmental Toxicology and Chemistry 9:13-19.

Longbottom, J. E., and J. J. Leichtenberg, editors. 1982. Methods for organic chemical analysis of municipal and industrial wastewater. U. S. Environmental Protection Agency, Report 600/4-82-057, National Technical Information Service, Springfield, Virginia.

McCauley, R. W. 1977. Laboratory methods for determining temperature preference. Journal of the Fisheries Research Board of Canada 34:749-752.

Mills, K. H., S. M. Chalanchuk, L. C. Mohr, and I. J. Davies. 1987. Responses of fish populations in Lake 223 to 8 years of experimental acidification. Canadian Journal of Fisheries and Aquatic Sciences 44 (Supplement 1):114-125.

Montgomery, J. C. 1989. Lateral line detection of planktonic prey. Pages 561-574 in S. Coombs et al., editors. The mechanosensory lateral line. Springer-Verlag, New York.

Mount, D. I. 1973. Chronic effect of low pH on fathead minnow survival, growth and reproduction. Water Research 7:987-993.

NAS-NAE (National Academy of Sciences -

National Academy of Engineering) Committee on Water Quality Criteria. 1973. Water quality criteria 1972. U. S. Government Printing Office, Washington, District of Columbia.

Pedder, S. C., and E. J. Maly. 1985. The effect of lethal copper solutions on the behavior of rainbow trout. Archives of Environmental Contamination and Toxicology 14:501-507.

Sprague, J. B. 1964. Avoidance of copper-zinc solutions by young salmon in the laboratory. Journal of the Water Pollution Control Federation 36:990-1004.

Sprague, J. B. 1968. Avoidance reactions of rainbow trout to zinc sulphate solutions. Water Research 2:367-372.

Sprague, J. B., and D. E. Drury. 1969. Avoidance reactions of salmonid fish to representative pollutants. Pages 169-179 *in* S. H. Jenkins, editor. Advances in water pollution research. Pergamon Press, New York.

Sprague, J. B., P. F. Edson, and R. L. Saunders. 1965. Sublethal copper-zinc pollution in a salmon river-a field and laboratory study. International Journal of Air and Water Pollution 9:531-543.

USEPA (U. S. Environmental Protection Agency). 1979. Methods for chemical analysis of water and wastes. EPA, Report 625/5-74-003 (Revised 1983), Washington, District of Columbia.

USEPA (U. S. Environmental Protection Agency). 1988. Ambient water quality criteria for aluminum - 1988. EPA, Report 440/5-86-008, Washington, District of Columbia.

Weber, C. I., and thirteen coauthors. 1989. Short-term methods for estimating the chronic toxicity of effluents and receiving waters to freshwater organisms. Second edition. U. S. Environmental Protection Agency, Report 600/4-89/001, Environmental Monitoring Systems Laboratory, Cincinnati, Ohio.

Weir, P. A., and C. H. Hine. 1970. Effects of various metals on behavior of conditioned goldfish. Archives of Environmental Health 20:45-51.

Wildish, D. J., H. Akagi, and N. J. Poole. 1977. Avoidance by herring of dissolved components in pulpmill effluents. Bulletin of Environmental Contamination and Toxicology 18:521.

Wilson, D. W. 1973. The ability of herring and plaice larvae to avoid concentrations of oil dispersants. Pages 589-602 *in* J. H. S. Blaxter, editor. The early life history of fish. Springer-Verlag, New York.

Yoshii, K., and K. Kurihara. 1983. Role of cations in olfactory reception. Brain Research 274:239-248.

Zischke, J. A., J. W. Arthur, K. J. Nordlie, R. O. Hermanutz, D. A. Standen, and T. P. Henry, 1983. Acidification effects on macroinvertebrates and fathead minnows (*Pimephales promelas*) in outdoor experimental channels. Water Research 17:47-63.

Methodology

Behavioral Methods for Assessing Impacts of Contaminants on Early Life Stage Fishes

EDWARD E. LITTLE, JAMES F. FAIRCHILD, AND AARON J. DELONAY

U. S. Fish and Wildlife Service
National Fisheries Contaminant Research Center
4200 New Haven Road, Columbia, Missouri 65201, USA

Abstract.— Laboratory studies with early life stage fishes show that these stages are typically less tolerant of environmental contamination than juveniles or adults. Laboratory observations of contaminant effects in early life stage fishes may not always correlate with the response of natural populations to contaminants because the more pervasive and subtle effects of sublethal exposure that diminish long-term survival may not be detected. Many behavioral responses are objective and readily quantifiable and they can be measured effectively in a variety of fish species to characterize the consequences of sublethal exposure to a range of chemical stressors. Methods that measure behavioral responses during toxicity tests may be useful in the assessment of toxic substances. Changes in responses such as swimming activity or respiratory rate are often effective indicators of changes in water quality, however the biological significance of such changes may be unclear. Alterations in feeding or predator avoidance behavior indicate injurious behavioral aberrations that would limit survival of the individual. In addition to demonstrating contaminant effect and injury, symptomatic behavioral responses may aid in the identification of classes of unknown contaminants. Behavioral data may also be predictive of impacts to populations. Impairment of critical behavioral functions may be reflected in population-level responses such as reduced population size, species numbers, biomass, and year-class success.

Behavior consists of a series of observable activities that operate through the central nervous system and enable an animal and its progeny to exist in an environment beneficial to its survival (Keenleyside 1979; Beitinger 1990). Early developmental stages of fishes are marked not only by transitions in morphological development, but also by the emergence of behavioral functions that will help ensure survival (Brown 1985; Huntingford 1986). Behavioral and morphological development are so intimately linked in early life stage fishes that certain developmental stages are often identified by alterations in, or the appearance of, behavioral functions, such as side plough and plough locomotor postures of Atlantic salmon *(Salmo salar)* alevins (Dill 1977). Behavior of early life stage fishes is therefore not a random sequence of activities, but is structured and predictable, of significant adaptive value, and essential to the fish's existence.

Behavioral responses are important to survival because they are necessary to perform essential life functions such as habitat selection, competition, predator avoidance, prey selection, and reproduction (Little et al. 1985). For example, selection of a habitat suitable for survival and reproductive success requires that the fish be capable of responding to appropriate environmental stimuli to locate beneficial habitats and to avoid less favorable environmental conditions. Alteration of the competitive interaction among species and individuals by impairing one competitor's ability to exploit resources or diminishing its capacity to ensure access to those resources may have significant consequences for fish populations and communities (Schoener 1974). Changes in the ability of a fish to detect, pursue, capture, and consume prey will have a considerable influence on its growth and survival (Brown et al. 1987). Conversely, impaired ability of fish to detect and respond appropriately to predators can increase predation-induced mortality.

Since appropriate behavioral function is obviously crucial to the survival of early life stage fishes, behavioral investigations are particularly pertinent when evaluating the effects of environmental contaminants on fishes. A diversity of quantifiable behavioral responses can be affected by changes in water quality (Table 1). These responses vary in complexity, ranging from reflexive responses to complex social reactions. Behavioral responses

TABLE 1.—Behavioral responses of various fish life stages measured during aquatic toxicity tests.

Respiration	Learning
Coloration	Swimming
Equilibrium	Schooling/shoaling
Rhythmicity	Feeding/predation
Thermotaxis	Predator avoidance
Rheotaxis	Competition
Phototaxis	Aggression
Chemotaxis	Reproduction
Avoidance/attraction	Migration
Shelter seeking	

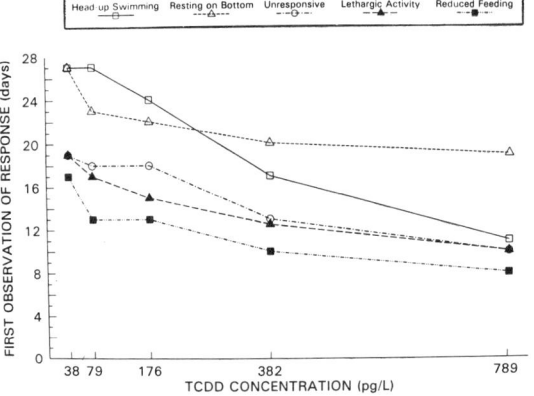

FIGURE 1.—Days of exposure to the dioxin TCDD to induce behavioral changes in rainbow trout exposed as free-swimming juveniles for 28 d (from Mehrle et al. 1988).

most useful in contaminant assessment should be (1) well defined endpoints that are practical to measure, (2) well understood relative to sources of variation, (3) sensitive to a range of contaminants and adaptable to different species, and (4) ecologically relevant (Rand 1985). The accumulated evidence described in several reviews (Olla et al. 1980; Westlake 1984; Little et al. 1985; Rand 1985; Atchison et al. 1987; Beitinger 1990) indicates that behavioral functions are usually quite sensitive to contaminant exposure and are often the first responses exhibited by fishes exposed to toxic substances. In addition to being directly affected by an environmental stressor, behavioral responses, such as avoidance/attractance or feeding, may also significantly influence the level or duration of exposure actually experienced by an organism through contact with their environment or diet (Beitinger 1990).

A number of behavioral changes may occur during exposure to a contaminant, depending on the type of chemical, concentration, and duration of exposure. For example, during a 28-d exposure of juvenile rainbow trout (Oncorhynchus mykiss) to the dioxin, TCDD, five different behavioral responses were altered (Figure 1; Mehrle et al. 1988). Reduced feeding was evident after 7 d of exposure to 789 pg/L; hypoactivity and diminished responsiveness to external stimuli were evident by day 10 of exposure; abnormal swimming postures were noted by day 12, and severe lethargy occurred by day 19. The appearance of a particular abnormality varied with concentration. Reduced feeding appeared after 7 d of exposure at 789 pg/L, by day 14 at 176 pg/L, and by day 17 at 38 pg/L. Thus, as with other toxicological endpoints, the type of response or behavioral symptom and the degree of response or severity of dysfunction varied with the concentration and duration of exposure.

Regardless of their sensitivity, utility, or biological significance, behavioral responses have not been routinely included in the standard aquatic toxicity assessment programs (Little 1990). The development of water-quality criteria often relies heavily on measured concentrations of chemicals causing mortality and impaired growth in laboratory exposures, with limited data on the chronic effects of exposure, such as impaired growth, on aquatic organisms. Acute mortality, however, may not necessarily be the best predictor of survival or toxicant effect in the field, where more pervasive and subtle sublethal exposures are prevalent. Most natural fish populations are exposed to sublethal concentrations of toxicants that can diminish long-term adaptability and survival by altering or impairing critical behavioral functions (Kleerekoper 1976). Investigation of the sublethal behavioral effects of contaminants on early life stage fishes is particularly relevant to the toxicity assessment process, because these rapidly developing life stages are typically more sensitive to environmental contamination than juveniles or adults (Mayer and Ellersieck 1986) and are less tolerant of long-term sublethal alterations of critical organism function. Unfortunately, due to the difficulty of working with early life stage fishes and the complexity involved in quantifying responses among rapidly developing organisms, most of the data regarding the effects of contaminants on behavioral responses reported in the literature have been measured in juvenile and adult life stages. However, many of these responses can be successfully adapted to earlier life stages.

The inclusion of behavioral test methods, especially early life stage tests, in toxicity evaluations would likely increase the sensitivity, ecological relevance, and predictive capability of contaminant assessments. In developing behavioral test methods for contaminant assessment, responses used in behavioral toxicology must address one or more of three assessment objectives, including (1) how well does the response measure effect or injury arising from exposure, (2) does the response aid in the identification of the toxic agent, and (3) does the use of the behavioral response increase the capability of contaminant assessments to predict the ecological consequences of exposure. In the following discussion we will examine the status of behavioral toxicology relative to these three assessment objectives and will describe techniques and approaches that address each objective. Behavioral test results featuring the most broadly studied species, rainbow trout, will be used to summarize and illustrate each objective.

Measurements of Effect and Injury

In measurements of toxicant effect the behavioral responses might not be clearly understood in terms of the survival of the affected organism or of the impact that such changes may have on the population, instead the behavioral responses provide a signal for changes in water quality. For example, a 20% reduction in swimming activity may provide a consistent, quantitative indication of a decline in water quality, yet be benign in terms of long-term survival. Many behavioral and physiological responses fall into this category.

Swimming behavior is well studied as a measure of toxicant effect (Little and Finger 1990) and has become one of the most common behavioral measures. Swimming can be measured in many ways, including the frequency and duration of swimming movements, as well as the form and pattern of swimming. Descriptions of swimming form and pattern include frequency and angle of turns, distance of movement, linear and angular velocity, position in the water column, and swimming posture (Kleerekoper 1974). The measurement of these swimming responses is usually limited to the laboratory. In contrast, assessment of fishes at contaminated field sites currently isn't possible since species-typical responses have not been defined to permit assessment of the normality of behavioral function of fishes from contaminated sites, except for the most extreme aberration. In the laboratory, the subtle changes that arise from sublethal exposures may be confirmed through comparisons with controls or with responses observed during a pre-exposure period.

Measurement of swimming frequency and dura-

TABLE 2.—Effects of contaminants on the pattern or frequency of swimming behavior and survival of rainbow trout.

Toxicant	Lowest behaviorally effective concentration (mg/L)	Percentage of lethal exposure
Aluminum[a]	0.510	6.0
Chlordane[b]	0.001	2.5
Copper[c]	0.005	12.0
DEF[b]	0.100	15.0
2,4-DMA[b]	2.500	2.5
Diquat[d]	0.500	0.5
Lindane[e]	0.060	25.0
Methyl-parathion[b]	0.010	0.2
Penta-chlorophenol[b]	0.007	2.5

[a]Freeman and Everhart 1971. [b]Little et al. 1990.
[c]Waiwood and Beamish 1978. [d]Dodson and Mayfield 1979.
[e]Poels 1977.

tion is highly effective in detecting toxic chemicals during on-site effluent tests. Finger and Bulak (1989) studied sites along three rivers in South Carolina where recruitment failure of striped bass *(Morone saxatilis)* populations had occurred. These sites were located in areas receiving effluent from several paper mills. The test involved exposing individual larval striped bass to a dilution series of water from the three rivers for 96 h. In addition to monitoring survival, swimming behavior was measured each day by counting the number of times the fish moved during a 2-min period. Significant reduction in swimming activity was observed within 24 h of exposure to full strength water from the Wateree River, while significant mortality did not occur until after 72 h of exposure. Although water from the other rivers was not lethal over the 96-h exposure, swimming behavior was significantly reduced after 48 h among fish exposed to 100% and 50% river water. Subsequent analysis revealed the presence of pentachloroanisol in all three rivers, but the concentration in the Wateree River was twice that of the other two rivers.

Behavioral changes, such as altered swimming behavior, usually occur much earlier than mortality when fish are exposed to contaminants. A review by Little and Finger (1990) revealed that the lowest behaviorally effective toxicant concentration that induced changes in swimming behavior of fishes ranged from 0.1% to 5.0% of the LC50 (Table 2 summarizes results from tests with rainbow trout). When observations were made over time, behavioral changes commonly occurred 75% earlier than the onset of mortality. Swimming behavior is affected by many chemical stressors, including metals, organochlorines, and industrial contaminants. Concentrations that induced behavioral changes often were considerably lower than the concentrations that caused mortality (Little and Finger 1990). Measurements of swimming activity may also rival or surpass growth as an index of sublethal exposure because swimming is often affected before reductions in growth are detected. For example, swimming activity was significantly reduced in rainbow trout after a 96-h exposure to an organophosphorus defoliant at concentrations affecting growth only after 30 d (Little et al. 1990). Development of locomotor responses, frequency of movements, and duration of activity were significantly inhibited in brook trout *(Salvelinus fontinalis)* alevins at lower aluminum concentrations under acidic conditions than those that affected survival or growth (Cleveland et al. 1991). Swimming responses have been employed in automated biomonitoring systems because of their consistent sensitivity to numerous contaminants (Miller et al. 1982; Smith and Bailey 1988).

As with physiological or biochemical responses that may signal a change in water quality, the ecological consequences of aberrant swimming behavior are presumably manifested through the impairment of other adaptive behaviors such as migration, predation, or predator success. Hypoactivity and hyperactivity, as well as deviations in adaptive diurnal rhythmicity, may disrupt feeding and increase vulnerability to predation (Steele 1983). Laurence (1972) found that heightened activity may increase an organism's vulnerability to predation, whereas reductions in swimming activity lessen the chance of encountering prey by reducing search areas.

Behavioral measures of injuries include behavioral changes that clearly represent a harmful response that would directly limit the individual's survival and long-term viability. Predator-prey interactions measure the ability of fish to feed, as well as their ability to avoid being eaten by other organisms. Predator-prey interactions provide a good example of a behavioral measure of injury because deficits in either response will have immediate implications for the organism's survival (Brown et al. 1987). Feeding and prey vulnerability have been used to examine sublethal contamination because predator and prey may be behaviorally affected by toxicants (Sandheinrich and Atchison 1990).

Survival of early life stage fishes will decline if the development of foraging behavior is delayed or inhibited. Impaired feeding behavior also correlates with reduced growth, which may lengthen a fish's

TABLE 3.—Effects of contaminants on the feeding behavior of juvenile rainbow trout.

Toxicant	Lowest behaviorally effective concentration (mg/L)	Percentage of lethal exposure
Aluminum[a]	0.052	2.0
Chlordane[b]	0.001	2.5
Copper[c]	0.100	10.0
DEF[b]	0.056	8.0
2,4-DMA[b]	25.000	25.0
Methyl-parathion[b]	0.100	20.0
Penta-chlorophenol[b]	0.007	2.5

[a]Freeman and Everhart 1971. [b]Little et al. 1990.
[c]Waiwood and Beamish 1978.

period of vulnerability to predation (Werner and Hall 1974) and impair overwinter survival (Oliver et al. 1979). Because changes in the ability of a fish to detect, pursue, capture, and consume prey will affect growth and survival, these behavioral variables provide a measure of contaminant-induced injury. Contaminants have been shown to affect every aspect of the feeding sequence, including detection of prey (Lemly and Smith 1987), prey capture (Little et al. 1990), handling time, and ingestion of prey (Sandheinrich and Atchison 1990), as well as general motivation to feed (Little et al. 1990). Thus, several variables of feeding behavior can be measured in toxicity studies (Sandheinrich and Atchison 1990). The experimental environment for such studies can vary in complexity depending on the amount of ecological realism one hopes to achieve, but it can also be readily adapted to standard toxicity testing procedures (Mathers et al. 1985).

A number of studies have shown that feeding behavior is impaired by sublethal exposure to a diversity of contaminants. Responses summarized in Table 3 for rainbow trout are typical of those reported in the literature for other species. Reductions in feeding have been observed after sublethal exposure to metals (Atchison et al. 1987), organophosphates (Bull and McInerney 1974), dioxins (Mehrle et al. 1988), and petroleum hydrocarbons (Woodward et al. 1987). Feeding was sensitive in detecting no-effect concentrations of aluminum in acid-exposed brook trout *(Salvelinus fontinalis)* and was among the most sensitive indices of sublethal pH and aluminum exposure (Cleveland et al. 1989). The sensitivity of feeding behavior as an index of sublethal exposure among rainbow trout exposed to various agricultural chemicals ranged from < 0.3% to 50% of the LC50 (Little et al. 1990). Inhibited motivation to feed seemed to be the predominant effect at higher con-

TABLE 4.—Effects of contaminants on the vulnerability of early juvenile rainbow trout to predation.

Toxicant	Lowest behaviorally effective concentration (mg/L)	Percentage of lethal exposure
Carbaryl[a]	0.010	0.5
Chlordane[a]	0.02	5.0
DEF[a]	0.05	5.0
2,4-DMA[a]	50.0	50.0
Methyl-parathion[a]	0.1	2.7
Penta-chlorophenol[a]	0.0002	0.5

[a] Little et al. 1990.

centrations, whereas reduced feeding efficiency and reduced strike frequencies were the predominant impairments at lower concentrations.

Predation tests measure the ability of prey to escape predation. Practically all organisms are vulnerable to predation during some portion of their life cycle. Increased vulnerability to predation may occur when a toxicant alters the ability of fish to detect or respond to predators (Brown et al. 1985). The general method for measuring predation in the laboratory is to combine predators with contaminant-exposed and unexposed prey, then count the surviving control and exposed fish when approximately 50% of the prey have been captured (Little et al. 1985). Structural complexity of the experimental setting, including the addition of refuges for prey, is used to increase the environmental realism of the test. A range of contaminants has been shown to increase predation-induced mortality (Table 4). Little et al. (1990) reported that different types of toxicants at concentrations as low as 2% of the lethal concentration clearly affected the response. The response did not consistently follow a typical dose-response relationship, however, because the behavioral aberrations resulting from the exposure sometimes made the prey less conspicuous to the predators owing to inactivity or reduced mobility.

Contaminant exposure may also cause aberrations or inhibitions of other behaviors such as schooling, shelter seeking, and locomotor responses, which may influence predator-prey interactions. For example, schooling declined following exposure of coho salmon (*Oncorhynchus kisutch*) to DDT (Besch et al. 1977) and of fathead minnows (*Pimephales promelas*) to the herbicide, 2,4-D (Holecombe et al. 1980). Assessment of predator-prey behaviors during toxicity tests provides not only a sensitive measure of toxicant effect, but also a measure of injury, since growth and predation-induced mortality are affected by exposure.

Behavior as an Index for Identifying Toxicants

Although there are many ways behavior can be used to determine effects of exposure to toxicants, the use of behavior as an index for toxicant identification is somewhat more limited because there are no known behavioral responses that are indicative of exposure to specific toxic agents. However, there are two approaches that may be valuable in identifying general classes of contaminants.

In an approach developed by Drummond et al. (1986), 17 behavioral responses were monitored during acute exposures of fathead minnows to more than 400 single compounds. Statistical pattern recognition, based on the types of behavioral changes induced by exposure, was used to identify three general responses (Type I, II, III) which correlated with three classes of contaminants.

Type I responses were indicative of narcosis-producing chemicals such as ethers, alcohols, ketones, and phthalates. These chemicals depress central and peripheral nervous system activities. Exposed fish exhibit depressed locomotor activity, loss of startle responses, rapid and shallow opercular rates, darkened coloration, and tetany.

Type II responses were indicative of chemicals such as rotenone, benzene, and phenol, which disrupt metabolic activity. Exposed fish exhibit heightened locomotor activity, hyperactivity to stimulation, increased rate and amplitude of opercular activity, slight darkening, and edema.

Type III responses were indicative of neurotoxic chemicals such as carbamates, organophosphates, caffeine, and strychnine. Exposed fish exhibit depressed locomotor activity, hyperactivity to stimulation, convulsions, spasms, tetany, scoliosis, lordosis, or vertebral hemorrhage.

A second approach, developed by Diamond et al. (1990), examined frequency and amplitude of the bluegill (*Lepomis macrochirus*) opercular rhythms and cough responses to differentiate contaminant types (Figure 2). For example, zinc at 300 μg/L typically reduces the amplitude of the respiratory response. Dieldrin, an organochlorine insecticide, increases ventilatory frequency at concentrations above 24 μg/L and causes cough responses and erratic movements. This approach has proven effective in detecting the presence of metals in complex mixtures (Diamond, personal communication).

Behavior as a Predictor of Contaminant Effects in the Natural Environment

Impaired behavioral performance may be predic-

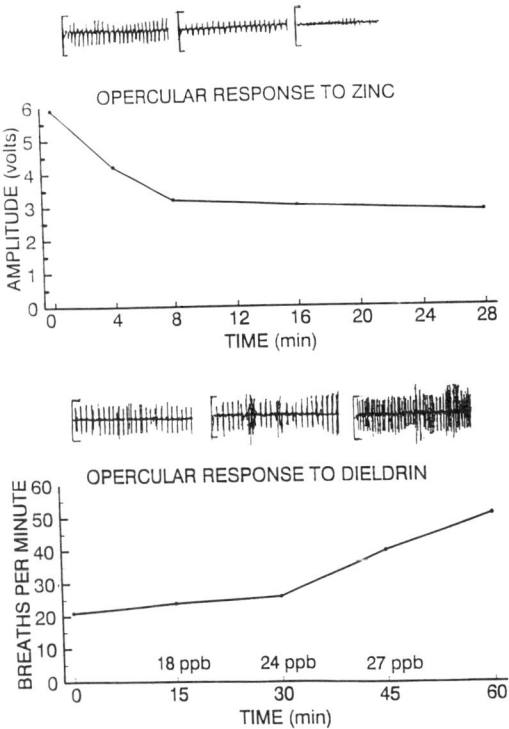

FIGURE 2.—Amplitude of opercular movements of juvenile bluegill during exposure to 300 ppb zinc and frequency of opercular movements induced in bluegill during exposure to three concentrations of dieldrin and a control. Oscillograph of opercular movements is shown above each graph, each vertical deflection represents one full respiratory cycle of gill opening and closing (from Diamond et al. 1990).

tive of contaminant effects in the field, when ecological consequences can be linked to impaired behavioral performance. Disruptions of essential functions such as habitat selection, competition, predator-prey relationships, or reproduction can become apparent ecologically through loss of population or changes in year-class strength when enough individuals are affected.

Avoidance reactions to contaminants provide an example of the type of behavioral measure useful as a predictor of ecological effects in contaminant assessment. The emigration of fishes from an affected area, resulting from an avoidance reaction to a chemical substance (Atchison et al. 1987), would be reflected in several ecological indices, including locally reduced biomass, loss of species diversity, diminished year-class strength, and loss of productivity and diversity in that area.

The first possible reaction that fishes may make in response to a toxicant is avoidance or attraction (Slobodkin 1968). Avoidance reactions provide short-term protection because they minimize exposure. In the long run, however, avoidance responses can be detrimental because organisms may be displaced from preferred habitats to suboptimal areas (Atchison et al. 1987), where they may face greater competition and predation pressure or inadequate resource availability. Preferences or aversions to contaminated space or food are useful in risk assessment because the probability of contaminant exposure may be behaviorally mediated as organisms detect and avoid contaminants in the field. Many contaminants induce avoidance responses and several reviews have been made of this extensive literature (Cherry and Cairns 1982; Giattina and Garton 1983; Hara et al. 1983; Beitinger 1990). Avoidance data from studies with rainbow trout are summarized in Table 5. The sensitivity of avoidance responses ranged from less than 3% of the LC50 for the herbicide, 2,4-D, to over 1,000 times the LC50 for the chlorinated biphenol, arochlor 1254 (Little et al. 1985). Similar ranges of responses to chemicals are seen in a number of species that have been examined in avoidance testing. Generalizations regarding the avoidance of aquatic contaminants by fishes are difficult to make because of the variety of species and experimental designs used to test behavioral responses, as well as variations in the modes and sites of action of the chemicals studied (Giattina and Garton 1983). In a recent review of published avoidance data for over 75 different chemicals, Beitinger (1990) found that roughly a third of the chemicals were avoided, whereas the others either failed to elicit a response or induced inconsistent responses. Generically similar organic compounds may not induce similar reactions, and closely related fishes may not respond similarly to the same compound (Little et al. 1985). Avoidance reactions are also influenced by variables such as flow and temperature (Kleerekoper 1976).

The avoidance response has been confirmed in a number of field studies. Sprague et al. (1965) diverted stream water to form a natural Y-maze and dosed the water with a mixture of copper and zinc. The avoidance thresholds observed in this study were even lower than those observed in the lab. Avoidance by a wild fish population was also confirmed in a field study by Geckler et al. (1976), who examined the effects of copper contamination in an intentionally dosed stream, Shyler Creek. This study concluded that among the toxicological and ecological responses observed, the avoidance reaction would have the greatest ecological impact, but it would not have been predicted from standard laboratory toxicity tests. Gray (1990), using telemetry, documented the avoidance of oil-contaminated water and gas-supersaturated water by free ranging fishes. Hartwell et al. (1987) determined that fathead minnows would avoid a 73.5 µg/L mixture of copper, chromium, arsenic, and selenium in a natural stream, and similarly would avoid 34.3 µg/L in an artificial stream.

Because the avoidance response has been verified as a response to contaminants in the field, it has been legally accepted as evidence of injury for Natural Resource Damage Assessment under proceedings of the Comprehensive Environmental Response, Compensation, and Liability Act of 1980 (NRDA 1986). However, with the exception of avoidance responses, the link between behavioral impairment and injurious population or community effects has not been well documented in the natural environment. It is this lack of well established causal links between behavioral responses and population effects, not the biological significance of the response, that limits the utility of behavioral data as predictive indices of field effects and measures of injury.

Verification of behavioral effects in the field is an important step in understanding the causal relation between observed behavioral changes and the impact of contaminants on natural populations and communities (Sandheinrich and Atchison 1990). However, relating behavioral change to higher ecological organization in the field poses significant technical challenges for behavioral toxicology. Experimental designs are difficult because of the mobility of the organisms, physical constraints of the environment, verification of exposure, and manpower required to accomplish such determinations (Little 1990).

Although most behavioral effects observed in laboratory studies have not yet been directly linked to ecological responses, they can still provide protective criteria which correspond with no-effect concentrations documented in the field. For example, in laboratory studies of bluegill exposed to the hydrocarbon, fluorene, the feeding-response was the most sensitive laboratory endpoint (Finger et al. 1985). This is particularly significant since these feeding-response data were the only laboratory observations that accurately predicted an adverse impact on growth and survival of bluegills in companion pond ecosystem studies. The behavioral response observed in the laboratory was not causally linked to an ecological impact, but it was predictive of field effects because it indicated a concentration that affected the mesocosm population.

Many of the ecological measurements obtained during field studies do not provide the resolution to link causal relationships between behavioral changes and the response of natural fish populations. Inhibited growth among fishes from field populations, for example, may be due to many factors, and this is where behavioral studies, even those conducted in the laboratory, could be useful in interpreting and defining the mechanisms of ecotoxicity.

Conclusions

Behavioral measurements may be useful as indicators of sublethal contamination because they frequently occur below concentrations that are chronically lethal and at lower concentrations than those that affect growth. Consequently, behavioral tests provide definable, interpretive endpoints that can be used for regulatory purposes in damage assessments and in the formulation of water-quality criteria.

Certain behavioral indices have been used to measure contaminant effects and injury, to provide diagnostic identification of contaminant classes, and to predict ecological impact. The extent to which a behavioral variable is used in these aspects of contaminant assessment depends upon the complexity of the behavioral variable and the extent to which the response is understood within different levels of biological organization. Certain elemental responses, such as frequency of movements, underlie more complex responses, such as feeding. Changes in the elemental responses may be excellent signals of change in environmental quality, yet be of little apparent significance to the organism's long-term survival. Likewise, growth inhibitions reflecting reduced feeding efficiency, although deleterious for the organisms, may not be directly evident in the status of the population or the community.

The challenges for behavioral toxicology are clear. More and better behavioral testing procedures, especially those using sensitive early life stage fishes, must be developed and refined into effective tools which can be readily and clearly applied in contaminant assessments. Secondly, more work must be done to establish links between behavioral effects observed in the laboratory and ecological effects observed in the field to aid in the development of contaminant assessments that are adequately predictive of population and community response. Finally, behavioral toxicologists must continue to work to dispel the erroneous paradigm prevalent in aquatic toxicology that suggests that behavioral responses are not suitable for inclusion in contaminant evaluation and assessment because they are either unquantifiable, too complex, too variable, extraordinarily difficult to measure, not biologically significant, or lacking in ecological relevance.

References

Atchison, G. J., M. G. Henry, and M. B. Sandheinrich. 1987. Effects of metals on fish behavior: a review. Environmental Biology of Fishes 18:11-25.

Beitinger, T. L. 1990. Behavioral reactions for the assessment of stress in fishes. Journal of Great Lakes Research 16:495-528.

Besch, W. K., A. Kemball, K. Meyer-Waarden, and B. Scharf. 1977. A biological monitoring system employing rheotaxis of fish. Pages 56-74 in J. Cairns, Jr., K. L. Dickson and G. F. Westlake, editors. Biological monitoring of water and effluent quality. American Society for Testing and Materials, STP 607, Philadelphia.

Brown, J. A. 1985. The adaptive significance of behavioral ontogeny in some centrarchid fishes. Environmental Biology of Fishes 13:25-34.

Brown, J. A., P. H. Johansen, P. W. Colgan, and R. A. Mathers. 1985. Changes in the predator-avoidance behavior of juvenile guppies (Poecilia reticulata) exposed to pentachlorophenol. Canadian Journal of Zoology 63:2001-2005.

Brown, J. A., P. H. Johansen, P. W. Colgan, and R. A. Mathers. 1987. Impairment of early feeding behavior of largemouth bass by pentachlorophenol exposure: a preliminary assessment. Transactions of the American Fisheries Society 116:71-78.

Bull, C. J., and J. E. McInerney. 1974. Behavior of juvenile coho salmon (Oncorhynchus kisutch) exposed to Sumithion (fenitrothion), an organophosphate insecticide. Journal of the Fisheries Research Board of Canada 31:1867-1872.

Cherry, D. S., and J. Cairns, Jr. 1982. Biological monitoring. V. Preference and avoidance studies. Water Research 16:263-301.

Cleveland, L., E. E. Little, C. G. Ingersoll, R. H. Wiedmeyer, and J. B. Hunn. 1991. Sensitivity of brook trout to low pH, low calcium and elevated aluminum concentrations during laboratory pulse exposures. Aquatic Toxicology 19:303-317.

Cleveland, L., E. E. Little, R. H. Wiedmeyer, and D. R. Buckler. 1989. Chronic no-observed-effect concentrations of aluminum for brook trout exposed in low calcium dilute acidic water. Pages 229-246 in T. E. Lewis, editor. Environmental chemistry and toxicology of aluminum. Lewis Publishers, Inc. Chelsea, Michigan.

Diamond, J. M., M. J. Parson, and D. Gruber. 1990. Rapid detection of sublethal toxicity using fish ventilatory behavior. Environmental Toxicology and Chemistry 9:3-12.

Dill, P. A. 1977. Development of behavior in alevins of Atlantic salmon, Salmo salar, and rainbow trout, S. gairdneri. Animal Behaviour 25:116-121.

Dodson, J. J., and C. I. Mayfield. 1979. Modification of the rheotropic response of rainbow trout (Salmo gairdneri) by sublethal doses of the aquatic herbicides diquat and simazine. Environmental Pollution 18:147-157.

Drummond, R. A., C. L. Russom, D. L. Geiger, and D. L. DeFoe. 1986. Behavioral and morphological changes in fathead minnows, Pimephales promelas, as diagnostic endpoints for screening chemicals according to modes of action. Pages 415-435 in T. M. Poston and R. Purdy, editors. Aquatic toxicology. 9th Aquatic Toxicity Symposium. American Society for Testing and Materials, STP 921, Philadelphia.

Finger, S. E., and J. S. Bulak. 1989. Toxicity of water from three South Carolina rivers to larval striped bass. Transactions of the American Fisheries Society 117:521-528.

Finger, S. E., E. E. Little, M. G. Henry, J. F. Fairchild, and T. P. Boyle. 1985. Comparison of laboratory and field assessment of fluorene: Part I. Effects of fluorene on the survival, growth, reproduction, and behavior of aquatic organisms in laboratory tests. Pages 120-133 in T.P. Boyle, editors. Validation and predictability of laboratory methods for assessing the fate and effects of contaminants in aquatic ecosystems. American Society for Testing and Materials, STP 865, Philadelphia.

Folmar, L. C. 1976. Overt avoidance reaction of rainbow trout fry to nine herbicides. Bulletin of Environmental Contamination and

Toxicology 15:509-514.

Freeman, R. A., and W. H. Everhart. 1971. Toxicity of aluminum hydroxide complexes in neutral and basic media to rainbow trout. Transactions of the American Fisheries Society 100:644-658.

Giattina, J. D., and R. R. Garton. 1983. A review of the preference-avoidance responses of fishes to aquatic contaminants. Residue Reviews 87:43-90.

Geckler, J. R., W. B. Horning, T. M. Neiheisel, Q. H. Pickering, E. L. Robinson, and C. E. Stephan. 1976. Validity of laboratory tests for predicting copper toxicity in streams. U. S. Environmental Protection Agency, EPA-600/3-76-116, Environmental Research Laboratory, Duluth, Minnesota.

Gray, R. H. 1990. Fish behavior and environmental assessment. Environmental Toxicology and Chemistry 9:53-68.

Hara, T. J., S. B. Brown, and R. E. Evans. 1983. Pollutants and chemoreception in aquatic organisms. Pages 248-306 in J. O. Nriagu, editor. Aquatic toxicology, John Wiley and Sons, New York.

Hartwell, S. I., D. S. Cherry, and J. Cairns, Jr. 1987. Field validation of avoidance of elevated metals by fathead minnows (Pimephales promelas) following in situ acclimation. Environmental Contamination and Toxicology 6:189-200.

Holcombe, G. W., J. T. Fiandt, and G. L. Phipps. 1980. Effects of pH increases and sodium chloride additions on the acute toxicity of 2,4-dichlorophenol to the fathead minnow. Water Research 14:1073-1077.

Huntingford, F. A. 1986. Development of behavior in fish. Pages 47-68 in T. A. Pitcher, editor. The behavior of teleost fishes. Johns Hopkins University Press, Baltimore, Maryland.

Keenleyside, M. H. A. 1979. Diversity and adaptation in fish behavior. Zoophysiology volume 11. Springer-Verlag, Berlin.

Kleerekoper, H. 1974. Effects of exposure to a subacute concentration of parathion on the interaction between chemoreception and water flow in fish. Pages 237-245 in F. J. Vernberg and W. B. Vernberg, editors. Pollution and physiology of marine organisms. Academic Press, New York.

Kleerekoper, H. 1976. Effects of sublethal concentrations of pollutants on behavior of fish. Journal of the Fisheries Research Board of Canada 33:2036-2039.

Laurence, G. C. 1972. Comparative swimming abilities of fed and starved larval largemouth bass (Micropterus salmoides). Journal of Fish Biology 4:73-78.

Lemly, A. D., and R. J. F. Smith. 1987. Effects of chronic exposure to acidified water on chemoreception of feeding stimuli by fathead minnows, Pimephales promelas: mechanisms and ecological implications. Environmental Toxicology and Chemistry 6:225-238.

Little, E. E. 1990. Behavioral toxicology: stimulating challenges for a growing discipline. Environmental Toxicology and Chemistry 9:1-2.

Little, E. E., R. D. Archeski, B. A. Flerov, and V. I. Kozlovskaya. 1990. Behavioral indicators of sublethal toxicity in rainbow trout. Archives of Environmental Contamination and Toxicology 19:380-385.

Little, E. E., and S. E. Finger. 1990. Swimming behavior as an indicator of sublethal toxicity in fish. Environmental Toxicology and Chemistry 9:13-19.

Little, E. E., B. A. Flerov, and N. N. Ruzhinskaya. 1985. Behavioral approaches in aquatic toxicity: a review. Pages 72-98 in P. M. Mehrle, Jr., R. H. Gray, and R. L. Kendall, editors. Toxic substances in the aquatic environment: an international aspect. American Fisheries Society, Water Quality Section, Bethesda, Maryland.

Mathers, R. A., J. A. Brown, and P. H. Johansen. 1985. The growth and feeding behavior responses of largemouth bass (Micropterus salmoides) exposed to PCP. Aquatic Toxicology 6:157-164.

Mayer, F. L., Jr., and M. R. Ellersieck. 1986. Manual of acute toxicity: interpretation and data base for 410 chemicals and 66 species of freshwater animals. United States Fish and Wildlife Service Resource Publication 160.

Mehrle, P. M., D. R. Buckler, E. E. Little, L. M. Smith, J. D. Petty, P. H. Peterman, D. L. Stalling, G. M. DeGraeve, J. J. Coyle, and W. J. Adams. 1988. Toxicity and bioconcentration of 2,3,7,8-tetrachlorodibenzodioxin and tetrachlorodibenzofuran in rainbow trout. Environmental Toxicology and Chemistry 7:47-62.

Miller, D. C., W. H. Lang, J. O. B. Graeves, and R. S. Wilson. 1982. Investigations in aquatic behavioral toxicology using a computerized video quantification system. Pages 206-220 *in* J. G. Pearson, R. B. Foster, and W. E. Bishop, editors. Aquatic toxicology and hazard assessment: fifth conference. American Society for Testing and Materials, STP 766, Philadelphia.

Natural Resource Damage Assessments: Final Rule. 1986. Federal Register 51:27674-27753.

Oliver, J. D., G. D. Holeton, and K. E. Chua. 1979. Overwintering mortality of fingerling smallmouth bass in relation to size relative energy stores and environmental temperature. Transactions of the American Fisheries Society 108:130-136.

Olla, B. L., W. H. Pearson, and A. L. Studholme. 1980. Applicability of behavioral measures in environmental stress assessment. Rapports et Procès-Verbaux des Réunions Commission Internationale pour l'Exploration Scientifique de la Mer Mediterranee Monaco 179:162-173.

Poels, C. L. M. 1977. An automatic system for rapid detection of acute high concentrations of toxic substances in surface water using trout. American Society for Testing and Materials, STP 607:85-95. Philadelphia.

Rand, G. M. 1985. Behavior. *in* G. M. Rand and S. R. Petrocelli, editors. Fundamentals of aquatic toxicology: methods and applications. Hemisphere Publishing, New York.

Sandheinrich, M. B., and G. J. Atchison. 1990. Sublethal toxicant effects on fish foraging behavior: empirical vs mechanistic approaches. Environmental Toxicology and Chemistry 9:107-120.

Schoener, T. W. 1974. Resource partitioning in ecological communities. Science (Washington, District of Columbia) 185:27-39.

Schumacker, P. D., and J. J. Ney. 1980. Avoidance response of rainbow trout *(Salmo gairdneri)* to single dose chlorination in a power plant discharge canal. Water Research 14:651-655.

Slobodkin, L. B. 1968. Towards a predictive theory of evolution. Pages 187-203 *in* R. C. Lewontin, editor. Population biology and evolution. Syracuse University Press, Syracuse, New York.

Smith, E. H., and H. C. Bailey. 1988. Development of a system for continuous biomonitoring of a domestic water source for early warning of contaminants. Pages 182-205 *in* D. S. Gruber and J. M. Diamond, editors. Automated biomonitoring: living sensors as environmental monitors. Ellis Horwood. Chichester, England.

Sprague, J. B. 1964a. Avoidance of copper-zinc solutions by young salmon in the laboratory. Journal of the Water Pollution Control Federation 36:990-1004.

Sprague, J. B. 1964b. Lethal concentrations of copper and zinc for young Atlantic salmon. Journal of the Fisheries Research Board of Canada 21:17-23.

Sprague, J. B. 1968. Avoidance reactions of rainbow trout to zinc sulfate solutions. Water Research 2:367-372.

Sprague, J. B., P. F. Elson, and R. L. Saunders. 1965. Sublethal copper-zinc pollution in a salmon river-a field and laboratory study. International Journal of Air and Water Pollution 9:531-539.

Steele, C. W. 1983. Effects of exposure to sublethal copper on the locomotor behavior of the sea catfish, *Arius felis*. Aquatic Toxicology 4:83-93.

Waiwood, K. G., and F. W. H. Beamish. 1978. The effects of copper, pH, and hardness on the critical swimming performance of rainbow trout, *Salmo gairdneri*. Water Research 12:611-619.

Werner, E. E., and D. J. Hall. 1974. Optimal foraging and the size selection of prey of the bluegill sunfish *(Lepomis macrochirus)*. Ecology 55:1042-1052.

Westlake, G. F. 1984. Behavioral effects of industrial chemicals in aquatic animals. Pages 233-250 *in* J. Saxena, editor. Hazard assessment of chemicals: current developments, Vol 3, Academic Press, Orlando, Florida.

Woodward, D. F., E. E. Little, and L. M. Smith. 1987. Toxicity of five shale oils to fish and aquatic invertebrates. Archives of Environmental Contamination and Toxicology 16:239-246.

The Teratological and Pathological Effects of Contaminants on Embryonic and Larval Fishes Exposed as Embryos: a Brief Review

JOEL E. BODAMMER

*URI/NOAA Cooperative Marine Education & Research Program,
Fisheries, Animal and Veterinary Science Department,
College of Resource Development, University of Rhode Island,
Kingston, Rhode Island 02881, USA*

Abstract.—Teratological abnormalities in embryonic and larval fishes have long been associated with polluted waters. Despite this lengthy relationship, little is presently understood about the precise developmental events that result in the formation of these morphological aberrations. This brief review is restricted to terata that result from exposure of embryos to contaminants. Recent reviews of this subject render any attempt to cover all the pertinent literature redundant; therefore, only general information is considered for work published prior to 1988. The teratogenic response of fish embryos to toxicants is non-specific. However, when carefully studied by quantitative methods in environments stressed by pollution and overfishing, a knowledge of the prevalence of embryos bearing these abnormalities may be useful in studies of recruitment. In addition to terata, pathological lesions in embryos and newly hatched larvae, resulting from contaminant exposure during embryogeny, are also reviewed.

Morphological aberrations (terata) in larval fishes have captured the attention of biologists since the 15th century and have long been associated with anthropogenically altered environments (Laale and Lerner 1981). The scope of this paper is restricted to some of the salient, but more general, information and problems regarding terata resulting from the sublethal exposure of embryos to contaminants, as previously reviewed by Rosenthal and Alderdice (1976), Laale and Lerner (1981), von Westernhagen (1988), and Weis and Weis (1989). To avoid redundancy with these reviews, only selected references from the period before 1988 will be considered, as well as results reported subsequent to that date. A slightly more detailed account of lesions in fish embryos exposed to contaminants is provided, plus a consideration of some of the ways in which embryonic-larval abnormalities are being used to evaluate the effects of contaminants as they exist in the field.

Non-Anthropogenic Environmental Alterations

As emphasized by the above authors, terata have been observed in developing embryos, particularly those collected in the field, whose environment had been affected by abnormal physical conditions rather than anthropogenic ones. For example, severe morphological aberrations (e.g., distorted notochords, cranio-facial defects, aberrant eye development) have been found in embryos whose normal environments have been altered by low oxygen levels and/or extreme salinity or temperature variations (Rosenthal and Alderdice 1976; von Westernhagen 1988; Wiegand et al. 1989). That physical and hydrological factors might influence the development of fish embryos in the natural environment is suggested by the recent study of Purcell et al. (1990) who observed malformed jaws in large numbers of moribund, starving Pacific herring *(Clupea pallasi)* larvae collected from Kulleet Bay (British Columbia) during a period of unseasonably warm weather accompanied by lower than normal midday tides. Having no reason to suspect that contaminants were involved, these authors attributed their findings to the unusual environmental conditions present at the time their observations were made. Therefore, as von Westernhagen et al. (1988) have cautioned, great care must be taken to measure and evaluate all hydrographic conditions and ancillary information on pollutants when examining contaminant effects on embryos and larvae in the field.

Non-Anthropogenic Genetic Considerations

As Laale and Lerner (1981) discussed, little consideration has been given to the problem of terata resulting from the influence of genetic factors on the development of fishes. Like modifications in the physical environment, genetic control of developmental processes may result in aberrant craniofacial development, vertebral abnormalities, and a reduction in eye size. Therefore, it is incumbant upon investigators to have adequate information regarding the frequency of morphological aberrations that may result from genetic factors for fish species they wish to evaluate for contaminant effects either in the laboratory or field. Lonning (1977), for example, indicated that up to 30% of Atlantic cod *(Gadus morhua)* embryos cultured under controlled conditions may undergo aberrant development, leading to death during embryogeny (i.e., cleavage and gastrulation) and therefore, should not be used for studies on contaminant-induced teratogenesis.

Contaminant Responses

Perhaps the most significant problem that exists in evaluating the relationship between contaminants and terata, observed in late-stage embryos or newly hatched larvae, is the *similarity* in morphological defects observed for embryos exposed to the major classes of contaminants (i.e., heavy metals, chlorinated hydrocarbons, petroleum hydrocarbons, altered pH). As detailed in recent reviews (von Westernhagen 1988; Weis and Weis 1989), terata such as axial deformation (notochord abnormalities), cranio-facial defects, brain and eye defects (e.g., exopthalmia, micropthalmia, cyclopia), cardiovascular defects (tube heart), and spinal deformation (vertebral abnormalities) may be induced in developing embryos exposed to a variety of contaminants. Because of the lack of specificity in their responses to the various classes of contaminants, the species-dependent variability of these responses, and the vagaries inherent in the definition of "sublethal," the endpoints used in the vast majority of studies on contaminants have been mortality or hatching success (von Westernhagen 1988; Weis and Weis 1989). Neither of these deals with terata and the effects they have on the animals that possess them.

While precise details of the mechanisms by which contaminants influence the developing organism at various stages of embryogenesis are not known, studies on toxicant effects during the early stages of development (early cleavage to blastopore closure) in various fish species have shown clearly that disruption of early morphogenetic events may ultimately result in the appearance of terata at later stages of development (i.e., late-stage embryos and newly hatched larvae) (Lonning 1977; Struhsaker et al. 1974; Linden 1978; Weis and Weis 1979; Falk-Petersen and Lonning 1984; Kjorsvik 1986; Carls and Rice 1989). Typically, the cellular interactions occurring during early development in contaminant-exposed embryos are either not examined or are difficult to observe (von Westernhagen 1988; Weis and Weis 1989). However, studies on the embryos of a number of marine species have shown that abnormal cleavage patterns and cellular development result from exposure to petroleum hydrocarbons (Kuhnhold 1974; Wilson 1976; Longwell 1977; Lonning 1977; Hughes and Longwell 1981; Falk-Petersen et al. 1982; Stene and Lonning 1984; Kjorsvik 1986). Similar observations have also been made on cod (Dethlefsen 1977) and winter flounder *(Pleuronectes americanus)*, (Smith and Cole 1973) embryos exposed to organochlorine insecticides. Recent experimental studies by Perry et al. (1988) have demonstrated the important relationship between cytogenetic effects (depressed and/or abnormal mitosis) of methylmercury accompanying the formation of terata in *Fundulus heteroclitus*. These and other supportive data from studies on teratogenesis in higher vertebrates have led Laale and Lerner (1981), and more recently Weis and Weis (1989), to conclude that the formation of terata in anthropogenically affected fish may largely result from the partial or complete disruption befalling any of the exquisitely coordinated cellular events known to occur during embryogenesis, especially the early stages of organ formation.

Middaugh et al. (1988) reported on the teratological effects of 2,4-dinitrophenol (DNP), "produced water" (a multicontaminant, saline formation water; a by-product of oil and gas production), and napthalene on embryos of the inland silverside *(Menidia beryllina)*. Exposure of embryos to these different classes of teratogens during cleavage and blastulation until 8 days post-fertilization resulted in cranio-facial, cardiovascular, and skeletal defects which were scored with a severity index as proposed by Weis and Weis (1977). While all three compounds produced morphological abnormalities and demonstrated the stage-dependent sensitivity of *Menidia* embryos to the toxicants, only exposure to DNP at the 2- to 4-cell stage resulted in mortalities that were significantly different from controls.

Another recent study, which emphasized the importance of stage-dependent sensitivity to teratogens, is that by Marty et al. (1990) on the response of medaka *(Oryzias latipes)* embryos to several carcinogenic N-nitroso compounds (N-nitroso-N-methylurea, N-methyl-N'-nitro-N-nitrosoguanidine, N-nitrosodiethylamine) during early blastula stages, early organogenesis, and late stage embryos just before hatching. All three compounds resulted in mortality of embryos exposed as blastulas, were teratogenic for embryos exposed during early organogenesis, and had little effect on embryos exposed just before hatching. Terata recorded for all three compounds were anisopthalmia, tube heart, pericardial edema, and abnormal swim bladder development (failure to inflate).

Biochemical and Metabolic Considerations

Because of the early demonstrations of oxygen depletion and temperature effects on embryonic development, Rosenthal and Alderdice (1976) advanced the hypothesis that some contaminants exert their effect by blocking oxidative phosphorylation, thereby inhibiting the formation of adenosinetriphosphate (ATP) needed for normal metabolism. As these authors have reviewed, compensatory attempts to increase respiratory rates after exposure of Atlantic herring *(Clupea harengus)* to dinitrophenol (DNP) typically fail, and in the early embryo differentiation is arrested or dedifferentiation may occur. Also, exposure of older herring embryos to various levels of DNP results in animals demonstrating numerous types of malformations at hatching. Recent studies on *Menidia beryllina*, referred to in the preceding section (Middaugh et al. 1988), have reconfirmed the earlier observations made on herring by Rosenthal and Alderdice (1976). Similar results on the modification of basal metabolic processes have been reported for rainbow trout *(Oncorhynchus mykiss)* embryos exposed to pentachlorophenol (Hodson and Blunt 1981). As noted by several reviewers (Rosenthal and Alderdice 1976; von Westernhagen 1988), metabolic retardation and ineffective energy utilization have also been observed for fish embryos exposed to aromatic hydrocarbons, organochlorines, and heavy metals. Presently, however, it is not clear whether these toxic compounds act directly to depress energy-related processes or whether energy needed for normal metabolism and growth is diverted toward attempts to detoxify the compounds (von Westernhagen 1988).

As the reader will quickly note upon considering the more detailed reviews provided by others, particularly Rosenthal and Alderdice (1976) and von Westernhagen (1988), the preceding abbreviated treatment of energy-related processes in relation to contaminants represents only one of a myriad of influences that a contaminant or a class of contaminants (e.g., petroleum hydrocarbons, heavy metals, etc.) may exert on a developing fish embryo. For example, the initial treatment of the egg and continued exposure of the embryo to cadmium can: (1) modify the permeability of the egg membrane prior to and after fertilization; (2) disrupt gastrulation and axiation during the mid- to later stages of embryogenesis; (3) retard the growth, development, and organogenesis; (4) reduce embryonic heart rate; (5) reduce or modify embryonic motility; (6) decrease the activity of several biosynthetic enzymes in late-stage embryos; (7) disrupt normal osteogenesis, resulting in skeletal malformations; (8) reduce yolk-sac size via osmotic effects on perivitelline fluid, resulting from cadmium affected egg membranes; and (9) result in premature or delayed hatching (Rosenthal and Alderdice 1976; von Westernhagen 1988; Weis and Weis 1989).

As has been mentioned, it is beyond the intended scope of this brief review to provide a detailed consideration of the multiplicity of effects exerted by the various classes of contaminants (or their members) on developing fish embryos and their relationship to the occurrence of terata in newly hatched larvae. However, the remarkable similarity of these effects, independent of contaminant type, as exhibited by morphological, physiological, and/or behavioral abnormalties, led Rosenthal and Alderdice (1976) to propose that the embryo responds to these toxic insults with a generalized "stress" response that has as its principal objective to remain alive, whatever the cost. Therefore, in keeping with this objective, these similarities may be considered as the repetitive expression of the limited repertoire of stress responses available to them (i.e., the embryo) at this stage of development.

Contaminant-Related Lesions

Unlike the plethora of reports on contaminant-induced gross abnormalities in embryonic fishes, only modest numbers of histopathological or electron microscopical studies on toxicant effects have been conducted.

Studies by Somasundaram et al. (1984a,b) on the effects of zinc on Atlantic herring trunk muscle (1984a), brain (1984b), and epidermis (Somasundaram 1985) constitute the only reports of which I am aware on the cytopathic nature of heavy-metal exposure to fish embryos. In both muscle and brain, zinc exposure resulted in degenerative changes characterized by mitochondrial swelling, cytoplasmic vacuolization, swelling of the endoplasmic (including the perinuclear cistern) or sarcoplasmic reticulum, and other degenerative changes associated with cell death. In the case of the epidermis, frank necrosis and sloughing of epithelial cells was noted at zinc concentrations of 12 ppm; also chloride cells had fewer mitochondria with normal cristae, Golgi organelles were absent, and numerous profiles of congested, smooth endoplasmic reticulum were present. The authors concluded that exposure of embryos to zinc at concentrations of 6 to 12 ppm may seriously compromise the osmoregulatory (chloride cell), respiratory (epidermal cells), muscular, and neural capability of newly hatched larvae.

Owing to its environmental significance, the histopathological and cytopathological effects of petroleum hydrocarbons on embryos has received more attention than other types of contaminants. Ernst and Neff (1977) examined developing embryos of *Fundulus grandis* at 12 and 22 days post-fertilization which had been continuously exposed to 25 and 50 percent water soluble fractions of No. 2 fuel oil. They observed degenerative and necrotic tissues in the liver, gut, kidney, and central nervous system (retina, brain, and spinal cord), as shown by the presence of cellular vacuolation and large numbers of pycnotic nuclei. The extent of damage to each tissue was dose dependent. Utilizing both light and electron microscopy, Cameron and Smith (1980) examined the cytopathological effects of a water soluble fraction of Prudhoe Bay crude oil on Pacific herring embryos, and Hawkes and Stehr (1982) conducted a similar study on surf smelt *(Hypomesus pretiosus)* embryos exposed to seawater-accommodated fractions of oil from Cook Inlet. Both of these studies demonstrated cytopathological ultrastructural changes (e.g., increased extracellular space, dilated endoplasmic reticulum cisternae, vacuolation, and nuclear pycnosis) in the forebrain, with necrotic cells also reported for the retina of surf smelt and degenerative, swollen mitochondria found in the head and trunk muscle of the Pacific herring. In this instance, no gross abnormalities were observed for the embryos and newly hatched larvae of either species.

Histopathological studies on embryos of the flathead sole *(Hippoglossoides elassodon*; Hose et al. 1982) and rainbow trout (Hose et al. 1984), after exposure to the aromatic hydrocarbon benzo[a]pyrene resulted in nuclear pycnosis and vacuolation in the cells of the brain and eye in embryos of both species. The authors reported additional necrotic tissues (epidermis, connective tissues, kidneys, liver, and muscle) were found in the trout.

The results of the above histopathological investigations on petroleum-derived contaminant effects on developing embryos, indicate that a number of tissue types can be affected by the presence of these toxicants; however, the central nervous system and neural retina appear to be particularly sensitive.

Histopathological studies on the effects of embryonic exposure to pesticides have demonstrated lesions and dislocations in the notochords and vertebral columns of rainbow trout exposed to dithiocarbamates (van Leeuwen et al. 1986) and in the inland silverside exposed to the organophosphorous pesticide, terbufos (Middaugh et al. 1990). In the case of the rainbow trout, notochordal disruption and ectopic osteogenesis was accompanied by hemorrhagic lesions in the brain, biliary hyperplasia, and hepatocyte necrosis when high levels of dithiocarbamates were used. Histological study of inland silverside embryos exposed for 5 days to 50 µg/L of terbufos, transferred to clean water, and examined after hatching, demonstrated some of the lesions (vertebral wall hypertrophy, retention of cellular notochord, longitudinal compression of vertebrae, and hypertrophy of zygapophyseal processes containing proliferating fibroblast, osteoblasts, and osteocytes) that had been previously reported by Couch et al. (1979) for the sheepshead minnow *(Cyprinodon variegatus)* exposed to the herbicide trifluralin.

In an extensive histopathological study on the effects of lowered pH on Atlantic salmon *(Salmo salar)* embryos exposed from fertilization, Daye and Garside (1980) observed a wide variety of pH-dependent degenerative and/or hyperplastic changes, primarily in the surficial tissues of the developing animal which were in more direct contact with the toxicant. Among the most important of these were the degenerative changes observed in the developing ectoderm (presumptive epidermis), which would be involved in respiration. Effects on the developing ectoderm could endanger all subsequent phases of development.

An ultrastructural study (Bahgat et al. 1989) on the effects of lowered pH on muscle tissue of newly hatched (yolk-sac) Atlantic herring larvae incubated from the time of fertilization at pH 7.5, 7.0, 6.0, or 5.2 demonstrated degenerative muscle changes at all pH levels below 7.5. The cytopathology observed consisted of swollen elements of sarcoplasmic reticulum, mitochondrial degeneration (e.g., shrinkage and loss of cristae), and loss of muscle-fiber volume. The authors interpret these changes as potentially significant in their effect on mitochondrial energy metabolism and muscular activity, taking the form of impeded swimming and maintenance of equilibrium.

In recent years, several contaminants have caused great concern for environmental toxicologists. Two of these are dioxin and tributyltin. The following recent studies on the cytopathic effect of these extremely toxic compounds signifies the reason for concern.

Spitzbergen et al. (1991) have extensively studied the pathological consequences of exposing lake trout *(Salvelinus namayacush)* embryos to < 15, 40, 400 ng/L [^{3}H]-2,3,7,8-tetrachlorodibenzo-*p*-dioxin for 48 h and examined late-stage embryos by histopathological methods. Fertilized eggs exposed to dioxin (400‰) developed into embryos having severe hemorrhagic lesions and subcutaneous edema resulting in the cessation of blood flow throughout the animal. Subsequent necroses (lesions) involving the brain, retina, liver, and spinal cord were present in embryos just prior to hatching as a result of the generalized hypoxia and anemia caused by the animal's failed circulatory system. Exposure of fertilized eggs to the lower concentrations (< 15, 40‰) of dioxin resulted in an edematous syndrome comparable to that of blue-sac disease, which had been previously reported in developing lake trout.

The pathological effects of tributyltin chloride (TBT) recently have been reported for newly hatched larvae of the European minnow *(Phoxinus phoxinus)* by Fent and Meier (1992). Like dioxin, low levels (0.82 to 19.51 μg/L) of toxicant resulted in a serious pathological condition in yolk-sac fry exposed as embryos for 8 to 9 days after fertilization. Degenerative lesions with cells demonstrating nuclear pycnosis, karyolysis, and hydropic vacuolation were observed in a wide variety of tissues (skin, eye, muscle, kidney, and central nervous system).

Field-Related Studies and Recruitment

Despite the fact that laboratory studies have clearly shown the non-specific nature of embryonic responses to contaminants, the examination of field-collected embryos for terata is not without merit.

One of the first studies to employ cytogenetic and teratological observations on embryos from a contaminated area was that of Longwell and Hughes (1981) in their study of Atlantic mackerel *(Scomber scombrus)* embryos from the New York Bight. These authors measured a number of parameters regarding the health of these embryos which were correlated with contaminant concentrations of heavy metals and toxic hydrocarbon levels in plankton and sea-surface (microlayer) water samples at specific locations within the Bight. They showed significantly lower egg viability in highly impacted Bight areas (i.e., those near the coast or disposal sites) than in areas more distant from the coast. As reviewed by Longwell et al. (1992), cytogenetic abnormalities observed for different developmental stages of Atlantic mackerel embryos from the New York Bight are correlated with the concentration of certain classes of contaminants.

In order to overcome the problems and objections of examining and evaluating embryos obtained from potentially polluted areas in the field (von Westernhagen et al. 1988), a number of recent studies have been conducted in which embryos were exposed in the laboratory to whole or extracted sea-surface and subsurface waters (bulk water) and scored for a number of parameters, including terata. Such studies have examined: (1) effects of the sea-surface microlayer of Puget Sound, on sand sole *(Psettichthys melanostictus)* embryos (Hardy et al. 1987); (2) toxicity of hexane extracted sea-surface waters on cod and herring embryos from locations in the North and Baltic seas (Kocan et al. 1987; and 3) cytogenetic effects of the sea-surface microlayer near Los Angeles, on embryos of the kelp bass *(Paralabrax clathratus;* Cross et al. 1987). All of these studies demonstrated that sea-surface waters (microlayers) from areas known or suspected to be contaminated by industrial or other forms of pollution produced significantly higher morphological aberrations or chromosomal abnormalities in developing embryos than comparable waters from control areas.

Along another vein of inquiry, Perry et al. (1991) have shown that cytogenetic and cytopathological evaluations of winter flounder embryos from fish captured at polluted sites in Long Island Sound along the coast of Connecticut and Boston Harbor and spawned in the laboratory, yield embryos having varying degrees of chromosomal damage and other forms of abnormal cell development. Therefore, as von Westernhagen et al. (1988) have indicated, it would appear that despite an inability to categorize terata and cytogenetic aberrations with respect to specific contaminants, their presence at higher than normal frequencies in embryos from polluted environments, in which fish stocks are already depleted, may serve as a significant monitoring tool that can aid in understanding recruitment under those conditions.

Terata and minor pathological lesions in embryonic or larval fishes are traditionally thought of as sublethal when evaluated in laboratory experiments. As von Westernhagen et al. (1988) have clearly indicated, however, field data have shown that only a small percentage of larvae survive in the natural environment and it would be anticipated that a 100% mortality would occur for larvae compromised by such defects. Therefore, a better assessment of survivability will be needed for laboratory studies before these types of data can be utilized in the field for a realistic analysis of recruitment.

References

Bahgat, F. J., P. E. King, and S. E. Shackley. 1989. Ultrastructural changes in the muscle tissue of *Clupea harengus* L. larvae induced by acid pH. Journal of Fish Biology 34:25-30.

Cameron, J. A., and R. L. Smith. 1980. Ultrastructural effects of crude oil on early life stages of Pacific herring. Transactions of the American Fisheries Society 109:224-229.

Carls, M. G., and S. D. Rice. 1989. Abnormal development and growth reductions of pollock *Theragra chalcogramma* embryos exposed to water-soluble fractions of oil. Fishery Bulletin, U. S. 88:29-37.

Couch, J. A., J. T. Winstead, D. J. Hansen, and L. R. Goodman. 1979. Vertebral dysplasia in young fish exposed to the herbicide trifluralin. Journal of Fish Diseases 2:35-42.

Cross, J. N., and six coauthors. 1987. Contaminant concentrations and toxicity of sea-surface microlayer near Los Angeles, California. Marine Environmental Research 23:307-323.

Daye, P. G., and E. T. Garside. 1980. Structural alterations in embryos and alevins of the Atlantic salmon, *Salmo salar* L., induced by continuous or short-term exposure to acidic levels of pH. Canadian Journal Zoology 58:27-43.

Dethlefsen, V. 1977. The influence of DDT and DDE on the embryogenesis and the mortality of larvae of cod *(Gadus morhua* L.*)*. Berichte Deutscher Wissenshaftliche Kommission für Meersforshung 25:115-148.

Ernst, V. V., and J. M. Neff. 1977. The effects of the water-soluble fractions of No. 2 fuel oil on the early development of the estuarine fish, *Fundulus grandis* Baird and Girard. Environmental Pollution 14:25-35.

Falk-Petersen, I. B., and S. Lonning. 1984. Effects of hydrocarbons on eggs and larvae of marine organisms. Pages 197-217 in G. Persoone, E. Jaspers, and C. Claus, editors. Ecological testing for the marine environment. State University of Ghent and Institute of Marine Scientific Research, Bredene, Belgium.

Falk-Petersen, I. B., L. J. Saethre, and S. Lonning. 1982. Toxic effects of naphthalenes and methylnaphthalenes on marine plankton organisms. Sarsia 67:171-178.

Fent, K., and W. Meier. 1992. Tributyltin-induced effects on early life stages of minnows *Phoxinus phoxinus*. Archives of Environmental Contamination and Toxicology 22:428-438.

Hardy, J., S. Kiesser, L. Antrim, A. Stubin, R. Kocan, and J. Strand. 1987. The sea-surface microlayer of Puget Sound: Part I. Toxic effects on fish eggs and larvae. Marine Environmental Research 23:227-249.

Hawkes, J. W., and C. M. Stehr. 1982. Cytopathology of the brain and retina of embryonic surf smelt *(Hypomesus pretiosus)* exposed to crude oil. Environmental Research 27:164-178.

Hodson, P. V., and B. R. Blunt. 1981. Temperature-induced changes in pentachlorophenol chronic toxicity to early life stages of rainbow trout. Aquatic Toxicology 1:113-127.

Hose, J. E., J. B. Hannah, D. DiJulio, M. L. Landott, B. S. Miller, W. T. Iwaoka, and S. P. Felton. 1982. Effects of benzo[a]pyrene on early development of flatfish. Archives of Environmental Contamination and Toxicology 11:167-171.

Hose, J. E., J. B. Hannah, H. W. Puffer, and M. L. Landolt. 1984. Histologic and skeletal abnormalities in benzo[a]pyrene-treated rainbow trout alevins. Archives of Environmental Contamination and Toxicology 13:675-684.

Hughes, J. B., and A. C. Longwell. 1981. Cytological-cytogenetic analyses of fourbeard rockling and yellowtail flounder eggs from plankton at Ocean 250 gasoline spill. Pages 21-30 in C.A. Griswald, editor. The barge ocean 250 gasoline spill, NOAA Technical Report, NMFS SSRF-751.

Kjorsvik, E. 1986. Morphological and ultrastructural effects of xylenes upon the embryonic development of the cod (*Gadus morhua* L.). Sarsia 71:65-71.

Kocan, R. M., H. von Westernhagen, M. L. Landolt, and G. Furstenberg. 1987. Toxicity of sea-surface microlayer: effects of hexane extract on Baltic herring *(Clupea harengus)* and Atlantic cod *(Gadus morhua)* embryos. Marine Environmental Research 23:291-305.

Kuhnhold, W. W. 1974. Investigations on the toxicity of seawater-extracts of three different crude oils in eggs of cod *(Gadus morhua* L.*)*. Berichte Deutscher Wissenschaftliche Kommission für Meeresforschung 23:165-180.

Laale, H. W., and W. Lerner. 1981. Teratology and early fish development. American Zoologist 21:517-533.

Linden, O. 1978. Biological effects of oil on early development of the Baltic herring *Clupea harengus membras*. Marine Biology 45:273-283.

Longwell, A. C. 1977. A genetic look at fish eggs and oil. Oceanus 20:46-58.

Longwell, A. C., S. Chang, A. Hebert, J. B. Hughes, and D. Perry. 1992. Pollution and developmental abnormalities of Atlantic fishes. Environmental Biology of Fishes 35:1-21.

Longwell, A. C., and J. B. Hughes. 1981. Cytologic, cytogenetic, and embryologic state of Atlantic mackerel eggs from surface waters of the New York Bight in relation to pollution. Rapports et Procès-Verbaux des Réunions pour l'Exploration de la Mer 178:76-78.

Lonning, S. 1977. The effects of crude Ekofisk oil and oil products on marine fish larvae. Astarte 10:37-47.

Marty, G. D., J. M. Nunez, D. J. Lauren, and D. E. Hinton. 1990. Age-dependent changes in toxicity of N-nitroso compounds to Japanese medaka (*Oryzias latipes*) embryos. Aquatic Toxicology 17:45-62.

Middaugh, D. P., M. J. Hemmer, E. M. Lores. 1988. Teratological effects of 2,4-dinitrophenol, 'produced water' and naphthalene on embryos of the inland silverside *Menidia beryllina*. Diseases of Aquatic Organisms 4:53-65.

Perry, D. M., J. B. Hughes, A. T. Hebert. 1991. Sublethal abnormalities in embryos of winter flounder, *Pseudopleuronectes americanus*, from Long Island Sound. Estuaries 14:306-317.

Perry, D. M., J. S. Weis, and P. Weis. 1988. Cytogenetic effects of methylmercury in embryos of killifish, *Fundulus heteroclitus*. Archives of Environmental Contamination and Toxicology 17:569-574.

Purcell, J. E., D. Grosse, and J. J. Grover. 1990. Mass abundances of abnormal Pacific herring larvae at a spawning ground in British Columbia. Transactions of the American Fisheries Society 119:463-469.

Rosenthal, H., and D. F. Alderdice. 1976. Sublethal effects of environmental stressors, natural and pollutional, on marine fish eggs and larvae. Journal of the Fisheries Research Board of Canada 33:2047-2065.

Smith, R. M., and C. F. Cole. 1973. Effects of egg concentrations of DDT and dieldrin on development in winter flounder *(Pseudopleuronectes americanus)*. Journal of the Fisheries Research Board of Canada 30:1894-1898.

Somasundaram, B. 1985. Effects of zinc on epidermal ultrastructure in the larva of *Clupea harengus*. Marine Biology 85:199-207.

Somasundaram, B., P. E. King, and S. E. Shackley. 1984a. The effect of zinc on the ultrastructure of the trunk muscle of the larva of *Clupea harengus* L. Comparative Biochemistry and Physiology 79C:311-315.

Somasundaram, B., P. E. King, and S. E. Shackley. 1984b. The effects of zinc on the ultrastructure of the brain cells of the larvae of *Clupea harengus* L. Aquatic Toxicology 5:323-330.

Spitsbergen, J. M., M. K. Walker, J. R. Olson, and R. E. Peterson. 1991. Pathologic alterations in early life stages of lake trout, *Salvelinus namaycush*, exposed to 2,3,7,8-tetrochlorodibenzo-p-dioxin as fertilized eggs. Aquatic Toxicology 19:41-72.

Stene, A., and S. Lonning. 1984. Effects of 2-methylnaphthalene on eggs and larvae of six marine species. Sarsia 69:199-203.

Struhsaker, J. W., M. B. Eldridge, and T. Echeverria. 1974. Effect of benzene [a toxic component of crude oil] in eggs and larvae of Pacific herring and northern anchovy. Pages 253-284 *in* F.J. Vernberg and W.B. Vernberg, editors. Pollution and physiology of marine organisms, Academic Press, New York.

van Leeuwen, C. J., T. Helder, and W. Seinen. 1986. Aquatic toxicological aspects of dithiocarbamates and related compounds. IV. Teratogenecity and histopathology in rainbow trout (*Salmo gairdneri*). Aquatic Toxicology

9:147-159.

von Westernhagen, H. 1988. Sublethal effects of pollutants on fish eggs and larvae. Pages 253-346 *in* W. S. Hoar and D. J. Randall, editors. Fish Phsyiology, Vol. XI. The physiology of developing fish. Academic Press, New York.

von Westernhagen, H., V. Dethlefsen, P. Cameron, J. Berg, and G. Furstenberg. 1988. Developmental defects in pelagic fish embryos from the western Baltic. Helgoländer Meeresuntersuchungen 42:13-36.

Weis, P., and J. S. Weis. 1977. Effects of heavy metals on development of the killifish, *Fundulus heteroclitus*. Journal of Fish Biology 11:49-54.

Weis, P., and J. S. Weis. 1979. Congenital abnormalities in estuarine fishes produced by environmental contaminants. Pages 94-105 *in* Animals as monitors of environmental pollutants. Symposium Proceedings. National Academy of Sciences, Washington, District of Columbia.

Weis, J. S., and P. Weis. 1989. Effects of environmental pollutants on early fish development. Reviews of Aquatic Sciences 1:45-73.

Wiegand, M. D., J. M. Hataley, C. L. Kitchen, and L. G. Buchanan. 1989. Induction of developmental abnormalities in larval goldfish, *Carassius auratus* L., under cool incubation conditions. Journal of Fish Biology 35:85-95.

Wilson, K. W. 1976. Effects of oil dispersants on developing embryos of marine fish. Marine Biology 36:259-268.

Calibrating Starvation-Induced Stress in Larval Fish Using Flow Cytometry

GAIL H. THEILACKER

*NOAA, National Marine Fisheries Service, Alaska Fisheries Science Center,
7600 Sand Point Way N.E., Seattle, Washington 98115, USA*

W. SHEN

*Department of Pathology, Health Sciences SM-30,
University of Washington, Seattle, Washington 98195, USA*

Abstract.—Two indices of physiological state, nutritional condition and growth, were developed for larval walleye pollock, *Theragra chalcogramma*, raised in the laboratory. The indices, estimated simultaneously for individual fish, rely on measurements of RNA and DNA content by flow cytometry of single cells dissociated from the walleye pollock brain. Nutritional condition was estimated from the fraction of brain cells that exhibited distinct RNA activity among larvae that were always fed, always starved, and starved before feeding. Fast- and slow-growing larvae were discerned by differences in their DNA synthetic activity, the potential for replication. We describe a simple method for freezing the dissociated brain cells that ensures nucleic acid stability during storage, thus permitting sampling and storage at sea. We envision that in the field, with a flow cytometer onboard the research vessel, a quick and accurate assessment of individual fish condition could be obtained on site.

The ratio of RNA to DNA is an established indicator of larval fish growth and condition (Bulow 1971, 1987; Buckley 1979, 1980, 1982, 1984; Buckley et al. 1984; Buckley and Lough 1987; Clemmesen 1987; Westerman and Holt 1988; Canino et al. 1991). Poor nutrition results in slow growth, yielding a low ratio. The analyses for RNA and DNA content have evolved from rather insensitive methods, requiring pooled larval fish samples, to a sensitive assay for single, whole animal tissue homogenates (Buckley 1980; Barron and Adelman 1984; Bulow 1987; Clemmesen 1988; Westerman and Holt 1988; Canino et al. 1991). Yet, because larval fish tissues respond differently to starvation conditions, homogenizing a whole larva integrates tissue-specific RNA-DNA responses. Histopathology of northern anchovy *(Engraulis mordax)*, jack mackerel *(Trachurus symmetricus)*, and striped bass *(Morone saxatilis)* larvae revealed that the digestive tract and its associated glands were the first tissues to be affected visibly by starvation, followed by muscle tissue, and then by brain tissue (O'Connell 1976; Theilacker 1978; Martin and Malloy 1980). Thus, analyzing the RNA-DNA response of a single tissue to stress should yield a more precise index of physiological condition than a homogenate of a whole larva. Furthermore, relative sizes of organs vary with developmental stage due to allometric growth during early life. Combining variable, tissue-specific nucleic acid responses in whole animal homogenates undoubtedly complicates interpretation of the RNA-DNA ratio.

To refine nucleic acid indices, we developed a flow cytometer (FCM) technique to assay RNA and DNA in single brain cells of larval walleye pollock, *Theragra chalcogramma* (Theilacker and Shen 1993). We used brain tissue for our analyses because flow cytometry requires suspensions of single cells, and the brain relinquished clean cell suspensions while other tissues did not. Additionally nerve cells are among one of the highest in RNA content of all body cells (Singhal et al. 1989; Darzynkiewicz 1991). In a preliminary laboratory study (Theilacker and Shen 1993), we found that the amount of RNA in the brain cells was indicative of larval feeding history. Additionally, we noted that larvae starved before feeding had a higher content of RNA per cell than their continuously fed counterparts. We inferred from this result that larvae subjected to delayed feeding had a greater potential to synthesize protein. This poten-

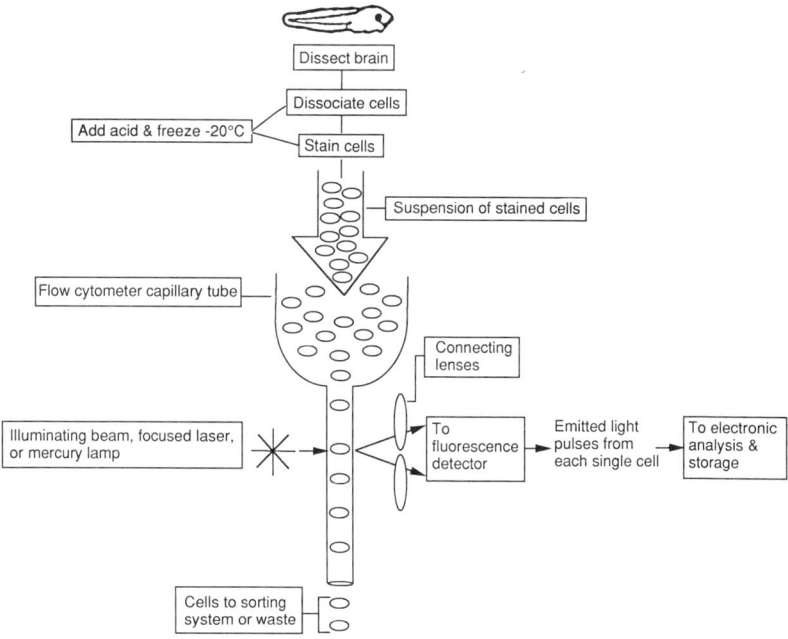

FIGURE 1.—Procedure for analysis of RNA and DNA in single larval pollock brain cells using flow cytometry. The figure from Theilacker and Shen (1993) includes the new acid-treatment and freezing step.

tial may be an adaptive response for larvae that are vulnerable to starvation during the early life-history transition from endogenous to exogenous feeding.

We analyzed relatively few samples in our original study because no suitable method was known for preserving the tissue samples. It was necessary to sample the fish and analyze the brain cells immediately to minimize loss of RNA. Here we describe a method for freezing dissociated brain cells and report on a similar investigation of larval walleye pollock condition that supports the results of our earlier study. The cell freezing method ensures nucleic acid stability during storage, thus permitting intensive sampling in the laboratory and field with subsequent flow cytometry analysis. We anticipate using flow cytometry to assess the condition of walleye pollock spawned in Shelikof Strait, Gulf of Alaska. Recruitment variability of the walleye pollock population in this area is being intensively studied (Schumacher and Kendall 1991).

Methods

Sample Collection and Larva Rearing

Walleye pollock eggs were spawned from adults collected by bottom trawl at 4°C in Shelikof Strait near Kodiak Island, Alaska, aboard the NOAA ship R/V *Miller Freeman* in April 1992. Fertilized eggs were maintained for several days aboard ship at 3°C and then flown to the Alaska Fisheries Science Center (AFSC), Seattle, Washington, USA. Eggs, at 5°C upon arrival, were incubated at 6°C and then stocked at 5/L into circular, 120-L black fiberglass tanks. Larvae were raised at 6°C on a 16-h daylight cycle. At the onset of feeding, 6 d after hatching, larvae were either fed, starved, or feeding was delayed for 1 or 2 d. Food was in the form of rotifers *(Brachionus plicatilis)* and nauplii and copepodite stages of copepods *(Acartia* sp.). The rotifers were raised on algal diets *(Isochrysis galbana* and *Pavlova lutheri)* high in unsaturated fatty acids (Nichols et al. 1989). Rotifer concentration in the rearing tanks was 10/ml, and copepod concentration was 1 to 3/ml.

Walleye pollock larvae were randomly sampled from the surface 8 to 10 h after the lights came on. Guts of fed larvae were full at this time. To estimate growth and brain cell condition, we sampled daily from hatching until 11 d after hatching. A final sample for growth was taken 19 d after hatching. Ten larvae from each feeding treatment were sampled for brain cell condition. Ontogenetic changes in condition will be presented elsewhere. Each larva was measured live, the yolk dimensions were taken, gut fullness noted, and then the brain was dissected.

Brain Cell Preparation and Storage

To dissect the brain, we placed an individual walleye pollock larva in seawater onto the left side of a microscope slide and placed a drop of cryoprotectant to the right. After measuring the larva at 120× magnification, we used forceps to grab the tail and dragged the larva into the cryoprotectant. We held the larva's head with the forceps and removed the brain by inserting a small dissecting needle under the posterior part of the brain and lifting it forward. The intact brain was transferred using a 100-μL pipette into a 1.5-mL centrifuge tube filled with 200 μL cryoprotectant on ice. The cryoprotectant was prepared by adding 1 mL fetal bovine serum and 1 mL dimethyl sulfoxide (DMSO) to 4 mL Eagles Minimum Essential Medium (MEM). The addition of cryoprotectants improves the recovery of whole, unruptured cells after freezing (Mazur 1970). We dissociated the brain cells by pipetting the brain 15 times in the cryoprotectant, using a 100-μL pipet, and then immediately added an equal volume of 0.08N HCl acid solution on ice to deactivate RNAase and to ensure RNA stability in the cell preparation. We limited the duration of these manipulations, from first invasion of the live fish to the addition of acid into the dissociated cell suspension, to 60 to 90 s. Although we found no differences in nucleic acid content between the anterior and posterior regions of the brain, we always dissected and used the whole brain, excluding olfactory organs, for the cell suspensions. Differences in nucleic acid content in separate regions of the brain of young Atlantic salmon, *Salmo salar*, have been documented (Nikonorov et al. 1985). The general procedure we followed to prepare the brain cells for assay by flow cytometry is given in Figure 1.

Assay Procedures and Data Analysis

Two-parameter flow cytometry was used to simultaneously determine RNA and DNA content per cell (Figure 1). All analyses were conducted using a mercury lamp flow cytometer at the University of Washington, Health Sciences Department.

For the two-parameter assay, we added the fluorochrome acridine orange to the quickly thawed brain cell suspension to differentially stain DNA and RNA in the single cells. The cells were treated with a detergent to make them permeable to acridine orange and to chelating agents used to selectively denature double-stranded and folded RNA (Darzynkiewicz 1991; Shapiro 1988 and references

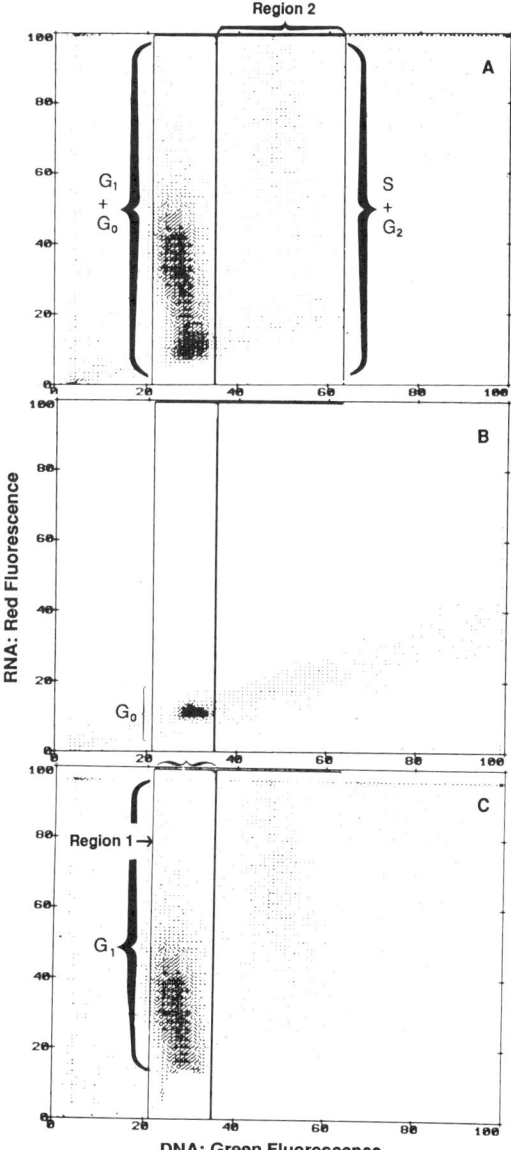

FIGURE 2.—A typical computer-display output. Panel A is a two-parameter histogram of relative cellular DNA and RNA content in 8,000 individual brain cells of a fed walleye pollock larva. The histogram is divided into two regions according to the intensity of green (DNA) fluorescence (see panels A and C). RNA and DNA fluorescence is in arbitrary units. Cells in region 1 are in G_0 and G_1 phases of the cell cycle; cells in region 2 are in G_2 and S. (See text for description of phases.) In region 1, to distinguish G_0 from G_1, RNAse was added and the sample was run again. Panel B is the two-parameter histogram of the RNAse-treated sample showing the elimination of RNA fluorescence and position of G_0. Panel C shows the histogram of region 1 after G_0 was subtracted, yielding G_1 cells only.

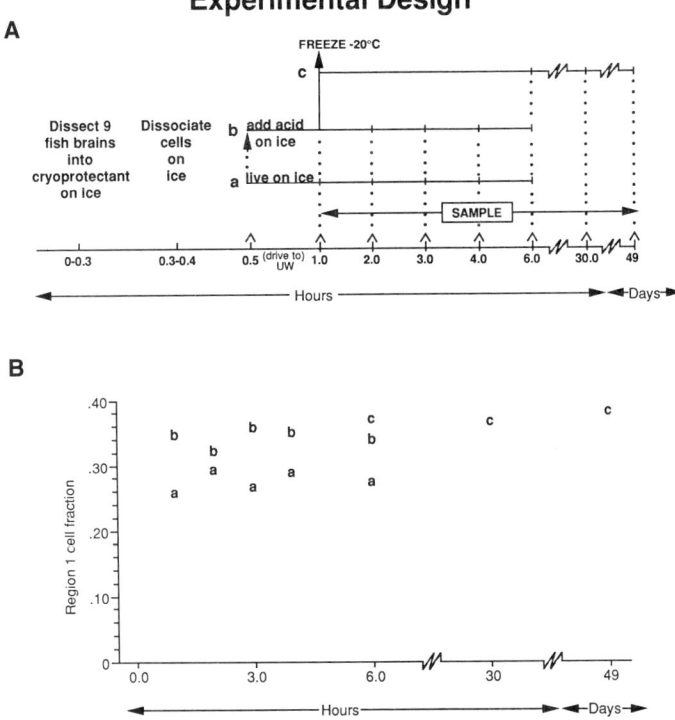

FIGURE 3.—Panel A illustrates the design of a time-course experiment to assess nucleic acid stability in larval walleye pollock brain cells during storage. Panel B compares the G_1 cell fraction in (a) fresh brain cells on ice, (b) acid-treated cells on ice, and (c) acid-treated cells frozen at -20°C

therein; Traganos et al. 1977). The emission wavelength for the fluorochrome was green at 530 nm, after intercalating and stacking into double-stranded DNA, and red at 640 nm, interacting electrostatically with the phosphates of single-stranded RNA (Darzynkiewicz 1991; Shapiro 1988 and references therein).

Fluorescence intensities of the emissions were visualized as a two-parameter histogram of relative cellular DNA and RNA content. The proportion of cells in the different phases of the cell cycle was calculated with a data analysis program, Multi2D (written by P. Rabinovitch and W. Shen and available from Phoenix Flow System, 11575 Sorrento Valley Road, Suite 202, San Diego, California, 92121). The eukaryotic cell cycle is divided into four phases: G_1, S, G_2, and M. The gaps before and after DNA synthesis during the S phase are G_1 and G_2, and M refers to mitosis. Phase length varies among species and between different developmental stages within species, although typically the duration of G_1 equals the duration of $G_2 + S + M$ (Lewin 1987). Cells in the G_1 phase may be in one of two stages of proliferation. Here we refer to the cells that are synthesizing RNA and proliferating as G_1 and to the low-RNA cells that are not proliferating as G_0.

We tested larvae individually, analyzing 8,000 brain cells per larva, and using the Multi2D program to count the number of cells belonging to phases of the cell cycle, which were compartmentalized into region 1 (G_0 and G_1 phases) and region 2 (G_2 and S phases) (Figure 2). We used the proportion of total cells in each of the two regions to establish two larval condition indices. Basically, region 1 represents RNA synthesis and region 2 represents DNA synthesis.

To distinguish the quiescent, non-proliferating, low-RNA cells called G_0 from the proliferating G_1 cells (region 1) that have the same DNA content as G_0 but elevated RNA content (Figure 2A), we added RNAse (1 mg/ml) to the cell suspension after the initial FCM analysis. The addition of this enzyme eliminated RNA fluorescence when the cell suspension was reanalyzed, distinguishing the division between G_0 and G_1 (Figure 2B). Then, using the Multi2D data analysis program, we mathematically aligned G_0 in the initial two-parameter his-

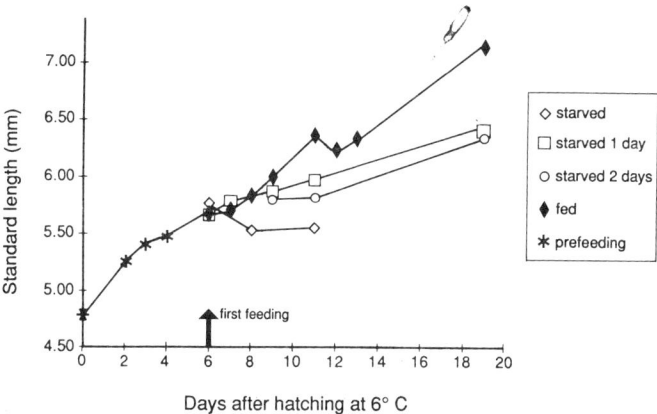

FIGURE 4.—Mean standard length of fed, starved, and delayed-fed walleye pollock larvae over time. Number sampled and standard deviations are given in Table 1.

togram (Figure 2A) with G_0 in the RNAse-treated histogram of the same sample (Figure 2B) and, using cumulative subtraction (Overton 1988), determined the number of proliferating G_1 cells in region 1 (Figure 2C).

We compared the fraction of cells in the two regions among larvae belonging to the different feeding treatments using an analysis of variance (ANOVA) procedure with feeding regime as the independent factor. Each pair of regimes was then compared using post-hoc Student-Newman-Keuls multiple comparisons. These analyses were performed using SYSTAT (Wilkinson 1988).

In our initial study, we were unable to store the cells and recover RNA, and all FCM measurements were made using fresh preparations (Theilacker and Shen 1993). In this study, all FCM measurements were made using frozen preparations. We determined that the cell preparation in HCl acid was stable when stored at -20°C by comparing the time course of the RNA content of fresh brain cells with acid-treated brain cells (Figure 3). To obtain sufficient cells for the multiple examination of RNA content, we pooled 9 brains and analyzed aliquots of fresh (a in Figure 3), acid-treated on ice (b in Figure 3), and acid-treated frozen cells (c in Figure 3) at intervals to 49 d. The RNA content of cells in acid-treated frozen samples was slightly higher than the acid-treated group on ice ($P < 0.05$) and both acid-treated groups were higher than fresh samples ($P < 0.05$) (Student-Newman-Keuls multiple comparisons). For this experiment, the fraction of high RNA cells at t_0 was unknown, but because the acid-treated frozen group was higher in RNA than the acid-treated on ice group, we assume there was some loss of RNA during the time the sample was

TABLE 1.—Live standard length (SL, mm) of walleye pollock raised under several feeding regimes at 6°C.

Days after hatching	Prefeeding and starved			Fed			Starved 1 d, then fed			Starved 2 d, then fed		
	N	Mean SL	SD	N	Mean SL	SD	N	Mean SL	SD	N	Mean SL	SD
0	14	4.78	0.13									
1												
2	10	5.23	0.11									
3	6	5.39	0.15									
4	10	5.46	0.17									
5												
6	4	5.76	0.23	10	5.66	0.16	6	5.64	0.11			
7				10	5.69	0.12	5	5.78	0.12			
8	10	5.53	0.13	5	5.82	0.19						
9				10	5.98	0.13	10	5.86	0.21	10	5.79	0.20
10												
11	9	5.56	0.21	10	6.35	0.13	10	5.96	0.22	10	5.80	0.13
12				10	6.23	0.27						
13				22	6.33	0.33						
19				17	7.12	0.42	20	6.39	0.53	15	6.34	0.53

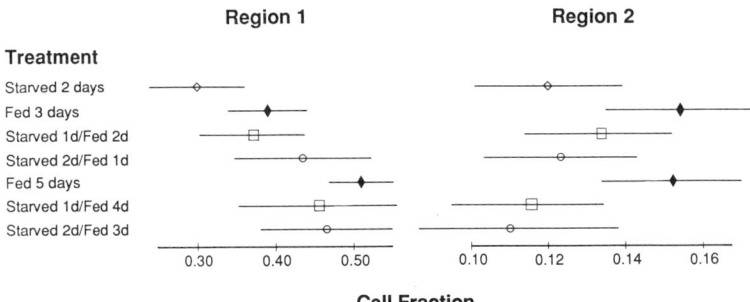

FIGURE 5.—Fraction of walleye pollock brain cells found in two regions of the cell cycle compared by feeding treatment. Symbols correspond to Figure 4 and represent treatment means, and the lengths of lines are ± 1 SD.

on ice. Using our current protocol, samples are acidified on ice in less than 90 s and frozen within 15 min.

Results

Growth characteristics

Walleye pollock larvae started eating 6 d after hatching. Growth was static for 1 d and then increased to 0.16 mm/d for the next 4 d which was the end of the condition experiment (11 d after hatching). Over the same period, larvae starved for 1 d before feeding grew at 0.06 mm/d, and larvae starved for 2 d before feeding grew at 0.03 mm/d (Figure 4; Table 1). The last length measurement for all treatments was taken 19 d after hatching. When growth was calculated over 19 d, fed larvae grew significantly faster, 0.14 mm/d ($P < 0.05$), than 1- and 2-d starved and fed larvae, which did not differ from each other and averaged 0.10 mm/d ($P > 0.05$). Thus, starving larval walleye pollock before feeding caused a decrement in growth rate after feeding resumed and a reduction in size at age (Figure 4; Table 1).

Distribution of Cells within Phases of the Cell Cycle

Cell proportions for the two established indices (regions 1 and 2) varied among the feeding treatments (Table 2; Figure 5). In region 1, the starved group differed from the group that fed continuously for 5 d ($P < 0.05$), while the delayed-fed groups together with the group fed 3 d were similar ($P > 0.05$). Thus, the proportion of G_1 cells in region 1 distinguished between starved and fed larvae but not between fed and delayed-fed larvae of the same age (Figure 5).

The fraction of S and G_2 cells in region 2 also discriminated among feeding treatments. Yet in this case, the delayed-fed groups were aligned more closely with the starved group than with the fed ones, as in region 1. In region 2, the 3- and 5-d fed groups were similar ($P > 0.05$) and separate from all other groups except starved 1 d/fed 2 d ($P < 0.05$; Figure 5).

Discussion

The time of first feeding for larval fishes is considered a critical period — yolk is being depleted and larvae must find and learn to eat prey. During this transitional period, larval fishes may experience high mortalities at sea due to starvation (see May 1974). Thus, reliable and sensitive proxies for condition of larval fishes are needed to estimate physiological state, predict survival probabilities, and refine estimates of recruitment of young fish into a fish stock.

Here we describe a technique that we are developing to identify the condition of larvae in the sea. This technique relies on two indices of parameters measured during phases of the cell cycle. Using these indices, a relative measure of the content of RNA and DNA in cells dissociated from the brain, starved walleye pollock were easily distinguished from fed ones. We recognize from studies of histopathology that larval fish brain tissue is highly defended (O'Connell 1976; Theilacker 1978; Martin and Malloy 1980). Additionally, growth of fish brains is slower than the increase in body size (Packard and Wainwright 1974; Bauchot et al. 1979; Bauchot et al. 1989; Brandstatter and Kotrschal 1990). Yet, here we show that brain cell proliferation and division is sensitive to nutritional levels. Cell division could stop without gross adverse effects to the brain tissue. In fact, neurons in culture remain healthy for long periods in the quiescent state of the cell cycle (Pardee 1989). Additionally, there is histological evidence for the cessation of cell division in brains of larval jack mackerel that were starved, with resumption of cell

TABLE 2.—Distribution of walleye pollock brain cells in two phases of the cell cycle compared to feeding regime. Region 1 includes the high-RNA fraction of proliferating diploid cells (G_1; Figure 3C). Region 2 includes the fraction of cells synthesizing DNA and tetraploids ($S + G_2$; Figure 3A).

			Fraction of cells			
			Region 1		Region 2	
Treatment	Age (d)	N	Mean	SD	Mean	SD
Starved 2 d	10	8	0.308	0.060	0.121	0.020
Fed 3 d	11	10	0.388	0.049	0.154	0.020
Starved 1 d, then fed 2 d	11	10	0.372	0.067	0.134	0.021
Starved 2 d, then fed 1 d	11	9	0.436	0.098	0.124	0.026
Fed 5 d	13	10	0.507	0.044	0.152	0.020
Starved 1 d/fed 4 d	13	10	0.455	0.133	0.115	0.024
Starved 2 d/fed 3 d	13	10	0.466	0.085	0.110	0.034

division upon renewed feeding (Theilacker 1986).

In our earlier study, we found a significantly higher fraction of G_1 cells in region 1 for delayed-fed larvae than for fed larvae of the same age (Theilacker and Shen 1993). Our current study confirms the high fraction of cells in region 1 for delayed-fed larvae (starved 2 d/fed 1 d), as compared to their fed counterpart (fed 3d) (Table 2; Figure 5). Then, after 2 more days of feeding (starved 2 d/fed 3 d), RNA levels returned to a lower level.

Cell culture experiments suggest that the status of cells in region 1 sets the stage for future growth (Pardee 1989 and references therein). The rate of increase of a population of cells in culture is primarily dependent on the fraction that are in cycle in G_1 as contrasted to those in G_0. Thus, we inferred that the high number of cells in G_1 found in delayed-fed larvae was an adaptive response to starvation, signifying a greater potential to synthesize protein after a short period of starvation (Theilacker and Shen 1993). In the present study, to determine whether the assumed greater potential to synthesize protein resulted in increased growth, we measured fish lengths for 1 week after feeding. Excess growth of the group of delayed-fed larval walleye pollock which exhibited a high fraction of cells in G_1 was not realized during the time period examined. The average growth rate for fed larvae 19 d after hatching was 0.14 mm/d, somewhat higher than other laboratory studies at 6°C water temperature (0.11 mm/d; Yamashita and Bailey 1989) and lower than field measurements (0.16 to 0.19 mm/d; Yoklavich and Bailey 1990). During this period, both groups of delayed-fed larvae grew at 0.10 mm/d, not achieving the greater than normal growth attained during a compensatory-growth response. Compensatory growth, noted in adults and juveniles of several fish species (Weatherley and Gill 1981; Dobson and Holmes 1984; Miglavs and Jobling 1989; Wieser et al. 1992; Zhang and Runham 1992) is defined as a rapid phase in growth following a period of undernutrition. Studies of larval fishes have presented inconsistent results. Although Pederson et al. (1990) found that growth of refed Atlantic herring (*Clupea harengus*) approached that of their fed counterparts, growth rates of Pacific herring *(C. pallasi)* declined (McGurk 1984) and results for coregonid larvae were mixed (Dabrowski et al. 1986). Because the refed herring attained a length similar to their fed counterparts 35 d after hatching at 8.5°C (Pederson et al. 1990), it would be informative to follow growth in length and weight of the larval walleye pollock groups over a longer period after the increase in ration.

We propose two explanations for high numbers of cells in G_1, region 1, of delayed-fed walleye pollock larvae: (a) a faster rate of entry of G_1 cells into S phase for fed larvae than for delayed-fed ones; and/or (b) a lower proportion of G_1 cells entering S phase for delayed-fed larvae than for fed ones. Our data show a higher fraction of cells in region 1 and a lower fraction in region 2 for delayed-fed larvae than for fed larvae (Table 2). Results with cultured cells are consistent with our data. Fibroblast cells in culture will move out of G_1 into G_0 after being placed in suboptimal conditions such as a low-nutrient medium, and then they will return to G_1 after the addition of nutrients. These returning cells require extra metabolism, thus more time to reach S phase than cells that have just completed dividing (Pardee 1989 and references therein). Thus, in these cell culture experiments, a higher fraction of cells would be in region 1 and a lower fraction in region 2.

Most studies of cycling cells are conducted with cells in culture where cycling can be synchronized. These studies show, as mentioned above, that external factors may affect events during G_1 that regulate cell proliferation (Pardee 1989). For example, serum stimulation of cell cultures in the G_1 phase increases cell proliferation and, conversely, suboptimal culture conditions depress proliferation (Pardee 1989). Our data suggest that the G_1 phase also may be regulated by nutrition in vivo. That is, fed walleye pollock larvae had a higher fraction of cells in G_1 than starved larvae. Additionally, feeding the starved larvae stimulated cell proliferation in G_1. Thus, the G_1 cell fraction may be indicative of feeding history and, potentially, the future state of the larva. The index may indicate short-term changes in nutrition. Timing the response to transient feeding conditions will depend in part on determining the length of the cell cycle and the response of G_1 cells to a meal.

In cell cultures, cell-cycle events are independent of external factors in S and G_2 phases, region 2, and dependent on intracellular controls to continue toward cell division (Hartwell and Weinert 1989; Murray and Kirschner 1989). Thus in cell cultures, the fraction of cells in the S and G_2 phases is an indicator of the number of cells that will divide independently of the current nutritional conditions. If cells react similarly in vivo, then the total number of cells in region 2 would seem to be the best indicator of brain-cell division, and thus by inference, of potential tissue growth.

Our brain-cell flow data for region 2 showed that the S and G_2 cell fraction for walleye pollock correlated well with growth rate. At 11 d after hatching, the always-fed group had a high $S + G_2$ fraction (0.152) that corresponded to a growth rate of 0.16 mm/d. The longer the fish were exposed to starvation, the lower the $S + G_2$ fraction (0.115 for 1 d starvation and 0.110 for 2 d starvation) and the slower the growth rate (0.06 mm/d and 0.03 mm/d).

We have partitioned the measure of RNA and DNA into two regions of the cell cycle: one depicting RNA synthetic capability (region 1), and the other illustrating the potential for replication, DNA synthesis (region 2). Measured simultaneously on an individual fish, these indices may yield information on both current condition and growth. For example, the region 1 analysis of delayed-fed fish shows nutrition is similar to always-fed fish and region 2 analysis shows that their growth parallels starved fish (Figure 5). If a fish sampled from the field had a similar signature, it would suggest current feeding but slow growth due to missing some meals in the past.

In summary, it appears that starvation has somehow damaged the walleye pollock brain cells or the cells may require extra nutrition to move from G_1 into region 2 and synthesize DNA. Even though cells were not prevented from entering G_1 through RNA synthesis and the initial feeding after starvation overstimulated the cells (starved 2 d/fed 1 d *versus* fed 3 d), subsequent stimulation from feeding did not achieve the same rate of DNA synthesis and growth attained by always-fed fish. Thus, previous starvation appeared to slow or prevent DNA synthesis, resulting in slower growth. The flow data showed that the cells recovered to some extent from the damage, but were not fully recovered over the duration of this experiment. Whether a prolonged feeding will achieve complete recovery is not known by flow data, but size information indicates that complete recovery did not occur within 19 d.

We do not know whether the mechanisms for the non-synchronous cell cycles in vivo that we are studying correspond to those described for synchronous cycles in culture (Hartwell and Weinert 1989; Murray and Kirschner 1989; Pardee 1989). Comparisons of cells in culture to cells in vivo may be too simplistic, yet the causal relationships appear to agree. Independently of the mechanisms involved, the two indices of RNA-DNA content we describe distinguish between starving and feeding walleye pollock larvae. Further studies of these indices may yield signatures of current feeding conditions, physiological state, and potential somatic growth for individual larvae. This sensitive and accurate analysis can be used in the field. With a flow cytometer onboard the research vessel, a quick, on-site nutritional assessment of larval fishes could be made.

More studies are needed to determine how RNA level responds to feeding rate and meal size over time, to describe the ontogenetic pattern of cycling cells, to determine the duration of the cell cycle, and to explore effects of food limitation after an initial period of good feeding. Additionally, the analyses need to be calibrated for temperature and fish size.

Acknowledgments

We thank Kevin Bailey for valuable discussions and comments on the first draft, Howard Browman for his advice, Art Kendall and Gary Stauffer for comments on the manuscript, Trish Brown for the statistical analyses, and Jerry Hardman for typing the tables. Steve Porter's unflagging effort during the rearing phase of this study was sincerely appreciated. This is FOCI Contribution No. 0168.

References

Barron, M. G., and I. R. Adelman. 1984. Nucleic acid, protein content, and growth of larval fish sublethally exposed to various toxicants. Canadian Journal of Fisheries and Aquatic Sciences 41:141-150.

Bauchot, R., M. Diagne, and J. M. Ridet. 1979. Post-hatching growth and allometry of the teleost brain. Journal für Hirnforschung 20:29-34.

Bauchot, R., J. E. Randall, J. M. Ridet, and M. L. Bauchot. 1989. Encephalization in tropical teleost fishes and comparison with their mode of life. Journal für Hirnforschung 30:645-669.

Brandstatter, R., and K. Kotrschal. 1990. Brain growth patterns in four European cyprinid fish species (Cyprinidae, Teleostei): roach *(Rutilus rutilis)*, bream *(Abramis brama)*, common carp *(Cyprinus carpio)* and sabre carp *(Pelecus cultratus)*. Brain, Behaviour and Evolution 35:195-211.

Buckley, L. J. 1979. Relationship between RNA-DNA ratio, prey density, and growth rate in the Atlantic cod *(Gadus morhua)* larvae. Journal of the Fisheries Research Board of Canada 36:1497-1502.

Buckley, L. J. 1980. Changes in ribonucleic acid, deoxyribonucleic acid, and protein content during ontogenesis in winter flounder, *Pseudopleuronectes americanus*, and effect of starvation. Fishery Bulletin, U. S. 77:703-708.

Buckley, L. J. 1982. Effects of temperature on growth and biochemical composition of larval winter flounder *Pseudopleuronectes americanus*. Marine Ecology Progress Series 8:181-186.

Buckley, L. J. 1984. RNA-DNA ratio: an index of larval fish growth in the sea. Marine Biology 80:291-298.

Buckley, L. J., and R. C. Lough. 1987. Recent growth, biochemical composition, and prey field of larval haddock *(Melanogrammus aeglefinus)* and the Atlantic cod *(Gadus morhua)* on the Georges bank. Canadian Journal of Fisheries and Aquatic Sciences 44:14-25.

Buckley, L. J., S. I. Turner, T. A. Halavik, A. S. Smigielski, S. M. Drew, and G. C. Laurence. 1984. Effects of temperature and food availability on growth, survival, and RNA-DNA ratio of larval sand lance *(Ammodytes americanus)*. Marine Ecology Progress Series 15:91-97.

Bulow, F. J. 1971. Selection of suitable tissues for use in the RNA-DNA ratio technique of assessing recent growth rate of a fish. Iowa State Journal of Science 46:71-78.

Bulow, F. J. 1987. RNA-DNA ratios as indicators of growth in fish: a review. Pages 45-64 *in* R. C. Summerfelt and G. E. Hall, editors. Age and growth in fish. The Iowa State University Press. Ames.

Canino, M. F., K. M. Bailey, and L. S. Incze. 1991. Temporal and geographic differences in feeding and nutritional condition of walleye pollock larvae *Theragra chalcogramma* in Shelikof Strait, Gulf of Alaska. Marine Ecology Progress Series 79:27-35.

Clemmesen, C. M. 1987. Laboratory studies on RNA/DNA ratios of starved and fed herring *(Clupea harengus)* and turbot *(Scophthalmus maximus)* larvae. Journal du Conseil International pour l' Exploration de la Mer 3:122-128.

Clemmesen, C. M. 1988. A RNA and DNA fluorescence technique to evaluate the nutritional condition of individual fish larvae. Meeresforschung 32:134-143.

Dabrowski, K., F. Takashima, and C. Strussmann. 1986. Does recovery growth occur in larval fish? Bulletin of Japanese Society of Scientific Fisheries 52:1869.

Darzynkiewicz, A. 1991. Flow cytometric methods for RNA content analysis. Methods: Companion to Methods in Enzymology 2:200-206.

Dobson, S. H., and R. M. Holmes. 1984. Compensatory growth in the rainbow trout, *Salmo gairdneri* Richardson. Journal of Fish Biology 25:649-658.

Hartwell, L. H., and T. A. Weinert. 1989. Checkpoints: controls that ensure the order of cell cycle events. Science 246:629-634.

Lewin, B. 1987. Genes. John Wiley and Sons, Inc. New York.

Martin, F. D., and R. Malloy. 1980. Histologic and morphometric criteria for assessing nutritional state in larval striped bass, *Morone saxatilis*. Pages 157-166 *in* L. A. Fuiman, editor. Proceedings of the fourth annual larval fish conference. U. S. Fish and Wildlife Service FWS/OBS-80/43.

May, R. C. 1974. Larval mortality in marine fishes and the critical period concept. Pages 3-19 *in* J. H. S. Blaxter, editor. The early life history of fish. Springer-Verlag, New York.

Mazur, P. 1970. Cryobiology: the freezing of biological systems. Science 168:939-949.

McGurk, M. D. 1984. Effects of delayed feeding and temperature on the age of irreversible starvation and on the rates of growth and mortality of Pacific herring larvae. Marine Biology 84:13-26.

Miglavs, I., and M. Jobling. 1989. Effects of feeding regime on food consumption, growth rate and tissue nucleic acids in juvenile Arctic charr, *Salvelinus alpinus*, with particular respect to compensatory growth. Journal of Fish Biology 34:947-957.

Murray, A. W., and M. W. Kirschner. 1989. Dominoes and clocks: the union of two views of the cell cycle. Science 246:614-621.

Nichols, P. D., D. G. Holdsworth, J. K. Volkman, M. Daintith, and S. Allanson. 1989. High incorporation of essential fatty acids by the rotifer *Brachionus plicatilis* fed on the prymnesiophyte *Pavlova lutheri*. Journal of Marine Freshwater Resources 40:645-655.

Nikonorov, S. I., A. V. Pol'gin, M. Yu. Pichugin, and L. V. Vitvitskaya. 1985. Size and nucleic acid content of brain of juvenile *Salmo salar* L. Zhurnal Evolyutsionnoi Biokhimii i Fiziologii 21:549-554. [*Translation*].

O'Connell, C. P. 1976. Histological criteria for diagnosing the starving condition in early post yolk sac larvae of the northern anchovy, *Engraulis mordax* Girard. Journal of Experimental Marine Biology and Ecology 25:285-312.

Overton, W. R. 1988. Modified histogram subtraction technique for analysis of flow cytometry data. Cytometry 9:619-626.

Packard, A., and A. W. Wainwright. 1974. Brain growth of young herring and trout. Pages 499-507 *in* J. H. S. Blaxter, editor. The early life history of fish. Springer-Verlag, Berlin.

Pardee, A. B. 1989. G_1 events and regulation of cell proliferation. Science 246:603-608.

Pedersen, B. H., I. Ugelstad, and K. Hjelmeland. 1990. Effects of a transitory, low food supply in the early life of larval herring *(Clupea harengus)* on mortality, growth and digestive capacity. Marine Biology 107:61-66.

Schumacher, J. D., and A. W. Kendall, Jr. 1991. Some interactions between young walleye pollock and their environment in the western Gulf of Alaska. California Cooperative Oceanic Fisheries Investigations Report 32:22-40.

Shapiro, H. M. 1988. Practical flow cytometry. Alan R. Liss, Inc. New York. 286 p.

Singhal, R. N., H. B. Sarnat, and R. W. Davies. 1989. Effects of anoxia, hyperoxia, and salinity on the neurons in the leech *Nephelopsis obscura* (Erpobdellidea): RNA redistribution by fluorescence histochemistry. Journal of Invertebrate Pathology 53:93-101.

Theilacker, G. H. 1978. Effect of starvation on the histological and morphological characteristics of jack mackerel, *Trachurus symmetricus*, larvae. Fishery Bulletin, U. S. 76:403-414.

Theilacker, G. H., and W. Shen 1993. Fish larval condition using flow cytometry. *in* B.T. Walther and H.J. Fyhn, editors. Physiological and biochemical aspects of fish development. University of Bergen, Norway.

Traganos, F., Z. Darzynkiewicz, T. Sharpless, and M. R. Melamed. 1977. Simultaneous staining of ribonucleic deoxyribonucleic acids in unfixed cells using acridine orange in a flow cytofluorometric system. Journal of Histochemistry and Cytochemistry 25:46-56.

Weatherley, A. H., and H. S. Gill. 1981. Recovery growth following periods of restricted rations and starvation in rainbow trout, *Salmo gairdneri* Richardson. Journal of Fish Biology 18:295-208.

Westerman, M. E., and G. J. Holt. 1988. The RNA-DNA ratio: measurement of nucleic acids in larval *Sciaenops ocellatus*. Contributions in Marine Science 30 (supplement):117-124.

Wieser, W., G. Krumschnabel, and J. P. Ojwang-Okwor. 1992. The energetics of starvation and growth after refeeding in juveniles of three cyprinid species. Environmental Biology of Fishes 33:63-71.

Wilkinson, L. 1988. SYSTAT: the system for statistics. SYSTAT, Inc. Evanston, Illinois.

Yamashita, Y., and K. M. Bailey. 1989. A laboratory study of the bioenergetics of larval walleye pollock, *Theragra chalcogramma*. Fishery Bulletin, U. S. 87:525-536.

Yoklavich, M. M., and K. M. Bailey. 1990. Hatching period, growth and survival of young walleye pollock *Theragra chalcogramma* as determined from otolith analysis. Marine Ecology Progress Series 64:13-23.

Zhang, Z., and N. W. Runham. 1992. Effects of food ration and temperature level on the growth of *Oreochromis niloticus* (L.) and their otoliths. Journal of Fish Biology 40:341-349.

Use of Mesocosm Studies to Examine Direct and Indirect Impacts of Water Quality on Early Life Stages of Fishes

JAMES F. FAIRCHILD AND EDWARD E. LITTLE

U. S. Fish and Wildlife Service
National Fisheries Contaminant Research Center
4200 New Haven Road, Columbia, Missouri 65201, USA

Abstract.—Mesocosms are replicated experimental ecosystems designed to simulate natural environments. A variety of mesocosm designs have been used to simulate freshwater and marine systems, including plastic in situ enclosures, earthen-lined experimental ponds, and outdoor experimental streams. Mesocosms have been used to study both direct and indirect impacts of water quality on early life stages of fishes. Numerous studies have examined the direct impacts of toxicants on early life stages of fishes; these studies have primarily been used to validate predictions derived from laboratory data. Mesocosm studies have also been used to examine the indirect impacts of water quality on survival and growth of early life stages of fishes, including food reduction, food-chain mediated contaminant transfer, habitat alteration, changes in predation rates, and changes in competition. Most recently, mesocosm studies were incorporated within the pesticide registration process to negate a presumption of risk derived from laboratory data; however, this requirement was subsequently rescinded. Studies in aquatic mesocosms can offer significant advantages over studies in natural systems because of increases in experimental control and statistical inference.

Early life stages of fishes undergo a series of ontogenetic shifts through the egg, larval, and juvenile stages which must be surpassed to ensure survival of the individual or population (Houde 1987). Numerous factors control the survival of early life stages of fishes, including food availability (Hjort 1914; May 1974), predation (Hunter 1981), and local environmental conditions (e.g., temperature, dissolved oxygen, and ammonia; Houde 1987). Mortality of many species of fishes is significant during the early life stages; yet, populations persist because fishes have developed various reproductive and life-history strategies to survive these environmental challenges (Potts and Wootton 1984). These strategies have evolved over thousands of years to accommodate relatively predictable forms of natural environmental stress.

Over the last century fish have been exposed to additional forms of anthropogenic stress such as fishing pressure, physical habitat deterioration, and environmental contaminants. These stresses have been introduced abruptly compared with the evolutionary scale of natural events; therefore, one may expect that populations have reduced capabilities for adaptation or compensation. Further, anthropogenic stressors act both individually and in combination with other natural challenges. Consequently, many species of fishes have declined in numbers to threatened or endangered status; others have become extinct. Although numerous species are in decline, causal factors are difficult to determine in natural environments due to the complexity of interactions that occur.

Laboratory studies are often used to gather information concerning the impacts of water quality and other factors on survival and growth of early life stages of fishes (McKim 1985). Laboratory studies allow researchers to examine the impacts of single and multiple stressors under controlled, replicable, and repeatable conditions. Many acute and chronic toxicity tests have been standardized (ASTM 1988a, ASTM 1988b) and are the primary mechanisms to establish Water Quality Criteria under the Toxic Substances Control Act (TSCA) and to register pesticides under the Federal Insecticide, Fungicide, and Rodenticide Act (FIFRA). However, reliance on standardized laboratory single-species tests has been criticized for several reasons: (1) most water-quality studies are conducted using only a few laboratory surrogate species; (2) indirect effects, such as changes in habitat or biological interactions are not studied; (3) avoidance or attractance to the chemical are not considered; and (4) exposure regimes do not simulate true environmental conditions (Cairns 1983, 1986; Kimball and Levin 1985). Although results of laboratory tests are relatively precise, they are not always accurate in predicting the direction or degree of response of organisms to water-quality changes in natural environments.

Mesocosm studies offer a reasonable compromise between laboratory and field studies in assessment of water-quality impacts on early life stages of fishes. They are potentially more accurate than laboratory studies because multiple species can be studied in simulated natural environments exposed under realistic conditions. Indirect effects on biological interactions (e.g., competition and predation), habitat quality, and biomagnification can be studied. Further, mesocosm studies offer significant advantages over studies in natural systems because they can be replicated, manipulated, and sampled in a cost-effective, efficient manner. Although the term "mesocosm" is relatively new (Odum 1984), the basic concept has been used frequently in aquatic research in freshwater (Swingle 1947; Macek et al. 1972) and marine (Menzel and Case 1977; Pilson and Nixon 1980) studies for nearly 50 years.

Herein, we discuss the use of mesocosm studies to examine the direct and indirect impacts of water quality on the early life stages of fishes. We provide examples of research applications of mesocosms in two areas: direct impacts of contaminants on survival and growth of early life stages of fishes; and indirect impacts on survival and growth due to food-chain reductions, food-chain mediated contaminant transfer, changes in predation or competition, and habitat alteration. We also discuss the recent consideration of mesocosm tests in the pesticide registration process under the FIFRA. This information is presented to demonstrate the use of mesocosm studies as an intermediate step between laboratory and field studies for examining the direct and indirect impacts of water-quality factors on changes in survival and growth of early life stages of fishes.

Mesocosm Design and Construction

Several mesocosm designs have been used to simulate freshwater and marine systems, including in situ plastic limnocorrals (Menzel and Case 1977; Kaushik et al. 1986), fiberglass towers (Nixon et al. 1980), littoral corrals (Brazner et al. 1989), earthen-lined experimental ponds (Crossland and Wolff 1985, deNoyelles et al. 1989), and outdoor experimental streams (Zischke et al. 1985; Fairchild et al. 1987). Sanders (1985) provided an extensive review of the use of large enclosures for lentic perturbation studies, emphasizing design, operating

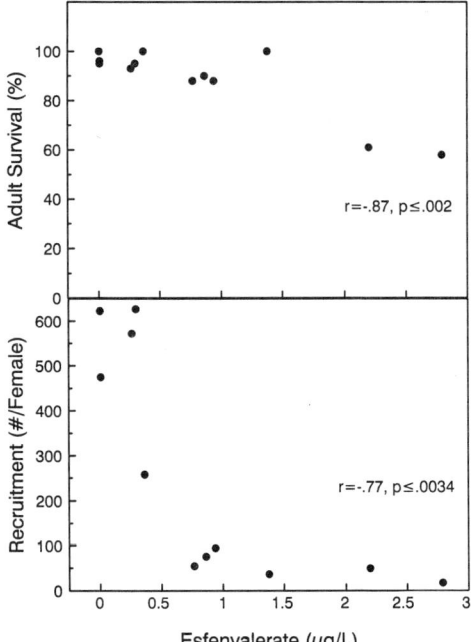

FIGURE 1.—Survival and reproduction of bluegill exposed to esfenvalerate in aquatic mesocosms. Each point represents results from one mesocosm. Figure adapted from Fairchild et al. (1992).

criteria, variability, and applications. Refer to Warren and Davis (1971) for approaches and constraints in using experimental streams for aquatic research.

Research Approaches for Use of Mesocosms

Direct impacts of water quality on early life stages of fishes.—Mesocosms have been used to study a variety of direct water-quality impacts on early life stages of fishes, including insecticides (Siefert et al. 1989; Fairchild et al. 1992), ammonia (Hermanutz et al. 1987), chlorinated phenols (Crossland and Wolff 1985; Zischke et al. 1985), polynucleated aromatic hydrocarbons (Boyle et al. 1985), metals (Sherman et al. 1987), acidification (Zischke et al. 1983), and temperature (Wrenn and Granneman 1980). These studies allowed researchers to compare laboratory results with field responses observed under conditions that accurately simulate the entry, partitioning, bioavailability, and fate of environmental contaminants or other factors.

Fairchild et al. (1992) conducted a mesocosm study in earthen-lined freshwater mesocosms (0.1 ha, 1.5 m mean depth, 700 m^3 volume) to examine the effects of pulsed dosing of a pyrethroid insecticide on adult and early life stages of bluegill

(Lepomis macrochirus). Mesocosms were stocked with adult bluegill in early April prior to reproduction. The mesocosms were treated at nominal concentrations of 0, 0.25, 0.67, or 1.71 µg/L esfenvalerate on six biweekly dates from early June through August. Treatment concentrations and intervals were selected to simulate a "worst-case" exposure scenario. In early October the ponds were drained to recover bluegill populations and determine mortality, growth, and reproductive rates.

Esfenvalerate concentrations decreased rapidly in water, with a calculated half-life of dissipation of 10 h. Esfenvalerate concentrations frequently exceeded expected nominal exposure concentrations, with initial levels averaging nearly 2.8 µg/L in one high treatment mesocosm (Figure 1). Bluegill survival and reproductive success were negatively related to measured concentrations of esfenvalerate (Figure 1). Early life stages of bluegill were more sensitive to esfenvalerate than adults. Recruitment of early life stages of bluegill was reduced at measured concentrations ≥ 0.5 µg/L; however, adult survival was not affected until average concentrations exceeded 2.0 µg/L (Figure 1). Esfenvalerate was acutely toxic to juvenile bluegill under standard laboratory conditions at 0.31 µg/L (96-h static LC50), indicating that laboratory tests were predictive of the nominal concentrations that would impact juvenile bluegill in simulated field environments. Laboratory results were accurate because the data were applied to a single species, pyrethroid insecticides are metabolized by fishes and do not produce cumulative effects, and the fate of the chemical was similar under both laboratory and field conditions.

Indirect impacts of water quality on early life stages of fishes.—Indirect effects occur when a primary, direct effect of a water-quality factor alters one system component, which then affects a second component. Numerous types of indirect water-quality impacts can occur in aquatic ecosystems, and can be transmitted via habitat-mediated, food-mediated, and interactive (competition, predation, etc.) pathways. Indirect impacts can be difficult to predict from laboratory studies because of constraints on system size and complexity. However, indirect effects can be studied in mesocosms and natural aquatic ecosystems exposed to contaminant stress.

HABITAT-MEDIATED INDIRECT EFFECTS. Quantity, quality, and stability of habitat are primary factors controlling the structure and persistence of animal populations (Crowder and Cooper 1982; Southwood 1988). Contaminants or other water-quality factors that alter habitat can indirectly impact those populations and communities that rely on macrophytes for food, shelter, and reproductive needs.

DeNoyelles et al. (1989) examined the impacts of the herbicide atrazine on plant, invertebrate, and fish communities in experimental pond mesocosms. Mesocosms (0.04 hectare) were treated annually with 0, 20, 100, 200, or 500 µg/L atrazine and observed for 3 years. Atrazine concentrations remained relatively stable over time and remained within 30% of nominal concentrations for 6 months following each treatment. Atrazine was not directly toxic to fishes or invertebrates. However, atrazine, a potent inhibitor of photosynthesis, reduced macrophyte biomass and periphyton growth at concentrations of 20 µg/L. Decreases in macrophyte biomass and periphyton productivity led to indirect decreases in numbers of grazing insects. Bluegill reproductive success was significantly reduced from 1,376 larvae/pond (controls) to 59 larvae/pond (20 µg/L treatment) (Kettle et al. 1987), even though chronic laboratory exposures had no effects on growth or development of young bluegill at concentrations as high as 95 µg/L (Macek et al. 1976). Decreases in bluegill reproductive success were caused by the combined indirect effects of food reduction and increased predation following macrophyte reductions. Thus, a mesocosm study revealed several indirect, habitat-mediated pathways through which water quality can impact early life stages of fishes.

Giesy et al. (1979) studied the direct and indirect impacts of cadmium on experimental stream mesocosms. Six outdoor experimental stream channels (90 m long, 0.5 m wide, 0.2 m deep; flowing at 1.3×10^{-3} m/s) were continuously dosed with 0, 5, or 10 µg/L cadmium for 10 months. In control channels crayfish dominated system structure and function. Crayfish grazing activity reduced macrophyte biomass and benthos, which subsequently reduced bluegill growth because of food limitation. However, crayfish were extremely sensitive to cadmium and were rapidly eliminated in treated channels. Aquatic vegetation, nearly absent in the highly grazed control channels, increased in the cadmium-treated streams. That is, cadmium benefitted vegetative habitat indirectly. Thus, paradoxically, early life stages of fishes, in addition to their prey, could indirectly benefit from contaminant exposure in some instances.

FOOD-MEDIATED INDIRECT IMPACTS. Contaminants can indirectly impact early life stages of fishes via two food-mediated mechanisms: food reduction

TABLE 1.—Average wet weight (mg) of fathead minnows prior to and after exposure to chlorpyrifos. Values are means ± SE, with sample sizes in parentheses. An asterisk denotes significant difference from the control ($P \leq 0.05$, ANOVA). Table and data adapted from Siefert et al. (1989).

Day	Control	20 µg/L
-2	4.64 ± 0.31 (49)	4.56 ± 0.22 (120)
7	9.29 ± 0.80 (52)	7.31 ± 0.59 (96)
15	31.54 ± 1.59 (57)	9.28 ± 0.80* (84)
31	52.76 ± 4.47 (38)	22.69 ± 3.91* (46)

TABLE 2.—Effects of selenium on hatchability, survival, and condition of larvae from adult bluegill exposed in lotic mesocosms for 365 days. Values are means ± SE, with sample sizes in parentheses. An asterisk denotes significant difference from the control ($P \leq 0.05$). Data from Hermanutz et al. (1992).

Endpoint	Control	10 µg/L
Embryos, percent hatch	93.9 ± 9.1 (7)	71.5 ± 22.5* (7)
Larvae, percent survival[a]	69.7 ± 13.9 (5)	28.8 ± 23.1* (6)
Larvae, percent edema[b]	0.1 ± 0.2 (6)	80.0 ± 1.0* (5)
Larvae, percent lordosis[b]	1.8 ± 2.6 (6)	11.6 ± 15.9[c] (7)
Larvae, percent hemorrhaging[b]	0.1 ± 0.3 (6)	28.5 ± 40.6* (5)

[a] Survival on the 4th d posthatching.
[b] Highest percent occurrence during a 4-d posthatching period.
[c] Significantly different from the control at $P \leq 0.10$.

and contaminant transfer. Mesocosm studies have been used to study the mechanisms of both of these indirect water-quality impacts.

Siefert et al. (1989) studied the direct and indirect impacts of the organophosphate insecticide chlorpyrifos on larval fathead minnows (Pimephales promelas) in littoral corrals. Twelve plastic-lined enclosures (5 m long, 10 m wide, 1 m deep) were treated with 0, 0.5, 5.0, or 20.0 µg/L chlorpyrifos (three replicates/treatment) in early June. Chlorpyrifos dissipated rapidly (half-life approximately 10 h) from the water column. Chlorpyrifos significantly reduced numbers and diversity of zooplankton and macroinvertebrates, but was not directly toxic to larval fathead minnows. Growth rates of larval fathead minnows decreased significantly in the corrals treated with 20 µg/L chlorpyrifos (Table 1). Dietary analysis of fish indicated that the reduced growth was primarily due to decreases in food availability. Thus, although chlorpyrifos was not directly toxic to larval fathead minnows, it reduced growth rates indirectly, via food reduction.

Mesocosm studies can also provide useful approaches for examining the potential for food-chain transfer and biological effects of environmental contaminants. Hermanutz et al. (1992) studied the food-mediated bioaccumulation and effects of selenium on early life stages of bluegill in experimental stream mesocosms. Selenium is a metalloid element frequently found in saline soils of the western United States and is a common constituent of fly-ash derived from coal-fired power plants. Standard laboratory acute and chronic data indicate that inorganic selenium is not toxic to aquatic organisms at concentrations less than 50 µg/L (USEPA 1987). However, field evidence has revealed that survival and recruitment of early life stages of fishes is greatly reduced in aquatic habitats containing selenium at ambient concentrations less than 10 µg/L (USEPA 1987). Research has indicated that inorganic selenium in aquatic systems is microbially methylated and substituted for sulfur in amino acids such as methionine (Bottino et al. 1984). Enzymes containing these altered amino acids lose normal function, which results in mortality of larval fishes (Gillespie and Baumann 1986). Feeding studies have indicated that dietary levels of organically incorporated, food-borne selenium $\geq$ 16 ppm resulted in total mortality of larval bluegill, even though spawning frequency, fecundity, and hatching success were not affected (Coyle et al. 1993). However, controlled ecosystem studies examining the reproductive effects of selenium in fish following actual food-web transfer and organic transformation of inorganic selenium had not previously been conducted.

Hermanutz (1992) continuously dosed six experimental stream mesocosms (520 m long; flowing at 0.53 to 0.76 m³/min) with 0, 10, and 30 µg/L sodium selenite (two replicates/treatment) for 356 days to determine a safe environmental concentration of inorganic selenium. Uptake, storage, and effects of selenium in larval and adult bluegill were determined. Survival of bluegill larvae was 41% lower in the exposure to 10 µg/L inorganic selenium (Table 2). Further, surviving larval bluegill exhibited significant increases in edema, lordosis, and hemorrhaging. These effects occurred at dissolved inorganic selenium concentrations that were five-fold lower than safe levels previously calculated using laboratory-based data for inorganic selenate and selenite (Hermanutz et al. 1992). Adverse impacts occurred due to organic conversion and transfer of selenium through the food web. Thus, a mesocosm study was successful in validating the toxicological model of selenium fate and effects in aquatic ecosystems, and it supports the current Water Quality Criterion Continuous Concentration of 5 µg/L selenium (USEPA 1987).

TABLE 3.—Effects of fluorene on largemouth bass and bluegill in fluorene-treated mesocosms. Means within the same row followed by a common letter are not significantly different ($P \leq 0.05$, ANOVA). Data from Boyle et al. (1985).

Species and variable	Fluorene concentrations (mg/L)			
	0	0.12	0.5	2.0
Largemouth bass adults				
Survival, %	100 z	82 y	51 x	2 w
Individual growth, g	146 z	59 z	120 z	351 z
Bluegill adults				
Survival, %	99 z	84 y	56 x	50 w
Individual growth, g	20 z	19 z	19 z	25 z
Bluegill recruits				
Number per adult	15 z	13 y	5 x	4 w
Growth, g	1.3 z	0.9 y	1.6 z	3.3 x

COMPETITION- OR PREDATION-MEDIATED INDIRECT EFFECTS. Predation and competition are two biological interactions that can influence the structure of fish populations. Mesocosms have been used to study the effects of both predation (Werner et al. 1983; Deutsch et al. 1992) and competition (Werner and Hall 1979) on the behavior, survival, and growth of fishes. However, few mesocosm studies have examined the impacts of water quality on predation rates or interspecific competition of fishes. Water quality can alter the survival, growth, and behavior of fishes which can ultimately be reflected in the numbers, size, or age distribution of fish populations. Numerous approaches exist to study the impacts of contaminants on intraspecific and interspecific interactions in the laboratory (see Little et al. in this volume). These interactions can also be studied in outdoor experimental ecosystems such as mesocosms.

Boyle et al. (1985) examined the direct and indirect effects of fluorene, a polynucleated aromatic hydrocarbon, on an assemblage of largemouth bass *(Micropterus salmoides)* and bluegill in 0.1 hectare mesocosms. Mesocosms were treated once with 0, 0.12, 0.5, 2.0, or 5.0 mg/L fluorene and then monitored over a 5-month period. Effects on survival, growth, and reproduction of largemouth bass and bluegill were measured. Results were compared to predictions derived from standard (mortality and growth) and non-standard (behavioral) laboratory endpoints (Finger et al. 1985).

Standard acute toxicity tests accurately predicted concentrations of fluorene causing mortality of largemouth bass and bluegill in the mesocosms. The laboratory 96-h LC50 values (0.82 to 0.91 mg/L) were similar to the concentration (0.5 mg/L) causing approximately 50% mortality of both species in the mesocosms (Table 3).

However, standard laboratory chronic tests were less accurate in predicting growth responses. In standard 30-d laboratory exposures (with unlimited feeding conditions) fluorene reduced growth of juvenile bluegill at concentrations of 0.5 mg/L. However, growth of juvenile bluegill decreased at fourfold lower concentrations (0.12 mg/L) in mesocosm exposures (Table 3). This response was predicted by non-standard laboratory behavioral tests, in which fluorene reduced feeding efficiency at 0.06 mg/L (Finger et al. 1985). Growth of juvenile bluegill decreased at 0.12 mg/L in the mesocosm study, even though species composition and numbers of invertebrates were not statistically different at concentrations ≤ 2.0 mg/L. Field exposures to fluorene may have reduced feeding efficiency of juvenile bluegill (as predicted by laboratory behavioral studies) even though food availability was not altered. Alternatively, growth rates of juvenile bluegill may have provided a more sensitive, functional indicator of differences in invertebrate numbers than standard invertebrate density estimates, which are inherently variable (Crossland 1988).

Growth rates of largemouth bass revealed strong, indirect density-dependent relationships in the fluorene study. Although survival of largemouth bass decreased linearly with increasing chemical concentrations, growth rates of bass actually increased in the highest concentration due to releases from intraspecific density-dependent food competition (Table 3). Largemouth bass grew the least in the low (0.12 mg/L) fluorene treatment where mortality was only 18%. However, as the numbers of surviving largemouth bass decreased in the 0.5 and 2.0 mg/L treatments, growth began to increase due to relative increases in available prey. Similar intraspecific responses in growth of juvenile bluegill were observed by Fairchild et al. (1992) following pulsed exposures of mesocosms to esfenvalerate. None of the laboratory tests predicted these indirect density-dependent responses.

Survival of juvenile and adult bluegill in the mesocosms decreased in a non-linear response to fluorene (Table 3). Survival and growth of bluegills in the 2.0-mg/L treatment was higher than predicted by a linear dose-response curve, perhaps due to the large decrease in largemouth bass survival and predation rates. Thus, bluegill populations exhibited potential numeric and functional (growth) compensation following changes in water quality, food availability, and predation rates. Although this interpretation is speculative, the study demonstrates that

TABLE 4.—Tiered hazard assessment procedure for aquatic risk assessment of pesticides required by the FIFRA. Adapted from Urban and Cook (1986).

Tier	Toxicity assessment	Exposure assessment
1	Acute fish and invertebrate tests	Simple application model
2	Fish early life-stage tests Invertebrate life-cycle tests Fish bioaccumulation tests	Simple runoff model
3	Fish life-cycle tests	Computer runoff and exposure models (SWRRB and EXAMS)
4	Simulated (i.e., mesocosms) or actual field tests of aquatic organisms	Field residue studies

homeostatic capabilities for fish population regulation can be examined in controlled ecosystems such as mesocosms.

Regulatory Use of Mesocosms: Direct and Indirect Impacts of Pesticides on Early Life Stages of Fishes

In 1986 mesocosm studies were introduced as fourth-tier tests for pesticide registrations (Touart 1988; Jenkins et al. 1989; Touart and Slimak 1989). The impact of pesticides on early life stages of fishes is a primary focus of these studies.

The U. S. Environmental Protection Agency (EPA) regulates the registration of pesticides within the domain of the FIFRA (Bedford 1984). A registrant submitting a pesticide for registration must provide the EPA with data on the toxicity of the candidate chemical to non-target aquatic organisms. Data on the environmental fate and persistence of the new chemical must also be provided. The EPA Office of Pesticide Programs (OPP) uses this information to conduct a hazard assessment in which the toxicity of a chemical is compared with the expected environmental concentration (EEC) resulting from the anticipated use pattern (Urban and Cook 1986; Jenkins et al. 1989).

The toxicity of pesticides to non-target aquatic species is determined using a stepwise hazard assessment procedure consisting of four tiers; toxicity and exposure potentials must be assessed (Urban and Cook 1986) (Table 4). Acute toxicity tests are required of all new chemicals. Sequential tests at higher tiers, including bioaccumulation tests, early life-stage tests, full life-cycle reproductive tests, and simulated field tests are required when established hazard criteria (i.e., a factor of the calculated EEC) are exceeded, at which point the registrant can decide to rescind the registration request or continue testing at the next higher tier.

Previous simulated field tests had been conducted in unreplicated ponds or aquatic habitats adjacent to chemically treated fields. However, the lack of experimental control and statistical power made this information difficult to interpret. These limitations led the EPA to develop guidelines for conducting simulated field tests in aquatic mesocosms to satisfy test requirements under tier four of the FIFRA (Touart 1988; Touart and Slimak 1989).

Mesocosms are defined within the guidelines as simulated aquatic ecosystems (e.g., constructed ponds with clay-lined earthen bottoms) with > 300 m^3 volume. Typically, 12 mesocosms are used in a registration test. An EEC of a proposed chemical is calculated using computer models for runoff (Simulator for Water Resources in Rural Water Basins [SWRRB]) and receiving waters (Exposure Analysis Modeling System [EXAMS]) based on characteristics of the pesticide, soils, rainfall, use patterns, and anticipated receiving systems (Jenkins et al. 1989). Mesocosms are typically exposed to 0, 0.1X, 1X, and 10X factors of the calculated EEC. Parameters monitored in the mesocosm study include survival, growth, and reproduction of fish (usually bluegill); numbers and species composition of zooplankton and macroinvertebrates; biomass of algae and macrophytes; water quality; and pesticide dynamics in water, sediment, and tissue.

In 1992 the OPP reviewed the results of numerous mesocosm studies conducted since 1986. The OPP determined that laboratory-based, single-species tests, in the absence of mesocosm studies, were adequate for registration purposes in most instances. Consequently, the mesocosm test requirement is expected to be rescinded in 1993, except in unusual regulatory circumstances (PTCN 1992). The OPP acknowledged that mesocosm studies provide valuable information on the indirect effects of pesticides in aquatic ecosystems; however, they felt that they do not currently have the capability to adequately evaluate these effects within the time and fiscal constraints of the pesticide registration process.

The OPP stressed that the long-term implications of the direct and indirect impacts of pesticides in aquatic ecosystems remains a high research priority within the EPA (PTCN 1992). Much of this research should focus on the direct and indirect impacts of water quality on the survival, growth, and recruitment of early life stages of fishes. Undoubtedly, mesocosm studies will continue to be a useful approach in these research efforts. However, the potential for use of mesocosm studies in water-quality regulation in the future is unknown.

References

ASTM (American Society for Testing and Materials). 1988a. Standard practice for conducting acute toxicity tests with fishes, macroinvertebrates, and amphibians. ASTM, Designation E 729-88, Philadelphia.

ASTM (American Society for Testing and Materials). 1988b. Standard guide for conducting early life-stage toxicity tests with fishes. American Society for Testing and Materials, Designation E 1241-88, Philadelphia.

Bedford, B. 1984. The role of ecosystem science in USEPA's mission to regulate toxic substances. Environmental Management 8:377-383.

Bottino, N. R., C. H. Banks, K. J. Irgolick, P. Micks, A. E. Wheeler, and R. A. Zingaro. 1984. Selenium-containing amino acids and proteins in marine algae. Phytochemistry 23:2445-2452.

Boyle, T. P., S. E. Finger, R. L. Paulsen, and C. F. Rabeni. 1985. Comparison of laboratory and field assessment of fluorene — Part II: effects on the ecological structure and function of experimental pond ecosystems. Pages 134-151 in T. P. Boyle, editor. Validation and predictability of laboratory methods for assessing the fate and effects of contaminants in aquatic ecosystems, American Society for Testing and Materials, STP 865, Philadelphia.

Brazner, J. C., L. J. Heinis, and D. A. Jensen. 1989. A littoral enclosure for replicated field experiments. Environmental Toxicology and Chemistry 8:1209-1216.

Cairns, J., Jr. 1983. Are single species toxicity tests alone adequate for estimating environmental hazard? Hydrobiologia 100:47-57.

Cairns, J., Jr. 1986. The myth of the most sensitive species. BioScience 36:670-672.

Coyle, J. J., D. R. Buckler, C. G. Ingersoll, J. F. Fairchild, and T. W. May. 1993. Effect of dietary selenium on the reproductive success of bluegills *(Lepomis macrochirus)*. Environmental Toxicology and Chemistry 12:551-565.

Crossland, N. O. 1988. A method for evaluating effects of toxic chemicals on the productivity of freshwater ecosystems. Ecotoxicology and Environmental Safety 16:279-292.

Crossland, N. O., and C. J. Wolff. 1985. Fate and biological effects of pentachlorophenol in outdoor ponds. Environmental Toxicology and Chemistry 4:73-86.

Crowder, L. B., and W. E. Cooper. 1982. Habitat-structured complexity and the interaction between bluegills and their prey. Ecology 63:1802-1813.

deNoyelles, F., W. D. Kettle, C. H. Fromm, M. F. Moffett, and S. L. Dewey. 1989. Use of experimental ponds to assess the effects of a pesticide on the aquatic environment. Pages 41-56 in J. R. Voshell, editor. Using mesocosms to assess the aquatic ecological risk of pesticides: theory and practice. Miscellaneous Publications of the Entomological Society of America 75, Lanham, Maryland.

Deutsch, W. G., E. C. Webber, D. R. Bayne, and C. W. Reed. 1992. Effects of largemouth bass stocking rate on fish populations in aquatic mesocosms used for pesticide research. Environmental Toxicology and Chemistry 11:5-10.

Fairchild, J. F., T. Boyle, W. R. English, and C. Rabeni. 1987. Effects of sediment and contaminated sediment on structural and functional components of experimental stream ecosystems. Water, Air, and Soil Pollution 36:271-293.

Fairchild, J. F., T. La Point, J. L. Zajicek, M. K. Nelson, F. J. Dwyer, and P. A. Lovely. 1992. Population, community, and ecosystem-level responses of aquatic mesocosms to pulsed doses of a pyrethroid insecticide. Environmental Toxicology and Chemistry 11:115-129.

Finger, S. E., E. E. Little, M. G. Henry, J. F. Fairchild, and T. P. Boyle. 1985. Comparison of laboratory and field assessment of fluorene — Part I: effects of fluorene on the survival, growth, reproduction, and behavior of aquatic organisms in laboratory tests. Pages 120-133 in T. P. Boyle, editor. Validation and predictability of laboratory methods for assessing the fate and effects of contaminants in aquatic ecosystems. American Society for Testing and Materials, STP 865, Philadelphia.

Giesy, J. P. Jr., H. J. Kania, J. W. Bowling, R. L. Knight, S. Mashburn, and S. Clarkin. 1979. Fate and biological effects of cadmium introduced into channel microcosms. U. S. Environmental Protection Agency, EPA 600/3-79-039, Athens, Georgia.

Gillespie, R. B., and P. C. Baumann. 1986. Effects of high tissue concentrations of selenium on reproduction by bluegills. Transactions of the American Fisheries Society 115:208-213.

Hermanutz, R. O., K. N. Allen, T. H. Roush, and S. F. Hedtke. 1992. Effects of elevated selenium concentrations on bluegills *(Lepomis macrochirus)* in outdoor experimental streams. Environmental Toxicology and Chemistry 11:217-224.

Hermanutz, R. O., S. F. Hedtke, J. W. Arthur, R. W. Andrew, K. N. Allen, and J. C. Helgen. 1987. Ammonia effects on microinvertebrates and fish in outdoor experimental streams. Environmental Pollution 47:249-283.

Hjort, J. 1914. Fluctuations in the great fisheries of northern Europe viewed in the light of biological research. Rapports et Procès-Verbaux des Réunions Conseil International pour l'Exploration de la Mer 20:1-228.

Houde, E. D. 1987. Fish early life dynamics and recruitment variability. American Fisheries Society Symposium 2:17-29.

Hunter, J. R. 1981. Feeding ecology and predation of marine fish larvae. Pages 33-77 in R. Lasker, editor. Marine fish larvae: morphology, ecology, and relation to fisheries. University of Washington Press, Seattle.

Jenkins, D. G., R. J. Layton, and A. L. Buikema. 1989. State of the art in aquatic ecological risk assessment. Pages 18-33 in J. R. Voshell, editor. Using mesocosms to assess the aquatic ecological risk of pesticides: theory and practice. Miscellaneous Publications of the Entomological Society of America 75, Lanham, Maryland.

Kaushik, N. K., K. R. Solomon, G. L. Stephenson, and K. E. Day. 1986. Use of limnocorrals in evaluating the effects of pesticides on zooplankton communities. Pages 269-290 in J. Cairns, Jr., editor. Community toxicity testing. American Society for Testing and Materials, STP 920, Philadelphia.

Kettle, W. D., F. deNoyelles, B. Heacock, and A. Kadoum. 1987. Diet and reproductive success of bluegill recovered from experimental ponds treated with atrazine. Bulletin of Environmental Contamination and Toxicology 38:57-52.

Kimball, K. D., and S. A. Levin. 1985. Limitations of laboratory bioassays: the need for ecosystem-level testing. BioScience 35(3) 165-171.

Macek, K. J., K. S. Buxton, S. Sauter, S. Guilka, and J. Dean. 1976. Chronic toxicity of atrazine to selected aquatic invertebrates and fishes. U. S. Environmental Protection Agency, EPA 600/3-76-047, Washington, District of Columbia.

Macek, K. J., D. F. Walsh, J. W. Hogan, and D. D. Holz. 1972. Toxicity of the insecticide dursban to fish and aquatic invertebrates in ponds. Transactions of the American Fisheries Society 101:420-427.

May, R. C. 1974. Larval mortality in marine fishes and the critical period concept. Pages 3-21 in J. H. S. Blaxter, editor. The early life history of fish. Springer-Verlag, New York.

McKim, J. M. 1985. Early life-stage tests. Pages 58-95 in G. M. Rand and S. R. Petrocelli, editors. Fundamentals of Aquatic Toxicology. Hemisphere Publishing Corporation, New York.

Menzel, D. W., and J. Case. 1977. Concept and design: controlled ecosystem pollution experiment. Bulletin of Marine Science 27:1-7.

Nixon, S. W., D. Alonso, M. E. Pilson, and B. A. Buckley. 1980. Turbulent mixing in aquatic microcosms. Pages 818-850 in J. P. Giesy, Jr., editor. Microcosms in ecological research. National Technical Information Service, CONF-781101, Springfield, Virginia.

Odum, E. P. 1984. The mesocosm. BioScience 34:558-562.

Pilson, M. E., and S. W. Nixon. 1980. Marine microcosms in ecological research. Pages 724-741 in J. P. Giesy, Jr., editor. Microcosms in ecological research. National Technical Information Service, CONF-781101, Springfield, Virginia.

Potts, G. W., and R. J. Wootton. 1984. Fish reproduction: strategies and tactics. Academic Press, New York.

PTCN (Pesticide and Toxic Chemical News). 1992. EPA, OPP ecological risk data needs reduced, revamped. Vol. 21:19-20, Nov. 24, 1992. CRC Press, Washington, District of Columbia.

Sanders, F. S. 1985. Use of large enclosures for perturbation experiments in lentic ecosystems: a review. Environmental Monitoring and Assessment 5:55-99.

Sherman, R. E., S. P. Gloss, and L. W. Lion. 1987. A comparison of toxicity tests conducted in the laboratory and in experimental ponds using cadmium and the fathead minnow *(Pimephales promelas)*. Water Research 21:317-323.

Siefert, R. E., S. J. Lozano, J. C. Brazner, and M. L. Knuth. 1989. Littoral enclosures for aquatic field testing of pesticides: effects of chlorpyrifos on a natural system. Pages 57-73 in J. R. Voshell, editor. Using mesocosms to assess the aquatic ecological risk of pesticides: theory and practice. Miscellaneous Publications of the Entomological Society of America 75,

Lanham, Maryland.

Southwood, T. R. S. 1988. Tactics, strategies, and templates. Oikos 52:3-18.

Swingle, H. S. 1947. Experiments on pond fertilization. University of Auburn Agricultural Experiment Station, Bulletin 264, Auburn, Alabama.

Touart, L. W. 1988. Aquatic mesocosm tests to support pesticide registrations. U. S. Environmental Protection Agency, EPA 540/09-88-035, Washington, District of Columbia.

Touart, L. W. and M. W. Slimak. 1989. Mesocosm approach for assessing the ecological risk of pesticides. Pages 33-40 in J. R. Voshell, editor. Using mesocosms to assess the aquatic ecological risk of pesticides: theory and practice. Miscellaneous Publications of the Entomological Society of America 75, Lanham, Maryland.

Urban, D. J., and N. J. Cook. 1986. Hazard evaluation division, standard evaluation procedure, ecological risk assessment. U. S. Environmental Protection Agency, EPA 540/9-85-001.

USEPA (U. S. Environmental Protection Agency), 1987. Ambient water quality criteria for selenium-1987. USEPA 440/5-87-006, Washington, District of Columbia.

Warren, C. E., and G. E. Davis. 1971. Laboratory stream research: objectives, possibilities, and constraints. Annual Review of Ecology and Systematics 2:111-144.

Werner, E. E., J. F. Gilliam, D. J. Hall, and G. G. Mittlebach. 1983. An experimental test of the effects of predation risk on habitat use in fish. Ecology 64:1540-1548.

Werner, E. E., and D. J. Hall. 1979. Foraging efficiency and habitat switching in competing sunfishes. Ecology 60:256-264.

Wrenn, W. B., and K. L. Grannemann. 1980. Effects of temperature on bluegill reproduction and young-of-the-year standing stocks in experimental ecosystems. Pages 703-714 in J. P. Giesy, Jr., editor. Microcosms in ecological research. National Technical Information Service, CONF-781101, Springfield, Virginia.

Zischke, J. A., J. W. Arthur, R. O. Hermanutz, S. F. Hedke, and J. C. Helgen. 1985. Effects of pentachlorophenol on invertebrates and fish in outdoor experimental channels. Aquatic Toxicology 7:37-58.

Zischke, J. A., J. W. Arthur, K. J. Nordlie, R. O. Hermanutz, D. A. Standen, and T. P. Henry. 1983. Acidification effects on macroinvertebrates and fathead minnows *(Pimephales promelas)* in outdoor experimental streams. Water Research 17:47-63.

Using Mesocosms to Assess the Influence of Food Resources and Toxic Materials on Larval Fish Growth and Survival

GRACE KLEIN-MACPHEE, BARBARA K. SULLIVAN, AND AIMEE A. KELLER

University of Rhode Island Graduate School of Oceanography,
Narragansett Bay Campus, Narragansett, Rhode Island 02882-1197, USA

Abstract.—Mesocosms allow observations of larva survival and growth at near natural densities while feeding occurs on natural particle assemblages. Simulation of the whole ecosystem increases the probability that exposure to toxics occurs through the route that will occur in situ, and any indirect effects of toxics will be observed. As a result, responses of the larvae to controlled alteration of food, nuisance algal blooms, or water chemistry (such as introducing toxic materials) can be extrapolated to natural systems with much greater confidence than for studies done in smaller laboratory systems.

Experiments in the 13-m^3 MERL mesocosms illustrate the utility of mesocosms for work with larval fishes. Very high survival rates of 0 to 1-month-old flounder indicated that food availability was not an obvious controlling variable for survival during this period and that physiological requirements of the larvae were met. Blooms of *Phaeocystis* sp. (a nuisance alga) had no observed negative effect on winter flounder larvae. The presence of a benthos did significantly reduce larva survival; the greatest differences were noted between tanks with sediments and those without (average mortality rates 15%/d versus 4%/d). We hypothesize that predators or competitors in the benthos caused the higher mortality in systems with sediment.

In a separate experiment, a starch blend material being developed as a biodegradable substitute for plastics was added to mesocosms to determine its fate and effects in the marine environment. Larval cunner and tautog were the most abundant species present. There was no difference in survival of larval fishes between treated tanks and the control even at the highest concentrations of starch blend material (50 mg/L); however, growth rates of both species were significantly reduced and inversely correlated with treatment level. Gut content analyses of cunner and tautog indicated that feeding on copepods was reduced in treated tanks, although copepods were more abundant in treated enclosures than in the control. Fish may have been ingesting the less nutritious starch blend particles in preference to copepods, their normal diet. These results demonstrate the utility of mesocosms for addressing a range of questions related to the growth and survival of fish larvae.

Large enclosures (mesocosms) recently have become an important tool in determining the relative importance of food limitation and predation to growth and survival of larval fishes (de Lafontaine and Leggett 1987; Bailey and Houde 1989; Cowan and Houde 1990). They also are one of the best tools for evaluating effects of toxic substances because of the following advantages: (1) larvae in mesocosms exhibit very high rates of survival that are hard to achieve in the laboratory; (2) larvae can be reared in an experimental environment that more closely resembles in situ conditions than can be achieved in smaller systems; (3) larvae experience realistic routes of exposure to toxic substances; and (4) there is potential to observe indirect effects on larvae.

This paper addresses the utility of large tank (13-m^3) mesocosms at the Marine Ecosystems Research Laboratory (MERL) for work with larval fishes. The systems are designed to model Narragansett Bay, an estuarine ecosystem along the northeast coast of the United States (Lambert and Oviatt 1986). The MERL mesocosms have several advantages: (1) being self contained, long-term experiments can be conducted and toxicants can be added without introduction to the environment; (2) there are enough tanks to be able to replicate experimental conditions; (3) a benthic community as well as a water column community can be included; and (4) temperature control and thermal or salinity stratification can be maintained in the water column.

Although many components of the marine environment have been studied in these mesocosms, there had been only two experiments involving fish prior to the work described here; one with introduced haddock larvae (Bergen et al. 1985), and the

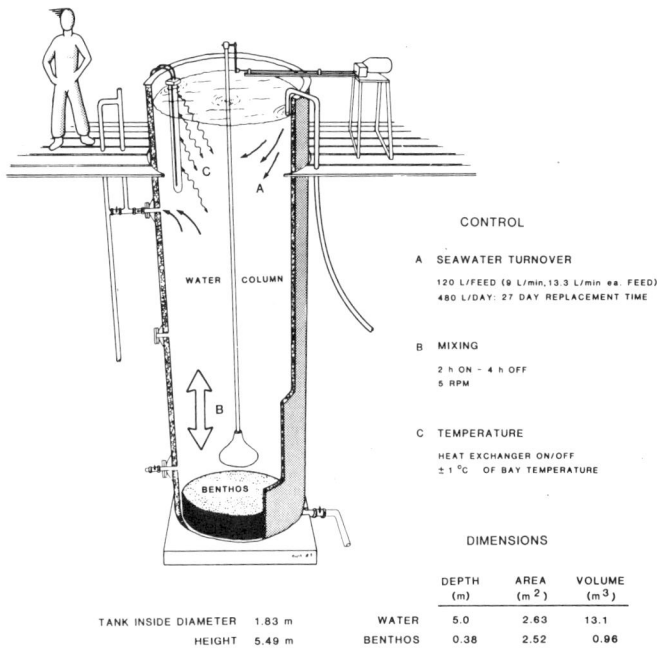

FIGURE 1.—Design of MERL mesocosms.

other an opportunistic experiment resulting from the introduction of menhaden eggs and larvae when the tanks were initially filled for a nutrient loading experiment (Keller et al. 1990). In these new experiments, we wished to utilize the system more fully for studying ecological and environmental effects on larval fishes. We describe our experiences with the MERL mesocosms and important findings concerning major factors affecting larval survival

In the first series of experiments we examined the effects of nutrient enrichment and sediments on growth and survival of larval winter flounder *Pleuronectes americanus*. The winter flounder was selected because it is one of the most important commercial and recreational marine species in the northeastern United States and has been experiencing a decline throughout its range for several years (NEFSC 1992). Larvae were reared at two nutrient levels in the MERL mesocosms at larva densities which one could encounter in the field with a natural plankton assemblage for food. The higher nutrient level represents conditions which exist in upper Narragansett Bay where winter flounder larvae are most abundant (Bourne and Govoni 1988), the control or unenriched nutrient level is found in lower Narragansett Bay where the MERL mesocosms are located. Experiments were run with and without sediments and their associated infauna. Winter flounder are demersal spawners that deposit their eggs in shallow water (Klein-MacPhee 1978). Eggs and newly hatched larvae are closely associated with sediments (Pearcy 1962; Klein-MacPhee 1978; Crawford and Carey 1985). Our hypothesis was that the best larval survival and growth would occur in nutrient enriched tanks with sediment. While conducting these experiments we had an opportunity to observe the effects of *Phaeocystis* sp., a potentially harmful alga (Rogers and Lockwood 1990), on larval winter flounder during blooms that developed in two mesocosms during spring of 1991.

The second set of experiments involved exposure of summer species of fishes including tautog *(Tautoga onitis)* and cunner *(Tautogolabrus adspersus)* to a potentially toxic material. We describe our experience here as an example of utility of mesocosms in applied research. The material studied was a starch blend (starch with vinyl alcohol copolymers but lacking any petroleum based plastic) that is being developed as a substitute for petroleum based, non-biodegradable plastics in packaging materials. The passage of international law to limit disposal of plastics at sea has created a need to develop new materials which effectively replace these products, producing less environmental damage.

Biodegradable blends of starch that are thermally processed with various copolymers make up one class of material currently being investigated. Harmful or undesirable effects of this type of material could theoretically come from several sources: (1) toxicity associated with either the blend or novel compounds created during thermal processing; (2) ingestion of particles of the degrading material by marine species; (3) physical effects of the particles that interfere with biological function; and (4) the biological oxygen demand created by decomposition of a carbon-based, biodegradable material. Predicting or documenting effects of materials for which complex or even unforeseen mechanisms such as those listed above are operating is a task that clearly benefits from test systems that retain much of the complexity of the natural environment. Experimental marine ecosystems (mesocosms) are such a tool.

We conducted an experiment documenting fate and effects of the starch blend on the water column community including phytoplankton, zooplankton, and fish in the MERL mesocosms (Sullivan et al. 1993). We concentrate here on additional data from that experiment describing the effect of this product on larval fishes.

Methods

The MERL mesocosms are 1.8 m in diameter and 5.5 m in depth (13.3 m^3) with a seawater delivery line and drain, heater, and plunger-type mixer (Figure 1). The 14 tanks are arranged along an access dock outdoors so that there is natural lighting and photoperiod. Mixing is periodic (2 h on, 2 h off) to resuspend sediments on a tidal period and to the same extent (3 to 5 ppm) as in Narragansett Bay. Temperature is maintained at the ambient value of Narragansett Bay water at the pumping station with glass heat exchangers suspended in the tanks (Figure 1). The area of the sediment (when used) is 2.52 m^2 and the depth 0.37 m.

The tanks were filled with unfiltered sea water from Narragansett Bay after which no water exchange occurred and therefore, no larvae were lost from the system except by mortality once the experiment began. Mesocosms in some experiments contained a sand-mud sediment and associated fauna collected from Greenwich Bay, a smaller segment of Narragansett Bay, in 1988 and from the Pawcatuck River Estuary in southern Rhode Island in 1990 and 1991. Nutrient-treated tanks received daily additions of nitrogen, phosphorus, and silica at a rate of 12 mmol NH$_4$Cl, 0.94 mmol KH$_2$PO$_4$ and 0.85 mmol Na$_2$SiO$_3$·9H$_2$O added as dissolved reagent-grade inorganic salts. This level was selected based on the response of phytoplankton and zooplankton observed in previous experiments (Sullivan and Ritacco 1985).

Winter Flounder Studies

Winter flounder larvae were obtained from broodstock collected in Narragansett Bay by otter trawl and maintained in the laboratory in tanks supplied with running seawater at ambient temperature and salinity. Eggs and larvae were obtained using techniques described in Klein-MacPhee et al. (1982). For each experiment, batches of eggs were obtained from two females fertilized with sperm from two males. Newly hatched larvae from the two groups of eggs were counted as they were gently poured into 4-L polycarbonate containers, combined in a 1:1 ratio and placed in the MERL tanks at a stocking density of 100/m^3, the maximum density of winter flounder larvae that occurs in Narragansett Bay (Bourne and Govoni 1988). During 1991, eggs were introduced into the mesocosms and initial numbers of larvae estimated from net tows. The natural community of zooplankton and phytoplankton in the tanks provided food for the larvae.

Handling mortality was calculated by floating two plastic 4-L containers with 202-μm screen bottoms in each MERL tank using styrofoam collars. Two hundred newly hatched flounder larvae were transferred into each container. Twenty-four hours later the containers were removed and the fish preserved and counted. Larvae preserved alive are clear and relatively straight, dead larvae are opaque, shrunken, and curved.

Temperature in the tanks was recorded daily. Salinity was measured weekly with a hand-held refractometer. In-vivo chlorophyll fluorescence was measured as an indication of phytoplankton abundance using a modification of the technique developed by Donaghay (1983). To ensure similarity of light history and mixing effects, samples were collected from well mixed tanks at the same time each day. All samples were measured using a Turner Design Model 10 Field Fluorometer equipped for chlorophyll a analysis.

Microplankton abundance (organisms retained on a 20-μm screen) was determined weekly by pumping 18 L of water from bottom, mid- and surface depths of the mesocosms. The water was filtered through a 20-μm screen, the filtrate preserved in 3% buffered formalin and the entire sample was

TABLE 1.—Design of experiments in MERL mesocosms, 1988-1991. N = number of mesocosms used. Nutrient-treated mesocosms are indicated by 4 × and had 4 times the ambient nutrient level (1×). Tanks with a benthos are indicated by +, those without by 0.

Date	Treatments		N	Duration (d)	Average temperature (°C)
	Benthos	Nutrients			
March 1988	+	1×	1	14	3.5
	+	4×	1	14	3.5
	0	1×	1	14	3.5
	0	4×	1	14	3.5
April 1988	0	1×	2	28	7.8
	0	4×	2	28	7.8
March 1990	0	1×	2	14	6.0
	+	1×	2	14	6.0
March 1991	0	1×	2	39	3.9
	+	1×	2	39	3.9

TABLE 2.—Average instantaneous daily growth rates of winter flounder larvae in mesocosms with two different nutrient levels during Experiment 2 (March 29 - April 27, 1988.

Treatment	Mean flourescense (relative units)	Growth rate (±SE)
No sediment control	5.85	0.022 (± 0.002)
No sediment control	6.74	0.021 (± 0.003)
No sediment high nutrient	11.39	0.019 (± 0.005)
No sediment high nutrient	19.35	0.020 (± 0.004)

counted under a dissecting microscope, except when small organisms such as tintinnids were abundant. These organisms were counted by resuspending the sample in 200 mL seawater and subsampling with a Stemple pipette. Macrozooplankton and fish larvae were collected weekly with three replicate samples using a 0.5-m diameter, 202-µm mesh plankton net equipped with a TSK flowmeter. The net was lowered to the bottom of the tank, then towed up by hand to the top of the tank. Water and organisms collected were briefly stored in a cold room in 4-L plastic containers until the larval fish could be removed and measured and weighed while alive. A subsample of larvae was preserved in 3% buffered formalin for gut content analysis. The remainder of the sample was filtered through a 200-µm screen and preserved in 3% buffered formalin. Fish lengths were obtained by anesthetizing 30 larvae with MS 222 and measuring them while alive using an image analyzer "image" measuring program, and compared to initial lengths taken from newly hatched larvae. Macrozooplankton abundances were determined from subsamples containing at least 200 individuals. Net tow capture efficiency of larvae was calculated by sampling the tanks with a 0.5-m net just prior to draining the entire tank. The number of larvae caught per tow was used to estimate the number of fish in the mesocosm. The total count of larvae obtained when the tank was completely drained plus the number removed by towing was then compared to the estimates to determine capture efficiency. Collections made by this method within the first 2 weeks of larval life had an 85 to 98% capture efficiency.

When experiments were terminated, a circular (1.75 m diameter) 333-µm mesh net was lowered to the bottom of the mesocosm on a vertical plane, and allowed to settle gently across the bottom. The water was then drained via a valve located 0.5 m from the bottom of the mesocosm through a 202-µm mesh plankton net suspended in a 30-L barrel. Larvae remaining in the tank were collected by raising the net through the remaining water in the mesocosm. Larvae from both nets were preserved in 3% buffered formalin. A final count for total surviving larvae was obtained by counting larvae collected in both the overflow net and the bottom net.

Experiments in 1988.—The first experiment in 1988 (Table 1) was designed to rapidly assess larva survival, and to determine whether daily addition of nutrients and the presence of a benthos affected survival. Four mesocosms were used for the experiment, two with sediment and two without. Nutrients were added to one of each of these treatments. No replicate treatments were established for this short-term (2-week) pilot study. The second 1988 experiment was conducted in four experimental tanks without benthos, two with nutrients added and two non-enriched controls to examine the effects on growth and survival. Sediments were not used because high mortality occurred in tanks containing a benthos in the pilot study. This experiment lasted 4 weeks and included two replicates per treatment (Table 1).

A series of experiments in 1990-1991 was designed to replicate observations on the sediment versus no-sediment effect and utilized two mesocosms with and two mesocosms without sediments for a period of 14 to 39 d (Table 1). Sediments from the Pawcatuck River estuary were used to ensure that larva mortality observed in 1988 was not the result of contaminants in the Greenwich Bay sediments, although we did not believe this was the case. The tanks were not enriched because long-term experiments in 1988 showed no significant effects of increased nutrients on larva survival and growth (Tables 2 and 3).

TABLE 3.—Instantaneous daily mortality rates of winter flounder in mesocosms with and without a benthos.

Treatment method	Days from start	Mortality rate (Z)	Collection
1988			
Expt. 1 (March 2-16)			
No benthos control	14	0.038	Total count
No benthos 4× nutrient	14	0.021	Total count
Benthos control	14	0.100	Total count
Benthos 4× nutrient	14	0.198	Estimate with net
Expt. 2 (March 31-April 27)			
Control (no benthos)	12	0.039	Estimate with net
Control (no benthos)	12	0.042	Estimate with net
4× Nutrient (no benthos)	12	0.066	Estimate with net
4× Nutrient (no benthos)	12	0.039	Estmate with net
Control (no benthos)	28	0.031	Total count
Control (no benthos)	28	0.025	Total count
4× Nutrient (no benthos)	28	0.025	Total count
4× Nutrient (no benthos)	28	0.058	Total count[a]
1990			
Expt. 1 (March 10-23)			
No benthos	14	0.088	Total count
No benthos	14	0.045	Total count
Expt. 2 (March 20-April 2)			
Benthos	14	0.131	Estimate with net
Benthos	14	0.193	Estimate with net
1991			
Expt. 1 (March 13-26)			
No benthos - (*Phaeocystis*)	14	0.002	Estimate with net
No benthos - (*Phaeocystis*)	14	0.020	Estimate with net
Benthos - (no *Phaeocystis*)	14	0.125	Estimate with net
Benthos - (no *Phaeocystis*)	14	0.148	Estimate with net

[a]Tank which underwent oxygen supersaturation.

Mathematical analysis.—For short-term experiments (14 d), daily instantaneous growth rates of larvae in the MERL experiments (assuming exponential growth) were estimated from the formula:

$$G = (\log_e SL_{t'} - \log_e SL_t)/ T \qquad (1)$$

where, G is daily larva growth rate, $SL_{t'}$ is the larva standard length (mm) at the end of the sampling interval, SL_t is larval standard length at the beginning, and T is the time interval in days. For longer term experiments (28 d), growth rate was determined by regression analysis using an instantaneous growth equation assuming exponential growth, where we regressed the natural log of length against day.

The average instantaneous daily larva mortality rate was calculated as:

$$Z = (\log_e S_t - \log_e S_{t'})/ T \qquad (2)$$

where, Z is the daily mortality rate, S_t is the number of winter flounder present at the start of the period, $S_{t'}$ is the number surviving at the end, and T is the time interval in days. For long-term experiments following Laurence (1974), S_t was corrected for potential survivors removed by sampling using the formula

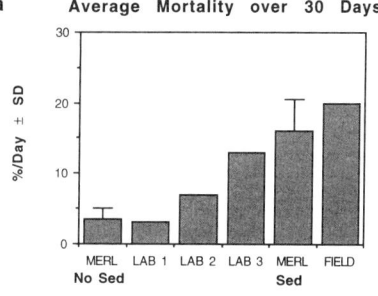

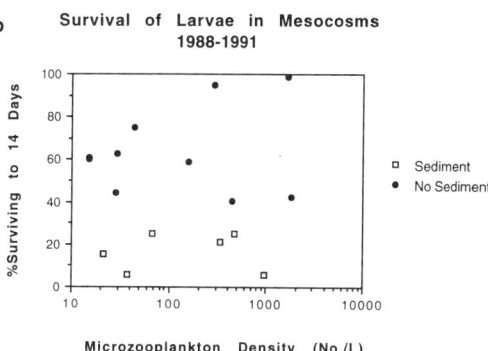

FIGURE 2.—(A) Rates of mortality for larval winter flounder compared for MERL mesocosms and other studies. "MERL Sed" refers to the 1988 MERL experiments with sediments. "MERL No Sed" refers to experiments conducted without sediments. LAB 1, Buckley et al. (1993) used 64-L containers stocked wth 50 to 1,000 prey/L. LAB 2, Buckley et al. (1993) used 36-L containers stocked with 1,000 prey/L. LAB 3, Laurence (1977) used 64-L containers stocked with 3,000 prey/L. "FIELD," refers to a study in the Mystic River Estuary, Connecticut (Pearcy 1962). (B) Relationship of survival of larval winter flounder to food availability during MERL experiments conducted with and without sediments.

$$\widehat{S_{t'}} = S_{t'} + N \sum_{K=1}^{W-1} e^{-7KZ} \qquad (3)$$

where, N is the number of individuals removed each week, W is the number of weeks of sampling, 7 is the number of days per week, and Z is the average instantaneous daily mortality rate. The initial Z value in equation 3 is the maximum mortality rate from equation 2, based on no survivors from samples taken and was used to calculate iteratively $S_{t'}$. S_t was then substituted for $S_{t'}$ in equation 2 to calculate an unbiased estimate of Z. Equation 3 was modified as needed for unequal sample intervals.

Summer Species Toxicology Studies

Water was pumped from Narragansett Bay at the dock of the University of Rhode Island's Graduate School of Oceanography and distributed among

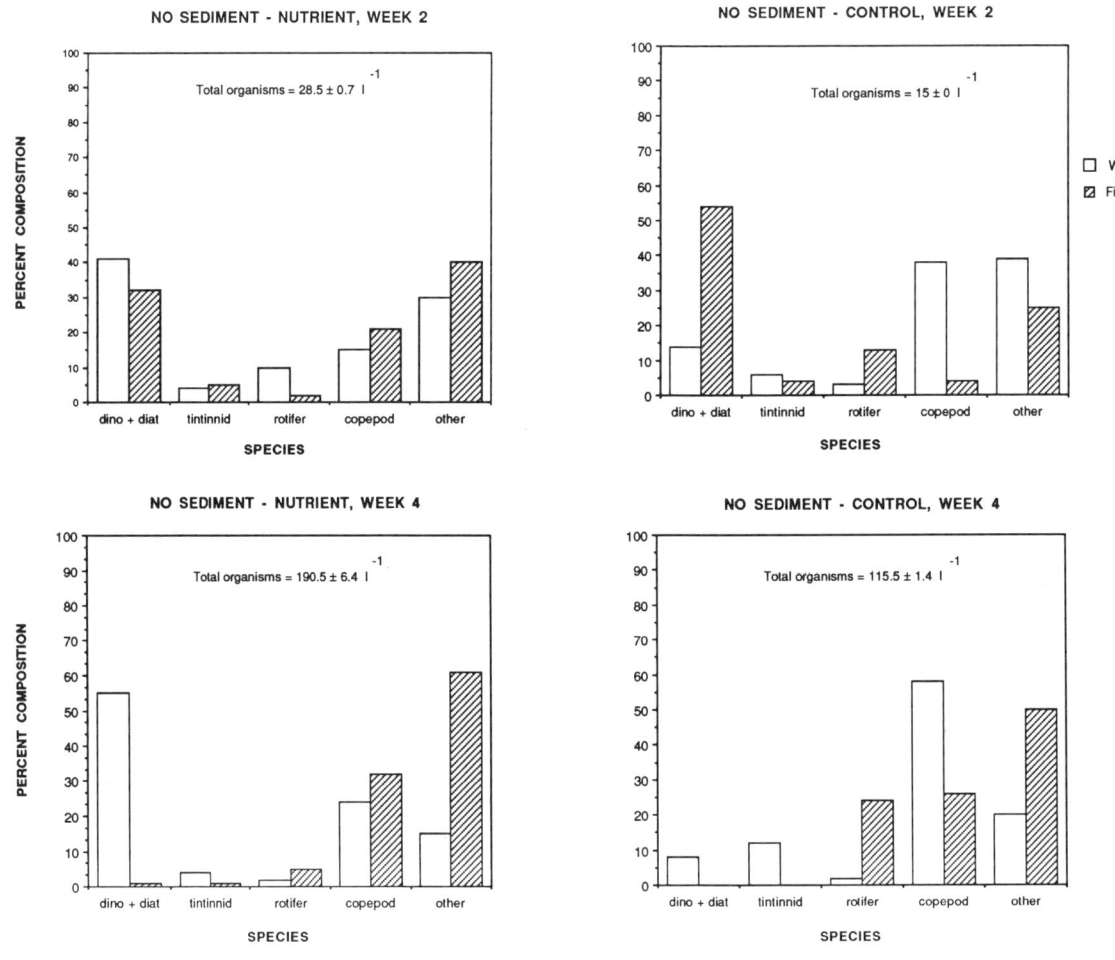

FIGURE 3.—Diet of larval winter flounder compared to microplankton available in mesocosms without sediments during 1988 (dino + diat refers to dinoflagellates plus diatoms). Open bars represent Water Column and shaded bars represent Fish Stomachs.

four separate enclosures at a rate of 10 L/min. Temperature was maintained at an average of 16.4°C. Chemical and biological responses to a single "spike" addition of the potentially toxic material were followed in three enclosures, with a fourth reserved as the control. Replicate samples within each mesocosm were used to determine differences between control and treatment mesocosms. Additions to the three experimental enclosures spanned an expected "no-effects" concentration (1 mg/L) to an expected "worst case effects" concentration (50 mg/L). The experiment was terminated after 1 month when oxygen concentration in the water column of the 50 mg/L treatment was reduced to levels known to be detrimental to marine plankton (2 mg/L).

Prior to addition to the enclosure, the plastic pellets were weighed, soaked in seawater for 12 h, and ground into small particles in a food mill to simulate disposal methods at sea. This resulted in many particles which either remained in the water column or were resuspended during mixing periods.

Biological parameters.—Dissolved oxygen was determined from Winkler titrations of samples withdrawn from 3 depths (1, 2, 3.4 m) on a dawn, dusk, dawn sampling schedule. Mesozooplankton were sampled with replicate net tows (0.25-m diameter, 64-μm mesh) taken from bottom to top of the enclosures. A flowmeter mounted inside the net determined the volume of water filtered. Subsamples containing at least 300 organisms were used to enumerate and identify each taxon. Microzooplankton (20 to 64 μm) were enumerated from replicate, integrated 18-L samples of water

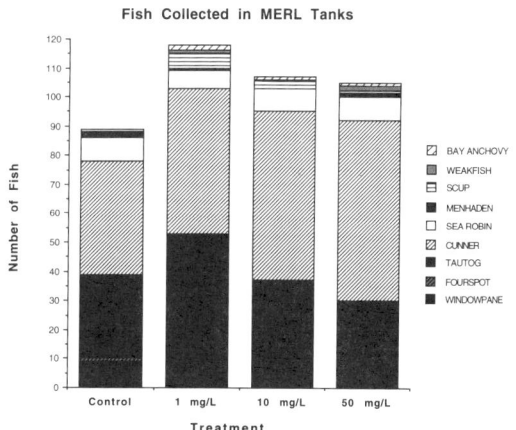

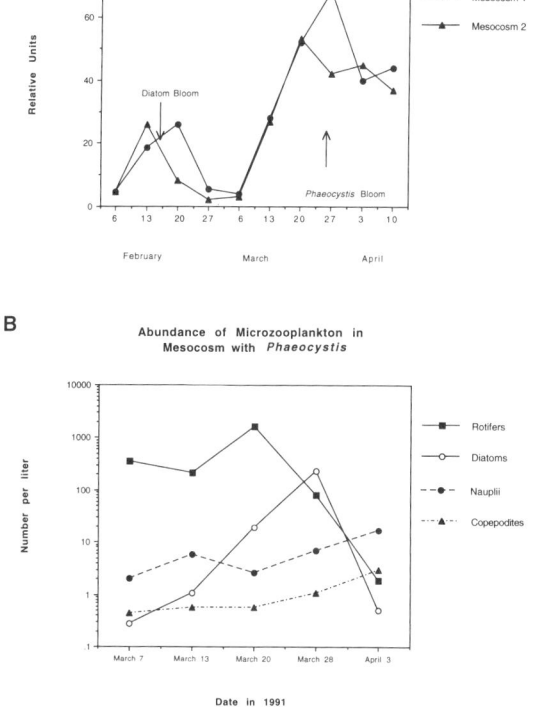

FIGURE 4.—(A) Flourescence and timing of phytoplankton blooms in two mesocosms in which blooms of *Phaeocystis* sp. were observed. The period of larval winter flounder hatching occurred from March 7 to 13. The *Phaeocystis* bloom extended from March 10 to April 10. (B) Abundance of diatoms and various zooplankton taxa in one of the mesocosms with *Phaeocystis* sp.

siphoned from 4 depths (4.5, 2, and 1 m, and surface). During the experiment, fish larvae were sampled with a 0.5-m diameter metered net with 202-μm mesh. To end the experiment, the entire contents of each mesocosm were drained through a 202-μm mesh net in order to collect all larval fish. All fish larvae were counted and measured to the nearest 0.01 mm. Gut contents of 20 larvae of the two most abundant species were recorded. Analysis of variance (using the GLM method for unbalanced designs where appropriate) was used to determine sources of variation (replicate versus treatment) and significance of results.

Results

Winter Flounder Study

Physical factors.—Temperatures in the mesocosms were similar to those in Narragansett Bay,

FIGURE 5.—Numbers of larval fishes collected from control and treatment mesocosms at the end of the 30-d experiment involving additions of biodegradable plastic material at concentrations shown at the bottom of each bar, July 1991. Species included: bay anchovy, *Anchoa mitchilli*; weakfish, *Cynoscion regalis*; scup, *Stenotomus chrysops*; menhaden, *Brevoortia tyrranus*; searobin, *Prionotus evolans*; cunner, *Tautogolabrus adspersus*; tautog, *Tautoga onitis*; fourspot, *Paralichthys oblongus*; windowpane, *Scophthalmus aquosus*.

fluctuating in a pattern which simulates natural temperature regimes. Average temperature increased slightly during the experiment, ranging from 3 to 8°C in 1988, 6 to 7.8°C in 1990 and 4°C in 1991. Salinities also reflected those in Narragansett Bay, ranging from 26 to 30‰. Handling mortality of larvae was low (1% or less) and consistent among replicates.

Effects of sediments and nutrients.—In mesocosms without sediment, mortality rates of larvae up to 30 d old were very low (Figure 2). In fact, they were comparable to the best rates achieved under laboratory conditions in which food is usually artificially concentrated. Furthermore, there was no significant difference in growth and survival among mesocosms without sediment despite nutrient enrichment of some tanks which stimulated phytoplankton growth (Table 2).

Using data available from shorter term experiments in 1990 and 1991, we examined the relationship between larva food abundance and survival (Figure 2). During these experiments microplankton consisted of dinoflagellates, diatoms, rotifers, tintinnids, and copepod nauplii and copepodites. Stomach contents of the fish showed they were feeding on a mixture of these organisms (Figure 3). Although there was high variability in survival, all

TABLE 4.—Average lengths (mm) of larval fishes at the end of experiment 1. Analysis of variance treatment effect was significant, $P \leq 0.0002$ for cunner, $P \leq 0.02$ for tautog.

Treatment	Cunner	Tautog
Control	18.22	16.41
1 mg/L	16.00	15.66
10 mg/L	15.31	14.40
50 mg/L	14.23	14.42

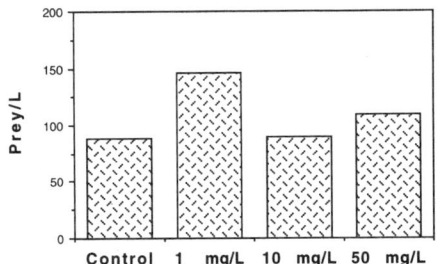

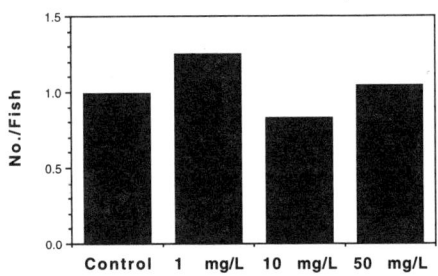

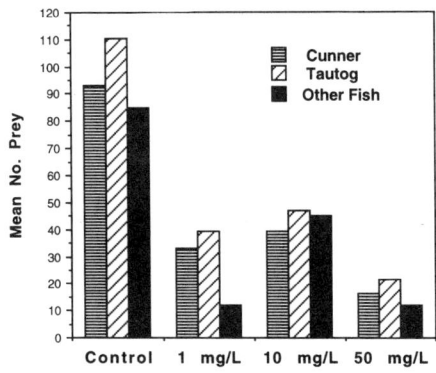

FIGURE 6.—Prey available and prey eaten per larval fish during the July, 1991 experiment with biodegradable plastice material.

low survival occurred in tanks with sediment. There was no obvious relationship between survival and food abundance among tanks without sediment (Figure 2, Table 3).

During experiments conducted in 1991, a *Phaeocystis* bloom occurred. Cells of *Phaeocystis* were observed in March following a diatom bloom in February (Figure 4). Winter flounder eggs incubated in 500-µm mesh baskets suspended in the mesocosms began to hatch in mid-March. During this period, abundance of *Phaeocystis* colonies ranged from 2 to 13.5/mL and colony diameters were 0.5 to 0.7 mm. Rotifers were extremely abundant during the first 2 weeks of winter flounder larva life and both diatoms and copepods, which were initially very rare, increased in abundance later in the experimental period (Figure 4). Survival of larvae over the next 40 d exceeded that seen in any of our other experiments.

Summer Species Toxicology Studies

Fate and biodegradation of material.— Qualitative observations indicated that particles of undegraded material were distributed throughout the water column and on the bottom of the mesocosm during the entire experimental period. Much of this material was resuspended during mixing. The amount of suspended particulate material was not quantified; however, its presence was evidence that the test material did not fully degrade during the month-long experiment.

Reduction in the oxygen concentration in mesocosms treated with the plastic provided evidence for progressive degradation of the material. The oxygen demand was so high in the 50 mg/L treatment that hypoxia (2 mg O_2/L) was observed 30 d after addition. Rate of degradation of the test material was approximated from the rate at which oxygen was consumed in the treated mesocosms relative to controls. There was a decrease of 1.5 g O_2/m^2/d. Assuming that it requires 1 g of O_2 to oxidize 1 g of carbohydrate, it would have required between 100 to 160 days to degrade all the material added to the 50-mg/L treatment if anoxia had not occurred.

*Biological effects.—*Larval forms of several species of fish typically found in lower Narragansett Bay in June (Klein-MacPhee unpublished data) were present in the enclosures during the experiment (Figure 5). The most abundant were cunner and tautog. Net tows made a week after the start of the experiment showed a relatively even distribution of species and numbers among the experimental tanks. Numbers of fish were somewhat higher in all treated tanks than in the control by the end of the experiment (Figure 5); however, growth rates of both cunner and tautog were significantly reduced in treated enclosures. Average length achieved by these two species was inversely correlated with treatment level (Table 4). Other species were not sufficiently abundant to test for length differences across treatments. The smaller size and reduced growth rates of fish in treated mesocosms was a surprising result because copepods, their primary food, were more abundant in treated enclosures so there was more food available per fish. However, gut content analyses indicated that feeding on copepods was reduced (Figure 6).

Discussion

Winter Flounder Study

Laboratory experiments have indicated that food availability is very important to growth and survival of larval winter flounder (Laurence 1975, 1977). However, the effects of natural variations in feeding conditions can be better understood through studies in large enclosures where food does not have to be artificially supplied. Laurence et al. (1979) showed that 4- to 6-week-old winter flounder had high rates of survival when reared in situ, in mesh chambers exposed to naturally occurring food organisms. Since food was not a limiting factor, the results reinforced the idea that predation may account for a large percentage of larva mortality.

Our studies on younger larvae (0 to 4 weeks) supply similar information for the first month of larval winter flounder life. During this critical stage when larvae begin to feed, food availability could be important. Lawrence (1977) found that prey levels of 10 and 100 organisms/L caused total mortality at the end of 2 weeks in laboratory studies. In 1988, the greatest differences in survival in the mesocosms were noted between tanks with sediments and those without (6 to 20% versus 60 to 74%). Unsedimented MERL mesocosms contained only 15 to 186 food organisms/L and had very high rates of larva survival. It appeared there was no relationship between food availability and survival in experiments conducted in 1988 to 1991, and that other factors were more important. Research in other mesocosms has demonstrated that larvae grow and survive with prey levels lower than laboratory estimates of required food amounts (Bailey and Houde 1989; Oiestad 1985; Cowan and Houde 1990). However, it should be noted that, since the microzooplankton community is so highly variable in species composition and over short time scales, any studies (including this one) that try to correlate larval fish success with food resources have several important limitations. Firstly, the exact nutritional value of the various microplankton groups (diatoms versus tintinnids for instance) are not well known; secondly, survival and growth processes probably depend on events occurring over days to weeks, during which time abundances of microzooplankton can shift dramatically; and thirdly, the importance of smaller microzooplankton such as ciliate protozoans to first feeding larvae may be of greater significance to larva survival then previously known.

Survival in mesocosms without sediment approximated rates more common for laboratory studies where there is no predation. In contrast, the lower survival rate in tanks with sediment more nearly approaches that determined by Pearcy (1962) for similar age winter flounder larvae in the field (natural mortality 20.7%/d for 9- to 26-d-old larvae) (Figure 2).

The effect of the benthos on larval fish has been demonstrated in another MERL experiment, where the presence of the benthos was observed to reduce abundance of larval menhaden and holozooplankton including copepods (Sullivan et al. 1991). The possible explanations for this effect include larvae feeding on non-nutritious sediment particles, competition with filter feeders in the benthos, or predation by other benthic forms. It has also been suggested that pollutants may arise from the sediments and poison the larvae, but this is unlikely as the sediments had been in place for at least 1 year before these experiments were conducted and they did not come from sites known to be contaminated.

If resuspended sediments interfered with feeding of the larvae, then one should see evidence of sediment particles in the gut contents. There was none. Thus, we believe it is more likely that the presence of competitors or predators in the benthos is responsible for the effect seen.

Filter-feeding benthic fauna may have prevented phytoplankton from blooming in the nutrient enriched sediment treatment, or may have removed smaller microzooplankton, such as ciliate protozoans, and thus limited food supply of larvae in a way not obvious from the microzooplankton data. The benthos has been suggested to be an important competitor of water-column species in several estuaries (Cloern 1982; Schwinghamer 1985; Daborn 1986), and protozoans are thought to be an important constituent of larval flounder diets (Scott Gallagher personal communication). There are also potential predators in the benthos including the mud anemone *(Ceriantheopsis americana)*, amphipods, and shrimp (Bailey and Stehr 1986; Bailey and Houde 1989).

The benthic effect could potentially be greater in the mesocosms than in nature because fish larvae share a smaller space in the enclosures with their predators or competitors. However, we hypothesize that the benthos plays at least a limited role in reducing larva growth and survival in shallow, well mixed estuaries that are typical of regions where winter flounder spawn.

In regards to the *Phaeocystis* bloom that developed in 1991, *Phaeocystis* sp. is a potentially harmful organism to larval fishes. Blooms of this haptophycean alga have been associated with negative effects on fishes and with low zooplankton abundances (Savage 1930; Admiral and Venekamp 1986; Rogers and Lockwood 1990), and it produces compounds that can inhibit feeding by other organisms (Estep et al. 1990). Reports of feeding inhibition in herbivorous copepods are inconsistent probably because of variations in physiological states of the alga that may occur (Estep et al. 1990).

The high rates of survival during this experiment are a clear indication that *Phaeocystis* had no negative effects on the fish larvae. There was no evidence of feeding inhibition; gut contents of larvae collected in early April indicated a diverse diet of diatoms, dinoflagellates, rotifers, and copepod nauplii. The larvae may have ingested *Phaeocystis* as well, but special techniques would be required to identify the "green slurry" which was also observed in most larva gut contents.

In summary, our experiments demonstrated the potential for very high survival rates of 0- to 1-month old winter flounder in mesocosms. Food availability was not an obvious controlling variable in survival during this period. *Phaeocystis* had no negative effect on winter flounder larvae. The presence of a benthos did significantly reduce larva survival.

Summer Species Toxicology Studies

There was no difference in survival of larval fishes between treated tanks and the control even at the highest concentrations of biodegradable plastics material, indicating the material was probably not toxic to larval fishes. There are several possible explanations for the reduction in feeding and growth by cunner and tautog in the tanks treated with plastic material. Firstly, low oxygen levels which existed at the time the fish were removed could have been inhibiting feeding. However, if this were the case, growth would not be expected to have slowed in the 1- or 10-mg/L treatments where oxygen was never very low. Secondly, fish may have been ingesting starch blend particles in preference to normal food. If so, a diet of starch is likely to be less nutritious than a diet of copepods. Digestibility of raw starch is low for most carnivorous fishes (NRC 1981). Most likely, the negative effect of the material was due to the starch base rather than to the additives; this was shown to be true for other planktonic organisms in a second experiment employing a pure starch control (Sullivan et al. 1993). Unfortunately that experiment did not include fishes. Nevertheless, our data indicate the effects of dumping the biodegradable product would be no greater than those from dumping similar amounts of food wastes (containing starch) at sea.

Conclusions

Mesocosms allow observations of larva survival and growth at near natural densities while feeding occurs on natural particle assemblages. High survival rates of larvae in mesocosms indicates that physiological needs of larvae are met by the experimental conditions. Simulation of the whole ecosystem increases the probability that exposure to toxics occurs through the route that will occur in situ and indirect effects through alterations in prey, predator, and competitor species will be observed. As a result, responses of the larvae to controlled alteration of food, predators, nuisance algal blooms, or water chemistry (such as introducing toxic materials) can be extrapolated to natural systems with much greater confidence than for studies done in smaller laboratory systems.

Acknowledgments

The research was funded by Sea Grant NA 85AA-D-SGO-94, University of Rhode Island, Graduate School of Oceanography, The Environmental Protection Agency CR812487-03, the Andrew W. W. Mellon Foundation, and The Department of Agriculture. We wish to thank the graduate students who worked on these projects, the MERL technical staff and Dr. Lawrence Buckley and an anonymous reviewer for their helpful suggestions.

References

Admiral, W., and L. A. H. Venekemp. 1986. Significance of tintinnid grazing during blooms of *Phaeocystis pouchetii* (Haptophycea) in Dutch coastal waters. Netherlands Journal of Sea Research 20:61-66.

Bailey, K. M., and E. D. Houde. 1989. Predation on eggs and larvae of marine fishes and the recruitment problem. Advances in Marine Biology 25:1-83.

Bailey, K. M., and C. L. Stehr. 1986. Laboratory studies on the early life history of the Walleye Pollock, *Theragra chalcogramma* (Pallas). Journal of Experimental Marine Biology and Ecology 99:233-246.

Bergen, R. H., G. C. Laurence, C. A. Oviatt, and L. J. Buckley. 1985. Effect of thermal stratification on the growth and survival of haddock larvae *(Melanogrammus aeglefinus)*. International Council for the Exploration of the Sea, C.M. 1985/Mini-Symposium/No. 6, Biological Oceanography Committee.

Bourne, D. W., and J. J. Govoni. 1988. Distribution of fish eggs and larvae and patterns of water circulation in Narragansett Bay 1972-1973. American Fisheries Society Symposium 3:132-148.

Buckley, L. J., A. S. Smigielski, T. A. Halavik, B. R. Burns, and G. C. Laurence. 1993. Growth and survival of the larvae of three temperate marine fish species at discrete prey densities 2. Cod, *Gadus morhua*; winter flounder, *Pseudopleuronectes americanus*; silver hake, *Merluccius bilinearis*. Pages 183-195 *in* B. T. Walther and H. J. Fyhn, editors. Physiological and biochemical aspects of fish development. University of Bergen, Norway.

Buckley, L. J., A. S. Smigielski, T. A. Halavik, E. M. Caldarone, B. R. Burns, and G. C. Laurence. 1991. Winter flounder *Pseudopleuronectes americanus* reproductive success. I. Among-location variability in size and survival of larvae reared in the laboratory. Marine Ecology Progress Series 74:117-124.

Buckley, L. J., A. S. Smigielski, T. A. Halavik, and G. C. Laurence. 1990. Effects of water temperature on size and biochemical composition of winter flounder *Pseudopleuronectes americanus* at hatching and feeding initiation. Fishery Bulletin, U. S. 88:419-428.

Cloern, J. E. 1982. Does the benthos control phytoplankton biomass in South San Francisco Bay? Marine Ecology Progress Series 9:191-202.

Cowan, J. H., and E. D. Houde. 1990. Growth and survival of bay anchovy *Anchoa mitchilli* larvae in mesocosms and enclosures. Marine Ecology Progress Series 68:47-57.

Crawford, R. E., and C. G. Carey. 1985. Retention of winter flounder larvae within a Rhode Island salt pond. Estuaries 8:217-227.

Daborn, G. R. 1986. Effects of tidal mixing on the plankton and benthos of estuarine regions of the Bay of Fundy. *in* M. J. Bowman, C. M. Yentsch and W. T. Peterson, editors. Tidal mixing and plankton dynamics. Lecture Notes Coastal and Estuarine Studies 17:390-413.

De Lafontaine, Y., and W. C. Leggett. 1987. Evaluation of *in situ* enclosures for larval fish studies. Canadian Journal of Fisheries and Aquatic Sciences 44:54-65.

Donaghay, P. 1983. *In vivo* fluorescence and DCMU fluorescence. *in* C. E. Lambert and C. A. Oviatt, editors. Manual of biological and geochemical techniques in coastal areas-MERL series I. pp 8-13. University of Rhode Island, Kingston.

Estep, K. W., J. C. Nejstgaard, H. R. Skyoldal, and F. Rey. 1990. Predation by copepods upon natural populations of *Phaeocystis pouchetii* as a function of the physiological state of the prey. Marine Ecology Progress Series 67:235-249.

Keller, A. A., P. H. Doering, S. P. Kelly, and B. K. Sullivan. 1990. Growth of juvenile Atlantic menhaden, *Brevoortia tyrannus* (Pisces: Clupeidae) in MERL mesocosms: effects of eutrophication. Limnology and Oceanography 35:109-122.

Klein-MacPhee, G. 1978. Synopsis of biological data for the winter flounder *Pseudopleuronectes americanus* (Walbaum). NOAA Technical Report, NMFS Circular 414. FAO Fisheries Synopsis 117, Washington, District of Columbia.

Klein-MacPhee, G., W. H. Howell, and A. D. Beck. 1982. Comparison of a reference strain and four geographical strains of Artemia as food for winter flounder, *Pseudopleuronectes americanus* larvae. Aquaculture 29:279-284.

Lambert, C. E., and C. A. Oviatt, editors. 1986. Manual of biological and geochemical techniques in coastal areas. MERL Series, Report 1, 2nd ed. Graduate School of Oceanography, University of Rhode Island, Narragansett.

Laurence, G. C. 1974. Growth and survival of haddock *(Melanogrammus aeglefinus)* larvae in relation to planktonic prey concentration. Journal of the Fisheries Research Board of Canada 31:1415-1419.

Laurence, G. C. 1975. Laboratory growth and

metabolism of the winter flounder *Pseudopleuronectes americanus* from hatching through metamorphosis at three temperatures. Marine Biology 32:223-229.

Laurence, G. C. 1977. A bioenergetic model for the analysis of feeding and survival potential of larval winter flounder *Pseudopleuronectes americanus* larvae during the period from hatching to metamorphosis. Fishery Bulletin, U. S. 75:519-528.

Laurence, G. C., T. A. Halavik, B. R. Burns, and A. S. Smigielski. 1979. An environment chamber for monitoring "in situ" growth and survival of larval fishes. Transactions of the American Fisheries Society 108:197-203.

NEFSC (Northeast Fisheries Science Center). 1992. Status of fishery resources off the Northeastern United States for 1992. NOAA Technical Memorandum NMFS-F/NEC-95.

NRC (National Research Council). 1981. Nutrient Requirements of Coldwater Fishes. Report 16. National Academy Press. Washington, District of Columbia. 63p.

Øiestad, V. 1985. Predation on fish larvae as a regulatory force, illustrated in mesocosm studies with large groups of larvae. North Atlantic Fisheries Organization Council Studies 8:25-32.

Pearcy, W. G. 1962. Ecology of an estuarine population of winter flounder, *Pseudopleuronectes americanus* (Walbaum). Bulletin of the Bingham Oceanographic Collection 18(1):1-78.

Rogers, S. I., and S. J. Lockwood. 1990. Observations on coastal fish fauna during a spring bloom of *Phaeocystis pouchetii* in the eastern Irish Sea. Journal of the Marine Biological Association of the United Kingdom 70:294-253.

Savage, R. E. 1930. The influence of *Phaeocystis* on the migration of herring. Fishery Investigations London (Series II) 12:5-14.

Schwinghamer, P. 1985. Observations on size structure and pelagic coupling of some shelf and abyssal benthic communities. Proceedings 19th European Marine Biology Symposium 347-359.

Sullivan, B. K., P. H. Doering, C. A. Oviatt, A. A. Keller, and J. B. Frithsen. 1991. Interactions with the benthos alter pelagic foodweb structure in coastal waters. Canadian Journal of Fisheries and Aquatic Sciences 48:2276-2284.

Sullivan, B. K., C. Oviatt, and G. Klein-MacPhee. 1993. Fate and effects of a starch based biodegradable plastic substitute in the marine environment. Pages 281-296. in C. Ching, D. L. Kaplan, and E. L. Thomas, editors. Biodegradable polymers and packaging. Technomic Publishing Company, Inc., Lancaster, Pennsylvania.

Sullivan, B. K., and P. J. Ritacco. 1985. The response of a dominant copepod species of food limitation in a coastal marine system. Archive für Hydrobiologie, Ergebnisse der Limnologie 21:407-418.

Contaminant Exposure and Population Growth of English Sole in Puget Sound: The Need for Better Early Life-History Data

JOHN T. LANDAHL AND LYNDAL L. JOHNSON

*Environmental Conservation Division, Northwest Fisheries Science Center
National Marine Fisheries Service, 2725 Montlake Boulevard, East Seattle, Washington 98112, USA*

Abstract.—The potential impact of contaminant-related mortality and reproductive impairment on the population growth rate in English sole *(Pleuronectes vetulus)* was examined using techniques of simulation modelling. A preliminary Leslie matrix population model was constructed for investigation of contaminant effects using the adult mortality rate for English sole in Puget Sound, Washington, estimated from recent historical data. Age-specific fecundity was determined from previously collected English sole ovary samples. Existing data on the effects of contaminants on reproduction, including impaired gonadal development, reduced spawning ability, and decreased egg and larva viability, were incorporated into the fecundity component of the model. Preliminary results suggest that declines in the fecundity component of the model, such as those observed in field studies in fish from contaminated sites in the Duwamish Waterway and at Eagle Harbor, could significantly decrease the population growth rate. To better evaluate the true magnitude of this effect, more accurate life-history data for larval and juvenile English sole are needed because the model is particularly sensitive to variation in mortality rates of these life stages.

Marine pollution in urban areas has been a source of increasing concern in recent years. Exposure to contaminants can affect individual fish in a variety of ways, including disease induction and reproductive impairment. Effects on individuals can in turn produce effects on the population. Documented effects of sediment-associated contaminants on the bottom-dwelling flatfish English sole *(Pleuronectes vetulus,* formerly *Parophrys vetulus)* in Puget Sound, Washington, include increased prevalences of neoplasms and other hepatic diseases (Malins et al. 1984; Rhodes et al. 1987; Myers et al. 1987, 1991; Landahl et al. 1990), and reproductive impairment (Johnson et al. 1988; Casillas et al. 1991).

Studies by state agencies have raised the question of whether or not the English sole population in Puget Sound is declining (e.g., Bargman 1988; Schmitt 1990). Commercial bottom-trawl catch of English sole in Puget Sound as a whole reached a peak in 1967 and has declined since then (see Schmitt 1990; Schmitt et al. 1991; Figure 1). Catch in the central region of Puget Sound (Admiralty Head to Blake Island), where contaminant "hot spots" have been identified in the sediments of adjacent areas, such as Elliott Bay and Eagle Harbor, has also declined since 1967 (see Schmitt 1990; Schmitt et al. 1991). Deleterious effects of contaminants on the health of English sole in the Duwamish Waterway were first described in 1977 (McCain et al. 1977). The question of whether the decline in the English sole population is real or apparent is best addressed through population studies, including trawl surveys and tagging studies. Unfortunately, population studies are relatively expensive to conduct and, to be definitive, should be carried out for a period of years.

An alternative approach to addressing the question of impacts of contaminants on the English sole population is to formulate and analyze a mathematical model of that population. While analyzing such a model constitutes a less direct approach than field

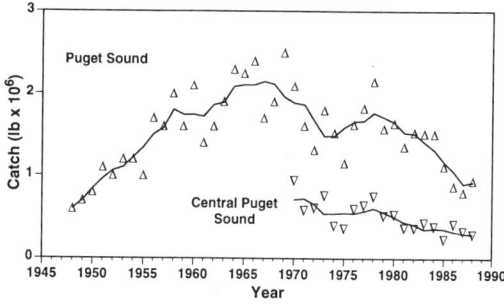

FIGURE 1.—Five-year moving average of English sole catch in Puget Sound, Washington, 1948 to 1987. Adapted from Schmitt (1990) and Schmitt et al. (1991).

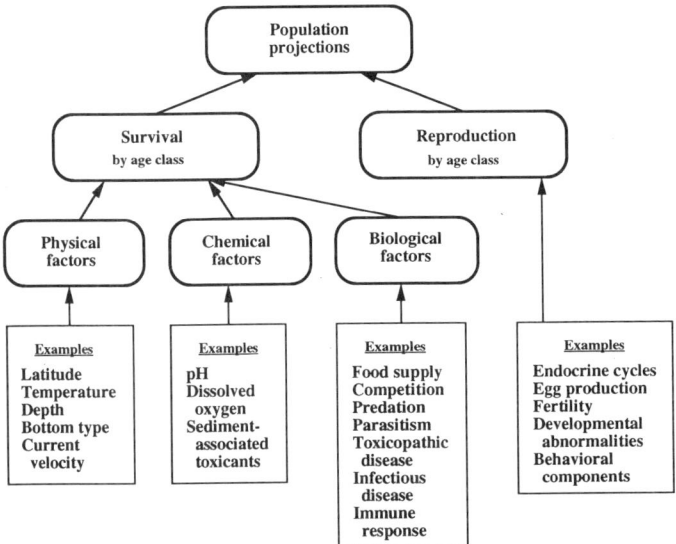

FIGURE 2.—Typical population projection model.

population studies, it has the dual advantages of being less expensive and of yielding results in a much shorter time. Moreover, a model permits the investigation of potential population fluctuations over much longer time scales than can feasibly be examined in field studies.

The type of model chosen depends on the modelling goal. Forecasting models are commonly used in fisheries management, but have the disadvantages of being relatively complex and of requiring the collection of considerable amounts of data. Projection models are simpler and require fewer data. However, because they do not include all variables needed for forecasting, their use is restricted primarily to organizing and planning research. A typical projection model (e.g., Figure 2) focusses on age-specific survival and reproduction, which may be affected by a variety of physical, chemical, and biological factors.

In recently completed preliminary studies, we have attempted to examine the effects of contaminant exposure on the English sole population. For this study, we have chosen a projection model rather than a forecasting model because it is more suited to the types of data presently available.

The model we present is an age-classified Leslie matrix model (Leslie 1945). Such models have been discussed in detail by Usher (1972) and Caswell (1989), among others, and their use in assessing effects of toxicants on populations of organisms is summarized by Barnthouse (1993). The preliminary model we have formulated is deterministic. Analysis of such a model can focus on a number of indices, including long-term population size and the intrinsic rate of natural increase (r), reproductive value or potential, population resilience and risk of population extinction, and sensitivity of the dominant eigenvalue of the matrix to changes in the model parameters (see Caswell 1989).

Methods

To formulate a Leslie matrix model for English sole, it was necessary to construct a life-cycle graph (Figure 3) and a life table specifying age-specific mortality and age-specific fecundity (see Caswell 1989). In theory, the age-specific mortality rate can be subdivided into three components: natural mortality, fishing mortality, and contaminant-related mortality. However, this preliminary model can consider only overall mortality, since no recent esti-

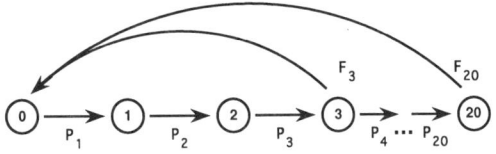

FIGURE 3.—Simplified age-classified life-cycle graph for English sole. The circles are age classes. Individuals have a probability (P) of surviving from one age class to the next. Older age classes contribute in varying degrees to the zero age class through reproduction (F).

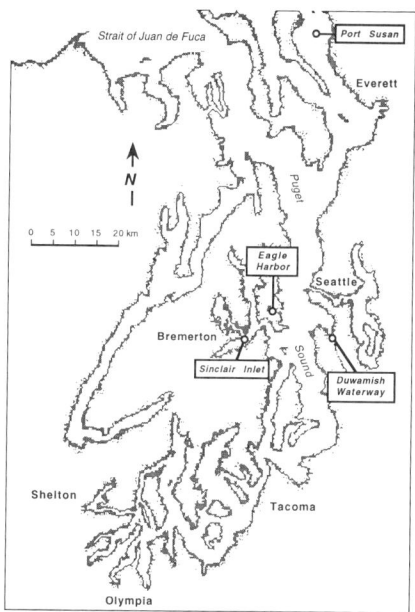

FIGURE 4.—Map of Puget Sound showing sites for which data on sediment contamination as well as fecundity and reproductive success estimates for English sole are available (Port Susan, Sinclair Inlet, Duwamish Waterway, and Eagle Harbor).

TABLE 1.—Relative levels of anthropogenic organic sediment contaminants at four well-studied Puget Sound sites.

Site	Type of contamination
Port Susan	Relatively uncontaminated[a]
Sinclair Inlet	Moderate PAHs, PCBs[a]
Duwamish Waterway	High PAHs, PCBs[a]
Eagle Harbor	Extremely high PAHs[b]

[a] Malins et al. (1984)
[b] Malins et al. (1985)

mates of natural or fishing mortality for the English sole population in central Puget Sound are available.

Our laboratory is concerned with the effects of complex mixtures of sediment-associated toxicants because this is what we have found in contaminated areas of Puget Sound, and for the present model we have used data on effects of contaminant mixtures from earlier studies involving field and laboratory exposures of English sole. Most of these data pertain to four sites (Port Susan, Sinclair Inlet, Duwamish Waterway, and Eagle Harbor; Figure 4) with varying degrees of sediment contamination (Table 1) which we have studied intensively for several years (Malins et al. 1984, 1985; Johnson et al. 1988; Casillas et al. 1991).

As part of our Puget Sound studies, we routinely collect and age otoliths from English sole collected for various research programs. Age is estimated by counting opaque zones of whole otoliths using a dissecting microscope (Chilton and Beamish 1982). Using catch-curve analysis of over 1,000 English sole collected in central Puget Sound between 1979 and 1985, we estimated an overall annual survival rate of 0.62 for fish 3 years old and older (Johnson and Landahl in press). Published annual survival rate estimates for other flatfish species range from 0.51 to 0.90; our estimate is thus comparable to those reported by other researchers.

To estimate age-specific fecundity (i.e., the number of eggs produced per female), we used fecundity estimates for 119 females (age range 3 to 12 years) collected at five sites in Puget Sound in 1986 and 1989 (the four sites listed in Table 1 and Yukon Harbor, for which no sediment contamination and reproductive success data are available, but which was included to increase sample size). Fecundity was estimated using the gravimetric method (Bagenal and Braum 1971). Age was estimated from length by constructing age-length curves for female English sole based on otolith age data collected in earlier studies, in many cases at the same sites. Our fecundity estimates might better be termed size-specific fecundity; since age is estimated from length, potential differences in size-at-age among our study sites cannot be assessed without obtaining otolith age determinations for the specimens involved.

Sensitivity analysis of the model was performed by calculating elasticities for the matrix elements. Elasticities help to standardize the effects of survival and fertility, quantitative estimates of which are very different, by measuring sensitivity on a proportional scale (see Kroon 1986).

TABLE 2.—Preliminary summary of fecundity study results for five Puget Sound sites for 1986 and 1989. Average fecundity at each site is given for the two age groups (5 and 6 years old) that were collected at all sites.

Statistic	Port Susan	Yukon Harbor	Sinclair Inlet	Duwamish Waterway	Eagle Harbor
Number of females	29	11	19	27	33
Minimum age (years)	3	4	5	4	3
Maximum age (years)	7	6	12	10	8
Average fecundity at age 5	422×10^3	382×10^3	485×10^3	446×10^3	422×10^3
Average fecundity at age 6	562×10^3	488×10^3	667×10^3	813×10^3	575×10^3

TABLE 3.—Sumary of reproductive success study results for four Puget Sound sites. Values are percentages, with sample sizes in parentheses.

Reproductive effect	Port Susan	Sinclair Inlet	Duwamish Waterway	Eagle Harbor
Vitellogenic females[a]	80 (50)	90 (96)	63 (101)	57 (97)
Females spawning[b]	90 (60)	75 (23)	54 (53)	35 (21)
Eggs fertilized[b]	52 (54)	35 (17)	44 (24)	24 (7)
Normal larvae[b]	74 (53)	54 (14)	59 (20)	68 (4)
Overall reproductive success	28	13	9	3

[a] Johnson et al. (1988); 1987 data for Port Susan, weighted average of 1986 and 1987 values for other sites
[b] Casillas et al. (1991); 1988 data for Sinclair Inlet and Eagle Harbor, weighted average of 1987 and 1988 values for other two sites

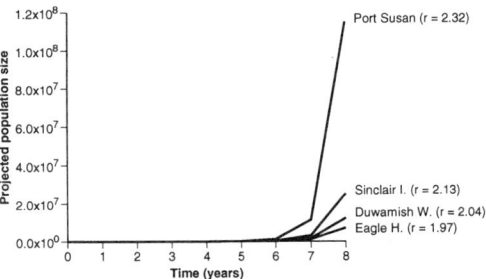

FIGURE 5.—Projected population trajectories over a period of 8 years for English sole at four sites in Puget Sound on the basis of r computed from the preliminary Leslie matrix model. Initial population size is one female adult.

Results

Detailed results of the fecundity study will be reported in a future publication; preliminary results are summarized in Table 2. As can be seen from this table, English sole usually become reproductively mature at about the age of 3 years. Young females produce about 300,000 eggs their first year. The oldest and largest females collected produced over 2 million eggs per year. For those females that spawned, the number of eggs initially produced at a given age was similar at all five of the sites studied.

However, our earlier field and laboratory studies had provided data on several aspects of reproductive success, namely gonadal development, spawning, fertilization success of eggs, and percent normal larvae at Port Susan, Sinclair Inlet, Duwamish Waterway, and Eagle Harbor. These studies found that some of these aspects of reproduction were impaired at the contaminated sites. Results of the reproductive success studies are summarized in Table 3. This summary is based on data for the years 1986, 1987, and 1988 reported by Johnson et al. (1988) and Casillas et al. (1991).

Reproductive impairment was most notable at the Duwamish Waterway, a site contaminated with both PAHs (polyaromatic hydrocarbons) and PCBs (polychlorinated biphenyls), and at Eagle Harbor, a site with very high levels of PAHs. At each site, the percentage of adult female English sole undergoing gonadal development was less than at Port Susan, which has little sediment contamination. Females at these two polluted sites were also less likely to spawn than at Port Susan, and their eggs were less likely to be fertilized. Those eggs which were fertilized were less likely to develop into normal larvae.

Having estimates for both fecundity and mortality, we were able to construct a generalized Leslie matrix model for Puget Sound as a whole. This matrix is shown in Table 4. Age-specific fecundity (before adjustment for reproductive impairment due to the effects of contaminant mixtures) is in the top row of the table, and the age-specific survival rate is along the subdiagonal. Lacking good mortality rate estimates for fish younger than 3 years old, we

TABLE 4.—Leslie matrix model for English sole in Puget Sound. Age-specific fecundity (in thousands of eggs produced, adjusted for a 50:50 sex ratio) is in the first row of the table, and the age-specific survival rate is along the subdiagonal.

Year	Year											
	0-1	1-2	2-3	3-4	4-5	5-6	6-7	7-8	8-9	9-10	10-11	11+
0-1	0	0	0	147,000	202,000	212,000	325,000	412,000	468,000	420,000	674,000	1,070,000
1-2	0.62											
2-3		0.62										
3-4			0.62									
4-5				0.62								
5-6					0.62							
6-7						0.62						
7-8							0.62					
8-9								0.62				
9-10									0.62			
10-11										0.62		
11+											0.62	0.62

TABLE 5.—Parameters with elasticity ≥ 0.20 for the preliminary model, indicating that the dominant eigenvalue of the Leslie matrix (and hence the estimate of r) is particularly sensitive to the accuracy of these estimates.

Parameter	Elasticity
Age 0 survival	0.24
Age 1 survival	0.24
Age 2 survival	0.24
Age 3 fecundity	0.20

used the mortality rate estimate for fish 3 years old and older for the younger age classes since it is the best information available at present (see Discussion). Fecundities were adjusted to take into account the 50:50 sex ratio of English sole.

We then used this generalized preliminary model to compare the intrinsic rate of natural increase (r) for English sole populations at the four sites for which reproductive success data were available. For all sites we used a mortality rate estimate based on fish 3 years old and older from Puget Sound as a whole (detailed mortality rate estimates are given in Johnson and Landahl in press). The fecundity estimates for Puget Sound as a whole were adjusted for reproductive impairment at each site. Based on the estimates of ovarian recrudescence, spawning, fertilization, and larval development, overall reproductive success was 28% for Port Susan, 13% for Sinclair Inlet, 9% for the Duwamish Waterway, and only 3% for Eagle Harbor (Table 3). These estimates of overall reproductive success are more accurate for Port Susan and the Duwamish Waterway than for the other two sites because they are based on 2 years of data rather than 1 (cf. Casillas et al. 1991).

These data permitted calculation of the dominant eigenvalue for the appropriate Leslie matrix (see Caswell 1989), and from that, r, which is 2.32 for the uncontaminated site and ranges from 2.13 to 1.97 for the contaminated sites. As can be seen from Figure 5, this difference in r results in a pronounced divergence of the projected population trajectories over a period of 8 years for fish from Port Susan and from contaminated sites. Values of this parameter decreased with increasing sediment PAH contamination, although these trajectories are probably not biologically realistic due to poor estimates of age 0 survival. It is the qualitative ranking of the estimates of r calculated using this preliminary model, rather than their precise values, which we wish to emphasize in this contribution.

Computation of elasticities for each element of the Leslie matrix (see Caswell 1989) indicates that the dominant eigenvalue (and hence the estimate for r) is particularly sensitive to the accuracy of the estimates for age 0, 1, and 2 survival, and to that for age 3 fecundity adjusted for overall reproductive success (Table 5).

Discussion

This preliminary analysis indicates that contaminant effects (particularly reproductive impairment) could substantially reduce the intrinsic rate of increase (r) of English sole populations from contaminated sites in Puget Sound. This could be a cause for concern, especially if fish inhabiting contaminated estuaries or embayments such as the Duwamish Waterway or Eagle Harbor contributed a significant proportion of recruits to the English sole population of Puget Sound prior to its decline beginning in 1967.

However, to obtain a more realistic assessment of the potential for contaminant-associated population decline, better life-history data are clearly needed for larvae and juveniles. The age 0 survival rate is several orders of magnitude too high to be biologically realistic for a species producing large numbers of eggs, and the sensitivity of the model to this parameter is high, at least as far as its quantitative projections are concerned. One possible approach to improving this estimate is to follow Schaaf et al. (1987) in using a value for age 0 survival which produces an r of 0.0 for fish from Port Susan in specifying the initial Leslie matrix. The utility of such an approach will be explored in a future contribution. Another approach would be to use survivorship probabilities estimated for larvae and juveniles of other flatfish species on the grounds that these would be better estimates than adult mortality rates.

Thus, although the preliminary model shows decreased r at the contaminated sites, the magnitude of this problem cannot be assessed without more accurate estimates of early age survival, and the first priority must be to refine the initial estimates. To refine the preliminary model further, field studies could be carried out to obtain better mortality rate estimates for larval and juvenile English sole. In addition, laboratory studies could be undertaken to further investigate effects of contaminant mixtures on survival, growth, and development of larval English sole, including susceptibility to predation. Studies of this nature are particularly important because it has been suggested from time to time that larva mortality rather than egg production is the primary determinant of recruitment success (see Magnuson 1991, who credits Hjort 1914 with originally advancing this idea). The model could also be refined by taking into account age-specific rather than overall effects of contaminants on reproductive success. Ultimately, the best test of the tentative conclusions drawn from analysis of the preliminary model would be provided by detailed stock assessment of English sole in this region.

Acknowledgments

We acknowledge helpful comments and discussion by B. B. McCain, S.-L. Chan, U. Varanasi, and T. Collier. Washington Deptartment of Fisheries staff, including C. Schmitt, W. Palsson, S. O'Neill, and H. L. Lai also provided helpful discussion in the early stages of this work. G. Nelson performed the egg counts for the fecundity estimates. We also thank two anonymous referees who provided a number of useful comments.

References

Bagenal, T. B., and E. Braum. 1971. Eggs and early life history. Pages 166-198 *in* W. E. Ricker, editor. Methods for assessment of fish production in fresh waters (Int. Biol. Programme, Handbook 3). Blackwell, Oxford, England.

Bargman, G. G. 1988. Trends in the abundance of economically important marine fish in Puget Sound. Pages 72-75 *in* P. S. W. Q. Authority, editor. Proceedings of the first annual meeting on Puget Sound research. Puget Sound Water Quality Authority, Seattle.

Barnthouse, L. W. 1993. Population-level effects. Pages 247-274 *in* G. W. Suter II, editor. Ecological risk assessment. Lewis Publishers, Boca Raton, Florida.

Casillas, E., D. A. Misitano, L. L. Johnson, L. D. Rhodes, T. K. Collier, J. E. Stein, B. B. McCain, and U. Varanasi. 1991. Inducibility of spawning and reproductive success of female English sole *(Parophrys vetulus)* from urban and nonurban areas of Puget Sound, Washington. Marine Environmental Research 31:99-122.

Caswell, H. 1989. Matrix population models: construction, analysis, and interpretation. Sinauer Associates, Inc., Sunderland, Massachusetts.

Chilton, D. E., and R. J. Beamish. 1982. Age determination methods for fish studies by the Groundfish Program at the Pacific Biological Station. Canadian Special Publication of Fisheries and Aquatic Sciences 60:1-54.

Hjort, J. 1914. Fluctuations in the great fisheries of northern Europe viewed in the light of biological research. Rapports et Procès-Verbaux des Réunions Conseil International pour l'Exploration de la Mer 20:1-228.

Johnson, L. L., E. Casillas, T. K. Collier, B. B. McCain, and U. Varanasi. 1988. Contaminant effects on ovarian development in English sole *(Parophrys vetulus)* from Puget Sound, Washington. Canadian Journal of Fisheries and Aquatic Sciences 45:2133-2146.

Johnson, L. L., and J. T. Landahl. In press. Chemical contaminants, liver disease, and mortality rates in English sole *(Pleuronectes vetulus)*. Ecological Applications.

Kroon, H. D., J. V. Groenendacl, A. Plaisier, and H. Caswell. 1986. Elasticity: the relative contribution of demographic parameters to population growth rate. Ecology 67:1427-1431.

Landahl, J. T., B. B. McCain, M. S. Myers, L. D. Rhodes, and D. W. Brown. 1990. Consistent associations between hepatic lesions in English sole *(Parophrys vetulus)* and polycyclic aromatic hydrocarbons in bottom sediment. Environmental Health Perspectives 89:195-203.

Leslie, P. H. 1945. On the use of matrices in certain population mathematics. Biometrika 33:183-212.

Magnuson, J. J. 1991. Fish and fisheries ecology. Ecological Applications 1:13-26.

Malins, D. C., M. M. Krahn, M. S. Myers, L. D. Rhodes, D. W. Brown, C. A. Krone, B. B. McCain, and S.-L. Chan. 1985. Toxic chemicals in sediments and biota from a creosote-polluted harbor: relationships with hepatic neoplasms and other hepatic lesions in English sole *(Parophrys vetulus)*. Carcinogenesis 6:1463-1469.

Malins, D. C., B. B. McCain, D. W. Brown, S.-L. Chan, M. S. Myers, J. T. Landahl, P. G. Prohaska, A. J. Friedman, L. D. Rhodes, D. G. Burrows, W. D. Gronlund, and H. O. Hodgins. 1984. Chemical pollutants in sediments and diseases in bottom-dwelling fish in Puget Sound, Washington. Environmental Science and Technology 18:705-713.

McCain, B. B., K. V. Pierce, S. R. Wellings, and B. S. Miller. 1977. Hepatomas in marine fish from an urban estuary. Bulletin of Environmental Contamination and Toxicology 18:1-2.

Myers, M. S., J. T. Landahl, M. M. Krahn, and B.

B. McCain. 1991. Relationships between hepatic neoplasms and related lesions and exposure to toxic chemicals in marine fish from the U. S. West Coast. Environmental Health Perspectives 90:7-15.

Myers, M. S., L. D. Rhodes, and B. B. McCain. 1987. Pathologic anatomy and patterns of occurrence of hepatic neoplasms, putative preneoplastic lesions, and other idiopathic hepatic conditions in English sole *(Parophrys vetulus)* from Puget Sound, Washington. Journal of the National Cancer Institute 78:333-363.

Rhodes, L. D., M. S. Myers, W. D. Gronlund, and B. B. McCain. 1987. Epizootic characteristics of hepatic and renal lesions in English sole *(Parophrys vetulus)* from Puget Sound. Journal of Fish Biolology 31:395-408.

Schaaf, W. E., D. S. Peters, D. S. Vaughan, L. Coston-Clements, and C. W. Krouse. 1987. Fish population responses to chronic and acute pollution: the influence of life history strategies. Estuaries 10:267-275.

Schmitt, C. 1990. Marine fish users and managers: how are we doing? Pages 118-141 *in* J. W. Armstrong and A. E. Copping, editors. Status and management of Puget Sound's biological resources. Environmental Protection Agency, Seattle.

Schmitt, C., S. Quinnell, M. Rickey, and M. Stanley. 1991. Groundfish statistics from commercial fisheries in Puget Sound. Department of Fisheries, State of Washington, 285.

Usher, M. B. 1972. Developments in the Leslie matrix model. Pages 29-60. *in* J. N. R. Jeffers, editor. Mathematical models in ecology. Blackwell Scientific Publications, Oxford, England.

Individual-Based Modelling of Environmental Quality Effects on Early Life Stages of Fishes: a Case Study Using Striped Bass

KENNETH A. ROSE

Environmental Sciences Division, Post Office Box 2008
Oak Ridge National Laboratory, Oak Ridge, Tennessee 37831-6036, USA

JAMES H. COWAN, JR.

Department of Marine Sciences
University of South Alabama, Mobile, Alabama 36688, USA

EDWARD D. HOUDE

Chesapeake Biological Laboratory
Post Office Box 38, Solomons, Maryland 20688, USA

CHARLES C. COUTANT

Environmental Sciences Division
Oak Ridge National Laboratory, Oak Ridge, Tennessee 37831-6036, USA

Abstract.—Environmental quality effects on early life stages can be important determinants of recruitment in many fish populations. We use a set of model simulations of age-0 striped bass *(Morone saxatilis)* to demonstrate the utility of an individual-based approach for assessing effects of changes in temperature, toxics, and liveable habitat at the population level. Temperature and toxics were varied to simulate chronic and episodic exposures; reduced liveable habitat was simulated by forcing juveniles to inhabit a smaller bottom area. Temperature affects mortality and development rates of eggs and yolk-sac larvae and the bioenergetics of feeding larvae. Effects of toxics were simulated as increased mortality and as the sublethal effects of reduced growth rate and reduced prey capture success. Increased mortality from chronic toxics exposure was implemented for egg and yolk-sac larva, feeding larva, and juvenile stages separately, and for all life stages simultaneously. Increased mortality was also implemented to be dependent on the degree of feeding success of the individual and on age to mimic a maternal route of toxics exposure. Model predictions of the number of age-1 survivors, and life-stage durations, growth rates, survival fractions, and mortality rates for each condition simulated were compared to those from baseline conditions. Toxics and chronic exposures generally had greater effects than temperature and episodic exposures; reduced liveable habitat had the least effect. Eggs, yolk-sac larvae and feeding larvae were more sensitive than juveniles. Whether mortality from chronic toxics exposure was increased equally for all individuals or increased based on their feeding success or age made little difference. Reduced growth rate and reduced prey capture success acted similarly, affecting larger (older) larvae more than smaller (younger) larvae. Priority research areas are estimation of exposure levels to define realistic effects for simulations and measurement of sub-lethal effects that can be explicitly related to growth and mortality. This information, in conjunction with further refinement of individual-based modelling, will improve our ability to assess effects of environmental quality on fish at the population level.

Environmental quality can play an important role in determining the growth and survival of early life stages of fishes. Early life stages are generally the most sensitive to environmental changes (e.g., McKim 1985) and recruitment level is believed to be established during the larval and juvenile stages of many fishes (Sissenwine 1984; Rothschild 1986; Houde 1987; Fogarty et al. 1991). Thus, changes in environmental quality experienced by early life stages can be a major determinant of year-class strength and ultimately the long-term dynamics of many fish populations.

Assessment of the consequences of changes in environmental quality on fishes is a difficult task because many factors affect environmental quality (e.g., temperature, dissolved oxygen, and contaminants). Each of these factors may cause effects that can impact one or more life stages. Effects of changes in environmental quality may be direct, as in increased mortality and decreased growth, or indirect, as in altered prey and predator dynamics. Often, several factors vary simultaneously in nature, introducing possible interactive effects. The number of possible combinations of factors that vary in nature far exceeds the ability to obtain information on interactive effects in controlled experimental studies. Further, comparison of the effects across factors is difficult because of the diverse nature of the accumulated database documenting factors and their effects on fishes. Experimental methods vary among different studies confounding comparison across studies. Finally, comparison of effects across life stages is difficult because life stages can be of different duration that exhibit different growth and mortality rates.

In this paper we demonstrate an individual-based approach to population modelling to evaluate environmental quality effects on early life stages of fishes. We believe that, regardless of the modelling approach, environmental quality effects ultimately must be evaluated at the population level. Determining population-level consequences of changes in environmental quality is critical because the population is the relevant endpoint of interest with respect to success of the species and its availability for harvest. Protecting individual populations, especially those deemed critical to the ecosystem, is the first step in protecting the integrity of the whole ecosystem. The population level offers a common metric upon which to compare among different environmental factors, effects, and life stages. We also believe that individual-based models offer advantages over more aggregated approaches to population simulation (Huston et al. 1988; DeAngelis and Gross 1992; DeAngelis et al. in press; Van Winkle et al. 1993). Individual-based modelling simulates population-level behavior by representing the population as an assemblage of interacting individuals. Each individual in a population is unique due to phenotypic and experiential differences. Furthermore, high mortality rates of early life stages of fishes (Bailey and Houde 1989; Houde and Zastrow in press) create a situation in which particular attributes, or combinations of attributes, may determine those small number of individuals that ultimately survive to recruitment (Crowder et al. 1992). Explicitly representing individuals in a population model permits straightforward incorporation of phenotypic and experiential differences (provided the information is available) and offers a conceptually simple approach to track many attributes of survivors. An individual-based approach also provides a conceptually simple mechanism to link effects of altered environmental quality to population responses. Most information on effects quantifies the responses of individuals to changes in the value of a factor such as temperature or a toxicant. Using an individual-based approach, these effects can be imposed upon individuals in the model and population-level consequences assessed.

We use an individual-based model of young-of-the-year striped bass to demonstrate the utility of this approach for assessing environmental quality effects on early life stages. A wide diversity of potential effects, including episodic and chronic changes in mortality at various life stages and sublethal effects, such as reduced ability to capture prey, are illustrated. Striped bass is a good species to use to demonstrate the individual-based approach because: (1) it is heavily exploited by commercial and recreational fisheries (NOAA 1991); (2) density-independent factors (including environmental quality factors) are primarily responsible for recruitment variability (Ulanowicz and Polgar 1980; Cooper and Polgar 1981; Uphoff 1989); (3) recent declines in many populations (Boreman and Austin 1985; Stevens et al. 1985) have provided copious information on the effects of various environmental quality factors on striped bass (e.g., Setzler-Hamilton et al. 1988; Rago et al. 1989; Coutant and Benson 1990; Hall 1991; Setzler-Hamilton and Hall 1991); and (4) it is illustrative of other species, especially other anadromous species that use estuaries as nursery areas.

Environmental Quality and Striped Bass

Overview

Assessment of the consequences of changes in environmental quality at the population level involves five major steps: (1) definition of the factor of interest; (2) determination of the life stage or stages affected; (3) specification of the route, temporal and spatial dimensions, and level of the exposure of the individuals to the factor in the system under study; (4) linkage of exposure level to effects on the individual; and (5) imposing effects on individuals in a model capable of producing realistic population-level responses. Once the factor of interest and the life stages affected are defined, aspects of how individuals are exposed must be delineated. Route of exposure may be direct from the environment inhabited by the individual, indirect from other individuals (e.g., ingestion of contaminated food), or via parental influences. The temporal dimension of exposure is viewed as either episodic or chronic, defined in relation to the duration of the affected life stage. Episodic exposure applies to changes in the factor over time scales substantially less than the duration of the life stage; chronic exposure applies to impacts experienced by individuals over most or all of the life stage. The spatial dimension of exposure is defined relative to the fraction of the population affected. Local exposures occur when only a portion of the population is affected. Global exposures imply that the entire population of individuals of that life stage are affected. Exposure level links laboratory data on effects, which are measured at a variety of exposure levels, to the likely magnitude of effects realized in nature. Finally, a quantitative relationship between the level of exposure and the effects on individuals is specified. Effects can be an increase in mortality or any number of sub-lethal effects such as reduced growth, increased metabolic rate, or altered behavior. Provided these steps are accomplished, and if the effects can be linked to processes represented in a population-level model, changes in a factor affecting environmental quality can be evaluated at the population level.

We define environmental quality factors operationally as those abiotic variables that exert a direct effect on individuals in the population. Water temperature, contaminants, turbidity, and liveable habitat all are examples of abiotic factors that affect environmental quality for fishes. Effects of changes in these factors, such as increased likelihood of dying, reduced growth rate, and impaired feeding ability, affect individuals directly. We have not considered biotic factors, such as size of the spawning stock and availability of prey, as environmental quality parameters. Nor have we considered the indirect effects of abiotic factors on components of the ecosystem other than the fish population of interest. For example, we have not included in our analysis indirect effects of toxics exposure that may cause changes in susceptibility of zooplankton to predation (e.g., Sullivan et al. 1983) or temperature changes that affect predators and consequently their predation rates (e.g., van der Veer and Bergman 1987).

We have considered how lethal and sub-lethal effects imposed on individuals can lead to population-level responses. For example, sublethal exposure to contaminants that cause decreased feeding ability (Drummond et al. 1973; Atchison et al. 1987; Morgan and Kiceniuk 1990) may result in increased likelihood of dying via reduced growth rates or from starvation. These types of sublethal effects, as well as lethal effects, at the individual level are an integral part of our assessment of environmental quality changes at the population level.

Factors Relevant to Striped Bass

Environmental quality can play a critical role in determining the health and dynamics of striped bass populations. Recruitment in striped bass is believed to be established by or before the juvenile stage (Goodyear 1985a; Stevens et al. 1985; Uphoff 1989; Houde and Rutherford 1992), and year class strength is dominated by density-independent factors (Ulanowizc and Polgar 1980; Cooper and Polgar 1981; Uphoff 1989).

The importance of environmental conditions to striped bass dynamics and the recent declines in Chesapeake Bay and Sacramento-San Joaquin River striped bass populations (Boreman and Austin 1985; Stevens et al. 1985) have promoted numerous studies of effects of environmental factors on early life stages of striped bass (CDFG 1987; Rago et al. 1989; Hall 1991; Setzler-Hamilton and Hall 1991). We have selected the factors of temperature, toxics, and liveable habitat to illustrate how environmental effects can be projected to the population level using individual-based modelling.

Temperature, toxics, and liveable habitat are good examples of factors that affect environmental quality because they affect different combinations of life stages, include episodic and chronic exposures, and involve both mortality and sub-lethal effects. Temperature may affect mortality of early life stages directly through thermal stress and is also an important determinant of development and growth rates, which can affect mortality rates indirectly (Houde and Rutherford 1992). In the model, we use toxics as a generic category that includes traditional contaminants (e.g., pesticides and metals), effects of low pH in conjunction with elevated aluminum (Hall 1987), and other anthropogenic agents that affect survival, growth, or behavior of early life stages (e.g., chlorine and ozone; Hall et

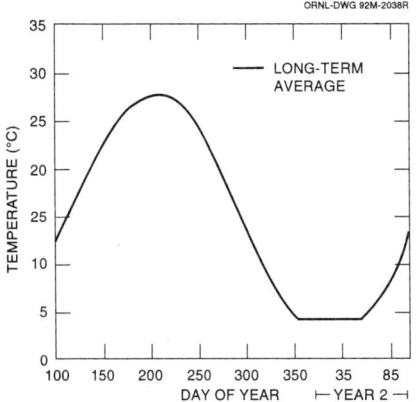

FIGURE 1.—Daily water temperatures used in model simulations.

al. 1981). Use of a generic toxicant for demonstration purposes permits great flexibility in the simulations that can be performed. Numerous bioassays for different toxicants have been performed to determine their mortality effects (Hall 1991) and sublethal effects (Brocksen and Bailey 1973; Eldridge et al. 1981; CNFRL 1983; Cech and Heath 1992) on striped bass. Reduction in liveable habitat causing individuals to be crowded into limited space also has been discussed as a factor affecting striped bass in estuarine and reservoir systems (Coutant 1985, 1987; Price et al. 1985; Coutant and Benson 1990).

To reduce the complexity of the analysis, we make some simplifying assumptions. We treat all effects as global (i.e., all individuals are assumed exposed), even though the effects of some of these factors could be more localized in nature. Some local effects could be easily incorporated by specifying effects on a subset of individuals in the model. Other localized effects that require spatial separation of individuals cannot be simulated in the present version of the model, which represents a single well-mixed compartment with all individuals sharing a common environment. We have not exhaustively investigated all combinations of factors, life stages, and all attributes of exposures. Instead, we have selected those combinations that are both realistic for striped bass and illustrative of the individual-based modelling approach to assessing environmental quality effects.

Description of the Striped Bass Model

Model Structure

The striped bass model is described in detail in Rose and Cowan (1993). Briefly, the model begins with the spawning of 100 individual females and simulates the growth and mortality of their progeny as they develop through the egg, yolk-sac larva, feeding larva, and juvenile stages to age-1. The model represents these dynamics on a daily time step in a single, well-mixed compartment (1,400 m × 1,400 m × 4 m deep). The environmental conditions considered in the compartment are daily average water temperature (Figure 1), fraction of the day there is daylight, and the daily densities of zooplankton and benthic prey types. Values for water temperature were determined from daily measurements from 1983 to 1987 at the Chain Bridge over the Potomac River. Fraction of the day that is daylight is from a regression equation fitted to observed values (Dalton 1987). Model simulations are for 1 year (365 d) beginning on calendar day 100 (April 10).

In the simulations individual females are all assumed to be the same size (750 mm TL, 6.44 kg wet weight) and are assigned to spawn in a predetermined pattern (see *Baseline Conditions* below). Each female produces 1.392×10^6 eggs (Setzler et al. 1980), with each egg weighing 0.27 mg dry weight (Zastrow et al. 1989). Each female's spawn is then followed as a cohort through the egg and yolk-sac larva stages. Egg and yolk-sac larva development and mortality rates are expressed as functions of temperature and are evaluated and updated each day. Total mortality is assumed to occur at less than 12°C for eggs and yolk-sac larvae. Upon initiation of first feeding, all larvae are assigned a length of 4.9 mm and a dry weight of 0.17 mg (Zastrow et al. 1989).

Larvae that survive from each female-cohort of eggs to initiate first feeding are followed as individuals in the model through the feeding-larva and juvenile stages. Length (total length in mm), weight (mg dry weight), age (d), life stage, and other attributes are tracked for each model individual. Feeding larvae develop into juveniles when their length exceeds 20 mm.

Feeding larvae and juveniles are followed day-by-day through growth and mortality processes. Daily growth in weight (mg dry weight) is represented with a difference form of a bioenergetics equation.

$$W_t = W_{t-1} + p \cdot C_{max} \cdot 0.7 - R_{tot} \qquad (1)$$

where p is the proportion of C_{max} realized, C_{max} is maximum consumption rate (mg/d), and R_{tot} is total metabolic rate (mg/d). The fraction of ingested food available for growth and metabolism is assumed constant at 0.7.

Maximum dry weight consumption rate (C_{max}) depends on an individual's weight and temperature, as follows:

$$C_{max} = \begin{cases} 0.55 \cdot W^{0.96} \cdot F(T) & \text{if } W \leq 63.7 \text{ mg} \\ 1.62 \cdot W^{0.70} \cdot F(T) & \text{if } W > 63.7 \text{ mg} \end{cases} \quad (2)$$

$$F(T) = \left[\frac{T_m - T}{T_m - T_0}\right]^X \cdot e^{X \cdot \left[1 - \left(\frac{T_m - T}{T_m - T_0}\right)\right]} \quad (3)$$

$$X = 0.0025 \cdot \log_e(\theta)^2 \cdot (T_m - T_0)^2 \cdot \left[1 + \sqrt{1 + \frac{40}{\log_e(\theta) \cdot (T_m - T_0 + 2)}}\right]^2 \quad (4)$$

Maximum consumptions as a function of weight are from laboratory data on early larvae (< 30 d posthatching or about 15 mm) reported by Tuncer (1988) and an equation developed by Moore (1988) based on age-1 and older fish; the two equations intersect at 63.7 mg dry weight (or 34 mm). The dependence of maximum consumption on temperature [F(T)] is a slowly rising function that reaches 1.0 at an optimal temperature and rapidly drops to zero at a maximum temperature (Hewett and Johnson 1987). Three parameters are required: the optimal temperature T_0 (22°C for larvae and 27°C for juveniles), θ that determines the rate of rise of the curve for temperatures below optimal (2.2 for larvae and 2.9 for juveniles), and the maximum temperature T_m (30°C for larvae and 35°C for juveniles). The θ parameter corresponds to a Q10 value (the factor by which rates increase for a 10-°C rise) for temperatures below optimal.

Metabolic dry weight losses (R_{tot}, mg/d) are determined as a routine component (R_r), which depends on larva weight and temperature, and an activity component, represented as a multiplier of routine metabolism for the period of feeding.

$$R_{tot} = [R_r + (ACT - 1) \cdot R_r \cdot FF] \cdot \delta \quad (5)$$

ACT is the activity multiplier of routine metabolism, FF is the fraction of day in which metabolism is considered active, and δ is an adjustment to metabolism under conditions of less-than-maintenance consumption. Routine metabolism is determined as a function of larva weight and a Q10 temperature relationship

$$R_r = \begin{cases} 0.096 \cdot W^{0.721} \cdot G(T) & \text{if } W \leq 179.6 \text{ mg} \\ 0.076 \cdot W^{0.766} \cdot G(T) & \text{if } W > 179.6 \text{ mg} \end{cases} \quad (6)$$

$$G(T) = e^{\frac{\log_e(Q_{10})}{10}(T - T_r)} \quad (7)$$

These equations for routine metabolism (R_r) are from Eldridge et al. (1982) for striped bass smaller than 2 mg dry weight and from Moore (1988) for primarily age-1 and older striped bass. Temperature-related parameters of T_r and Q10 are set to 20°C and 1.9 for larvae and to 22°C and 1.5 for juveniles. The fraction of a day during which active metabolism occurs (FF) is assumed to be the daylight fraction (DL) for larvae and the fraction of day required to achieve realized consumption for juveniles. If feeding by juveniles during all daylight hours would result in consumption exceeding maximum consumption, we assumed the fish achieved maximum consumption and only searched for the daylight hours needed to achive maximum consumption. An adjustment to the metabolic costs when consumption is less than maintenance (δ in equation 5) is needed to produce realistic larva weight loss (see Rose and Cowan 1993).

The proportion of C_{max} realized by an individual on a given day (p in equation 1) is computed based on random encounters of a feeding individual with different types of prey. Larvae feed on four zooplankton types (rotifers, *Eurytemora affinis* nauplii, *E. affinis* copepodites and adults, and cladocerans); juveniles are assumed to feed exclusively on benthos, which consists of 7 amphipod groups (*Gammarus* sp. and 6 size classes of *Leptocheirus plumulosus*), *Neomysis* sp., and 15 polychaete groups (15 size classes obtained by combining *Nereis succinea* and *Marenzelleria viridis*). Each prey type is defined by an average weight and length per individual which do not change over time, and specified equilibrium densities which are assumed to be constant for zooplankton, *Neomysis* sp. and *Gammarus* sp., and time-dependent for *Leptocheirus plumulosus* and polychaetes. Equilibrium densities for zooplankton were determined from Potomac River field data (Setzler-Hamilton et al. 1981; Houde et al. 1988). *Neomysis* sp. density was set to 16.6/m², the average observed in monthly surface, mid-depth, and bottom samples (expressed per square meter at a depth of 10 m) near Calvert Cliffs (Chesapeake Bay) between 1976 and 1980 (Olson 1987). Equilibrium densities of the remaining benthic types were determined from 7 years of approximately monthly samples collected with a 0.025-m² box corer (Holland et al. 1989) at three or four stations in the shallow portion of the Potomac River used by feeding juvenile striped bass (Boynton et al. 1981; Setzler-

Hamilton et al. 1981a).

Densities of each prey type (number/L for zooplankton and number/m² for benthos) are updated daily based upon the number of prey consumed by the striped bass (for zooplankton prey, by striped bass plus congeneric white perch larvae), by how close densities are to specified equilibrium densities, and by the turnover rate of the prey type. Density-dependence effects can result in model simulations if striped bass consumption exceeds the turnover rate of the prey causing prey densities to drop below equilibrium levels, thereby slowing down striped bass growth rates. Because one of the components of mortality decreases with increasing striped bass length, slower growth of striped bass results in higher mortality rates.

Values of p are determined stochastically by first determining mean encounter rates of individuals with each prey type. Mean encounter rates are based on the product of the volume (area for juveniles) searched and the density of each prey type. Search volume (area) is determined using the swimming speeds and the reactive distances of individual larvae, where reactive distance increases for larger individuals and longer prey. Realized encounter rates for each prey type are generated as a random deviate from a cumulative distribution function for larvae feeding on zooplankton and from a Poisson distribution for juveniles feeding on benthos. A fraction of the encountered prey of each type is "captured" and subsequently available for ingestion. Probability of successful capture increases with age for larvae feeding on rotifers and nauplii, and with length for all other individuals and prey types. Selection of which of the captured prey available for ingestion are actually ingested is based on general results from optimal foraging theory (Hughes 1980) that permit an individual to consume mostly preferred prey types when they are abundant and to switch to less preferred prey types when preferred types are rare. Prey types are added to the diet in decreasing order of preference. Prey types are included in the diet until the optimal foraging rules indicate that adding the next prey type does not increase the individual's average consumption rate or until consumption would exceed C_{max}.

Mortality of each feeding larva and juvenile depends upon an individual's weight and length. Weight-dependent mortality is based upon laboratory experiments without predators and is considered to be starvation-related. If the weight of an individual becomes less than some fraction (0.65 for feeding larvae and 0.5 for junveniles) of the weight expected for an individual of that length, it is assumed to die. Length-based mortality is based on a regression relating mortality rate (Z_L, 1/d) to length using field estimates of mortality rates of age-0 striped bass (see Rose and Cowan 1993). Probability of dying ($1-e^{-Z_L}$) is evaluated daily based on each individual's length

$$Z_L = 0.003 + 0.295 e^{-0.075 \cdot L} \qquad (8)$$

If a generated random number from a uniform distribution between 0 and 1 is less than the probability of dying, the individual is assumed to die.

To increase execution speed and reduce the computer memory requirements of the model, a sample of first feeding larvae, rather than every individual, is followed in simulations. A fraction of first-feeding larvae is taken from each female egg cohort inproportion to the number that survived to first-feeding in that cohort. Each model individual is assumed to represent some number of other identical individuals. For the model simulations presented in this paper, 200,000 first feeding larvae were followed, which resulted in greater than 40 individuals surviving to age 1. All simulations were performed on a 80486 personal computer.

Design and Rationale of Model Simulations

Baseline Conditions

Baseline conditions are defined such that spawning occurs throughout the season (day 101 [11 April] to day 140 [20 May]) with daily temperatures (Figure 1), day length, and prey densities set to values based upon Potomac River and Chesapeake Bay field data (Rose and Cowan 1993). Baseline simulations presented here differ from those previously reported (Rose and Cowan 1993; Cowan et al. 1993) in three ways. First, we use daily temperatures corresponding to the long-term averages observed in the Potomac River (Figure 1), but without stochastic variability. Second, simulations begin with 100 females in a 1,400 × 1,400 × 4 m compartment, which is double the number of spawners and double the compartment volume in Rose and Cowan (1993). Finally, all females are assigned identical lengths (750 mm) and identical days of spawning for all simulations. A fixed female size was chosen to eliminate effects of female size on size at first feeding (Zastrow et al. 1989). We impose a fixed temporal pattern of spawning because timing of spawning can influence the effects of environmental quality factors,

TABLE 1.—Simulations of the individual-based striped bass model to investigate the effects of temperature, toxics, and liveable habitat. Arrows indicate the direction of change in values.

Factor	Life stages affected	Exposure		Direct effects on individuals	Simulations
		Mode	Temporal		
Temperature	Eggs, yolk-sac larvae and feeding larvae present	Direct	Chronic	Increase or decrease daily temperature by 2°C	1) 2°C ↑ from 4/11-5/20 2) 2°C ↓ from 4/11-5/20
	Eggs, yolk-sac larvae and feeding larvae present	Direct	Episodic	Reduce daily temperatured by 4°C for a 4-d period	3) 2 episodes in early subperiod (4/11-30) 4) 1 episode in early / 1 in late (5/1-20) subperiod
Toxics	Eggs, yolk-sac larvae and feeding larvae present	Direct	Episodic	Kill all eggs and yolk-sac and add 0.5 to probability of dying of feeding larvae present on selected day	5) 2 episodes in early subperiod 6) 1 episode in early / 1 in late subperiod
	All eggs, yolk-sac larvae and feeding larvae present	Direct	Chronic	Add 0.2 to mortality rate of eggs and yolk-sac; increase probability of dying by 10% for larvae and 1% for juveniles	7) Add 0.2 to eggs and yolk-sac only 8) 10% ↑ applied to larvae only 9) 10% ↑ applied to juveniles only
	All feeding larvae	Direct	Chronic	Increase probability of dying by K, where $K = 0.5 - 0.5 \cdot p$, for feeding larvae	11) K ↑ in probability of dying
	All feeding larvae	Maternal	Chronic	Increase probability of dying by K', where $K' = 0.31 e^{-0.1 \cdot age}$, for feeding larvae	12) K' ↑ in probability of dying
	All feeding larvae	Direct	Chronic	Reduce growth rates by 10%	13) 10% ↓ in larval growth rates
	All feeding larvae	Direct	Chronic	Reduce capture success by 20%	14) 20% ↓ in capture success
Liveable habitat	All juveniles	Direct	Chronic	Change shape of compartment to squeeze juveniles	15) 50% of the baseline area 16) 10% of the baseline area

especially those imposed in an episodic manner during the spawning season. By fixing days of spawning of the 100 females and using the same pattern for all simulations, we eliminate spawning time as a variable in our analysis and focus on the comparative effects of environmental quality factors.

A spawning date was assigned to each of the 100 females to mimic typical temporal patterns of egg production in the Potomac River (Houde and Rutherford 1992). We assume that three major spawning events take place: day 105 (April 15), day 120 (May 1), and day 135 (May 15). Temperatures on these days are 13, 16.4, and 19.6°C, respectively. Twenty-one females are assumed to spawn on each of these 3 days; for each of the remaining 37 days between day 101 and day 140, a single female spawns. This results in a low level (1%) of constant egg production throughout the spawning season, in addition to the three evenly spaced peaks.

Model Output Variables

Model output variables include the number of age-1 survivors (day 100 of the second year), and life-stage durations, mortality rates, survival fractions, and growth rates. For selected simulations, additional model output variables are reported to facilitate interpretation of model predictions. Examples of additional model outputs are mean length and the length frequency distribution of age-1 survivors, and the attributes (e.g., mean length and growth rate) of a cohort of larvae (all spawned on the same day) as they grow.

Five replicate simulations were performed for each condition analyzed. Replicate simulations maintain all model inputs at identical values except for the initial seed value for the random number generator. Replicate simulations only differ by the random number sequence used in evaluation of prey encounter and capture and probability of dying. Predictions of individual replicate simulations, and the mean value and standard error of the mean of the five replicate simulations are reported. The standard error of the mean is computed as $\sigma/\sqrt{n}$, where σ is the standard deviation based on the 5 replicates and $n = 5$.

Simulations Assessing Environmental Quality Factors

Sixteen different conditions involving various combinations of the three factors (temperature, toxics, and liveable habitat) affecting different life stages via chronic or episodic exposures are simulated (Table 1).

Temperature.—Chronic changes in temperature are implemented by increasing (simulation 1) or decreasing (simulation 2) the average daily water temperatures by 2°C in the 11 April (day 101) to 20 May (day 140) period. Eggs, yolk-sac larvae, and some feeding larvae are present. Temperature effects are manifested through altered development rates of eggs and yolk-sac larvae and altered bioenergetics of feeding larvae.

Episodic temperature events are represented as either two events occurring early (simulation 3) or one event occurring early and one event occurring late (simulation 4) during the April 11 to May 20 spawning period. The early subperiod is from 11 to 30 April while the late subperiod is from 1 to 20 May. The starting day of each episode is randomly selected from the early or late subperiods. An episode is assumed to last 4 days with each day's daily mean temperature during this 4-day period reduced by 4°C. For episodic events in the early subperiod, no attempt was made to avoid overlap in time of the two events. Episodic temperature events can result in total mortality of eggs and yolk-sac larvae (100% mortality when temperatures fall below 12°) and short-term changes in development rates of eggs and yolk-sac larvae and bioenergetics of feeding larvae.

Toxics.—Episodic exposures to toxics (simulations 5 and 6) are imposed in a manner similar to episodic temperature simulations (i.e., two early events and one early plus one late event). For comparison with temperature episodes, the toxic events were imposed on the day that coincided with the beginning of each temperature episode in simulations 3 and 4. All individuals present on the day selected, regardless of life stage, are affected. No effects are imposed for any subsequent days unless there is another toxics event. All eggs and yolk sac larvae are killed and 0.5 is added to the probability of dying for each feeding larva. We added a constant to the probability of dying of feeding larvae under the assumption that the episodic exposure level is so high that any differences among individuals in their sensitivity to the toxicant are minor and can be ignored.

Chronic exposure to toxic stress is simulated by increasing the mortality rates of eggs and yolk-sac larvae by a constant and increasing the probability of dying of feeding larvae and juveniles by a percentage. Chronic exposures are imposed on egg and yolk-sac larva (simulation 7), feeding larva (simulation 8), and juvenile (simulation 9) stages separately, and on all life stages simultaneously (simulation 10). Mortality rates of eggs and yolk-sac larvae are increased by adding 0.2 to the daily rates predicted under baseline conditions; probability of dying for feeding larvae and juveniles is increased by 10% over baseline values. Probability of dying for feeding larvae is increased by 10% to result in a decreased larva survival rate intermediate to that caused by adding 0.2 to eggs and 0.2 to yolk-sac larvae. Adding 0.2 to egg mortality reduced the survival to 40% of baseline; for yolk-sac larvae, adding 0.2 to mortality reduced the survival fraction to 63% of baseline. Increasing daily probability of dying for feeding larvae by 10% reduced survival fraction to 52% of baseline. Because the probability of dying for feeding larvae and juveniles decreases with length, increasing the probability of dying by a percentage insured that susceptibility to death from toxic exposure decreased as length increased.

Chronic exposure to toxics also is simulated via increase in mortality of feeding larvae that is dependent upon feeding success (simulation 11). Eldridge et al. (1981) showed that benzene exposure caused increased mortality of starved striped bass larvae while having little effect on fed larvae. They suggested that larvae feeding on less than maximum ration potentially have higher mortalities. Simulation 11 mimics this scenario by increasing the daily probability of dying for feeding larvae by a factor $K = 0.5 - 0.5 \cdot p$, where p is the fraction of maximum consumption realized by a larva on each day. Larvae achieving maximum consumption ($p = 1$) are unaffected ($K = 0$), whereas larvae that realize the minimum designated value of $p = 0.5$ experience a 25% increase ($K = 0.25$) in the probability of dying. In baseline simulations p averages 0.8, resulting in an average K value of 0.1 in simulation 11, which is comparable to the 10% increase in probability of dying imposed on larvae under chronic toxics exposure (simulation 8).

Chronic exposure to toxics via a maternal mode of exposure (simulation 12) is simulated by increasing the mortality rate of feeding larvae by a percentage that decreases with age. Westin et al. (1985) found an inverse relationship between survival rates of striped bass larvae 30 d after fertilization (20 d of feeding) and initial organochlorine (e.g., PCB) concentrations in the eggs, which were inherited from parents. Organochlorine levels in the striped bass larvae declined with age from the initial egg levels (Westin et al. 1983, 1985). Assuming that the likelihood of dying increases with increasing toxic levels in the individual and that toxic levels decrease with age, simulation 12 increases the daily probability of dying for each feeding larva

TABLE 2.—Model predictions of life-stage durations, mortality rates, fraction surviving, and growth rates for five replicate simulations (different random number seeds) under baseline conditions. Standard errors of the mean are shown in parentheses.

Stage	Replicate simulation	Duration (d)	Mortality rate (Z, 1/d)	Fraction surviving (S)	Growth rate (mm/d)
Egg		2.90	1.04	0.049	–
Yolk-sac larva		7.35	0.26	0.153	–
Feeding larva	1	31.43	0.184	0.00308	0.50
	2	31.63	0.179	0.00348	0.50
	3	31.69	0.181	0.00323	0.50
	4	31.53	0.182	0.00322	0.50
	5	31.39	0.182	0.00330	0.50
	mean	31.53	0.182	0.00326	0.50
		(0.057)	(0.00081)	(0.000065)	(≈0.0)
Juvenile	1	303.35	0.0064	0.144	0.24
	2	302.70	0.0059	0.167	0.24
	3	304.13	0.0060	0.160	0.24
	4	303.69	0.0066	0.133	0.24
	5	304.80	0.0067	0.128	0.24
	mean	303.73	0.0063	0.146	0.24
		(0.354)	(0.00016)	(0.0075)	(≈0.0)

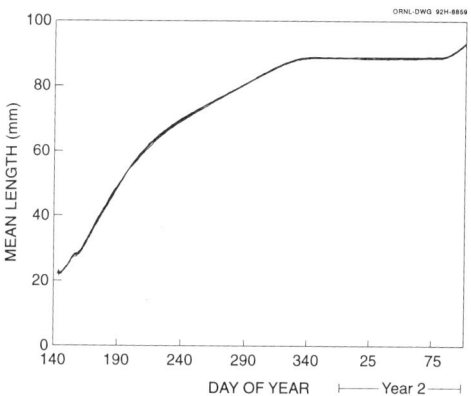

FIGURE 2.—Mean daily length of juveniles for the five replicate baseline simulations.

above baseline by a factor $K' = 0.31 \cdot e^{-0.1 \cdot \text{age}}$. The exponent coefficient was set to 0.1 based on reported PCB concentrations (μg/g) in striped bass larvae declining to about 12% of their initial levels in eggs, with most of the decline occurring during the 20 d of feeding (Westin et al. 1983). Daily probability of dying would be increased by 31% above baseline on its first day of feeding and by 1.4% at 31 d after first feeding. The average increased expectation of death for a surviving larva is approximately 10%, making simulation 12 comparable to the mortality imposed by the chronic exposures in simulation 8.

Sublethal effects of chronic exposure to toxics are simulated by reducing growth rates (simulation 13) and reducing capture success (simulation 14) of feeding larvae. Daily growth rates of feeding larvae are reduced by 10% from the baseline value; prey capture probabilities are reduced by 20% for each of the four zooplankton prey types. Sublethal exposure to toxics causing reduced growth rates (Collvin 1985; Mathers et al. 1985; von Westernhagen 1988) and reduced prey capture success (Morgan and Kiceniuk 1990; Mathers et al. 1985) has been documented for a variety of fishes.

Liveable habitat.—Effects of high temperature and low dissolved oxygen reducing liveable habitat for juveniles (simulations 15 and 16) are simulated by decreasing the bottom area-to-depth relationship of the compartment while maintaining its volume. Eggs and yolk-sac larvae, which are unaffected by the dimensions of the compartment, and feeding larvae, which search and encounter prey on a per-cubic-meter basis, are unaffected by changes in the bottom area-to-depth relationship. Reduced bottom area squeezes juveniles, which encounter prey on a per-square-meter basis, into a smaller area. Reduced liveable habitat is simulated by changing the dimensions of the compartment to result in 50% (simulation 15) and 10% (simulation 16) of the bottom area used for baseline conditions. Availability of suitable habitat for subadult striped bass in Chesapeake Bay is correlated with the juvenile index (Coutant and Benson 1990; Price et al. 1985).

Simulation Results

Baseline conditions

Model predictions under baseline conditions of life-stage durations, mortality rates (Z), fractions surviving (S), and growth rates are similar for the five replicate simulations (Table 2) and are generally realistic for the Potomac River (Rose and Cowan 1993). Because female size and day of spawning are fixed and unaffected by the random number sequence, egg and yolk-sac larva predictions are identical for all five baseline simulations. Egg stage duration is 2.9 d with $Z = 1.04$/d; yolk-sac larva stage duration is 7.35 d with $Z = 0.26$/d. Feeding-larva stage durations average 31.5 d with a growth rate of 0.5 mm/d and a mortality rate of 0.18/d. Fraction surviving for feeding larvae (first feeding to metamorphosis) ranges from 0.31% to 0.43%. The juvenile period, defined as metamorphosis to April 1 of the next year (i.e., age 1), have mean growth rates of 0.24 mm/d and Z values ranging from 0.0059 to 0.0067/d. Mean lengths of juveniles

TABLE 3.—Model predictions of mean life-stage durations, mortality rates, fraction surviving, and growth rates for baseline, 2-°C chronic increase, and 2-°C chronic decrease in temperature conditions. Values are the mean of five replicate simulations (different random numbers seeds). Standard errors of the mean are shown in parentheses.

Stage	Duration (d)			Mortality rate (Z, 1/d)			Fraction surviving (S)			Growth rate (mm/d)		
	Base	+2°C	−2°C	Base	+2°C	−2°C	Base	+2°C	−2°C	Base	+2°C	−2°C
Egg	2.90	2.36	3.29	1.04	1.21	1.07	0.049	0.058	0.029	–	–	–
Yolk-sac larva	7.35	6.85	8.01	0.25	0.28	0.23	0.153	0.143	0.153	–	–	–
Feeding larva	31.53	30.93	32.52	0.18	0.17	0.18	0.00326	0.00459	0.00302	0.50	0.51	0.48
	(0.057)	(0.038)	(0.101)	(0.00081)	(0.00042)	(0.0012)	(0.000065)	(0.000082)	(0.000083)	(≈0.0)	(0.00081)	(0.0015)

indicate that most juvenile growth is in the summer, with little or no growth during the winter months (Figure 2). Most survivors at age 1 (53 to 61%) originate from 1 of the 3 days of peak spawning when 63% of the total eggs were spawned. The middle peak of spawning (day 120), with 21% of the egg production, contributes 36 to 42% of the age 1 survivors, while the first and third peaks (days 105 and 135) contribute only 4.7 to 10.8%. Larvae from the three peak spawning events experience different temperatures, which result in different growth rates and mortality rates.

Temperature

The effects of a 2-°C chronic increase (simulation 1) and decrease (simulation 2) in daily water temperatures were similar in absolute magnitude but for different reasons. The temperature increase resulted in a mean ratio of age-1 survivors to baseline of 1.65 compared to a mean ratio of 0.54 for a temperature decrease (Figure 3). The major cause of increased survivorship at increased temperature was higher larva survival, whereas lower egg survival was the major cause of reduced survivorship for the temperature decrease (Table 3). Increased temperatures shortened egg and yolk-sac larva stage durations, while decreased temperatures lengthened them. For the increased temperature, reduced durations had little effect on the number of first feeding larvae produced (only 100,000 less than baseline) because increased survival of eggs (5.8% versus 2.9%) was offset by reduced survival of yolk-sac larvae (14.3% versus 15.3%). The

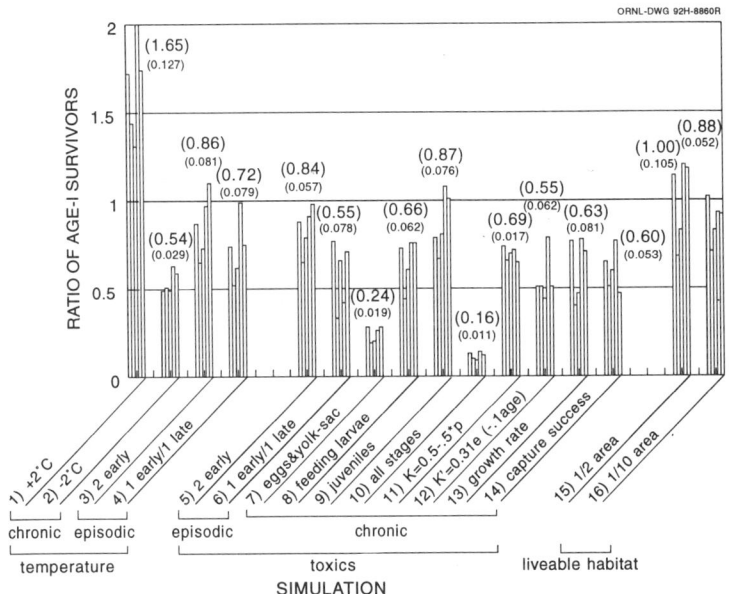

FIGURE 3.—Ratio of the number of age-1 survivors to the number under baseline conditions for each of five replicate simulations of the 16 different environmental quality effects simulations. See Table 1 for descriptions of the simulations. Mean value of the ratio (top number) and the standard error of the ratio (bottom number) based on the five replicate simulations are shown.

TABLE 4.—Model predictions of life-stage durations, mortality rates, and fraction surviving for five replicate simulations (different random number seeds) for simulations involving 1-early/1-late temperature episodes. Mean values for baseline conditions are shown in brackets. Standard errors of the mean are shown in parentheses.

Stage	Replicate simulation	Duration (d)	Mortality rate (Z, 1/d)	Fraction surviving (S)
Egg	1	2.98	1.031	0.0426
	2	2.97	1.047	0.0446
	3	3.20	1.048	0.0350
	4	2.97	1.054	0.0437
	5	3.01	1.037	0.0440
	mean	3.03 [2.90]	1.043 [1.04]	0.0420 [0.049]
Yolk-sac larva	1	7.65	0.266	0.131
	2	7.65	0.260	0.137
	3	7.40	0.254	0.152
	4	7.60	0.247	0.153
	5	7.45	0.261	0.143
	mean	7.55 [7.35]	0.258 [0.26]	0.143 [0.153]
Feeding larva	1	33.06	0.177	0.00292
	2	33.09	0.180	0.00258
	3	32.19	0.177	0.00330
	4	32.60	0.179	0.00295
	5	33.34	0.176	0.00281
	mean	32.86 [31.53]	0.178 [0.18]	0.00291 [0.00326]
		(0.205) (0.057)	(0.00074) (0.00081)	(0.00012) (0.00065)

major cause of increased age-1 survivorship at increased temperature was increased larva survival (0.459% versus 0.326% under baseline) due to slightly increased growth rates and consequently approximately 0.6 d shorter stage duration. In contrast, decreased egg survival was the major cause of reduced survivorship at decreased temperature. Egg survival was 2.9% at reduced temperature (versus 4.9% for baseline) resulting in 416,000 fewer first feeding larvae than baseline.

Episodic effects of a 4-°C temperature decline for both simulation 3 (2 episodes early) and simulation 4 (1-early/1-late episode) had relatively small effects on model predictions. Mean ratio of age-1 survivors was 0.94 for the 2 early episodes and 0.86 for 1-early/1-late episodes (Figure 3). At least one of the five replicate simulations for the 2-early and 1-early/1-late conditions produced a ratio of age-1 survivors to baseline ≥ 0.99. Losses of eggs and yolk-sac larvae, indicated by the fraction of individuals surviving to first feeding, roughly translated into proportional reductions in age-1 survivors. Two early episodes produced, on average, 87% of the numbers of first feeders as did baseline simulations and resulted, on average, in production of 86% as many age-1 survivors as did baseline conditions. Simulations involving 1-early/1-late episodes produced, on average, 81% of the baseline numbers of first feeders and 72% of the baseline numbers of age-1 survivors. Reductions in age-1 survivors due to episodic temperature events were spread among the egg, yolk-sac larva, and feeding-larva stages, resulting in slightly, but consistently, longer durations and lower survival compared to baseline values. Life-stage durations, mortality rates, and fractions surviving for the five replicate simulations involving 1-early/1-late temperature episodes illustrate the subtle changes observed (Table 4).

The effects of temperature episodes were most apparent on the percent of survivors that came from each spawning day. Because all simulations included either 1 or 2 early subperiod events during which episodic temperature drops result in lethal temperatures (i.e., less than 12°), the day-105 spawning peak was eliminated in all simulations. Effects of late subperiod temperature episodes were less dramatic (i.e., did not cause complete mortality of day-120 or day-135 spawning peaks). Ambient temperatures after day 118 (April 28) are high enough so that a 4-°C drop is not fatal to eggs and yolk-sac larvae in the model. Although 21% of the eggs are spawned on day 105, only 9% of the age-1 survivors result from this first spawning peak under baseline conditions. Thus, a complete loss of 21% of the eggs due to an episodic temperature event eliminating the day-105 spawning peak has less effect than might be expected considering only the percentage of total egg production that is lost.

Toxics

Episodic toxics simulations involving 2-early events (simulation 5) had effects similar to episodic temperature simulations. Two early toxic events resulted in an average ratio of age-1 survivors of 0.84 of baseline conditions, which is almost identical to the average age-1 ratio of 0.86 obtained for 2 early temperature events (Figure 3). As with episodic temperatures, reductions in the numbers of first feeders produced (average of 88.8% of baseline) roughly translated into proportional reductions in age-1 survivors (average age-1 ratio of 0.84). The primary response to the 2-early toxic events was a slight reduction in the surviving fraction of eggs (4.57% versus 4.93% for baseline) and yolk-sac larvae (14.7% versus 15.3% for baseline).

The 1-early/1-late episodic toxics simulation (6) had the greatest effect among the four episodic

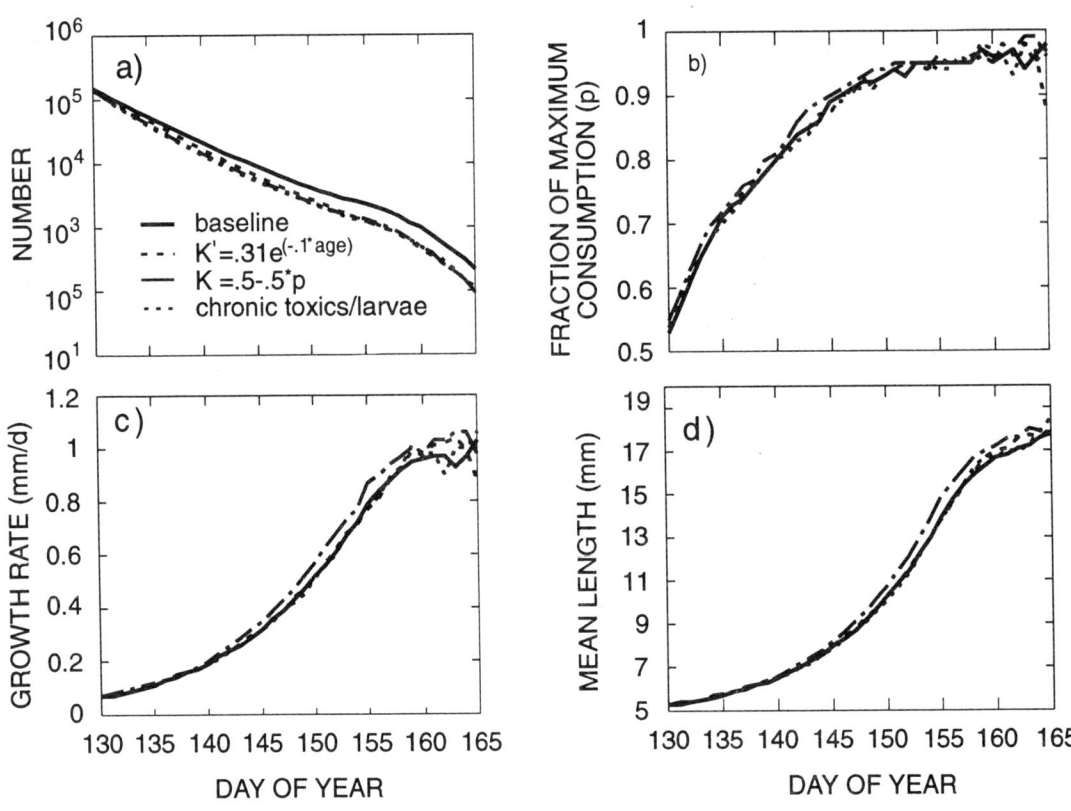

FIGURE 4.—Number surviving and mean values of attributes of individuals of the day-120 cohort (feeding initiated on day 130) for one of the replicate simulations under baseline conditions, and mortality of feeding larvae increased via chronic toxics exposure (10% applied to all larvae), feeding success (K = 0.5-0.5 · p), and age (K' = $0.31e^{-0.1 \cdot age}$). (a) number surviving, (b) p value, (c) growth rate, and (d) length.

events simulations. The 1-early/1-late toxics simulations produced the fewest first feeding larvae (64% of baseline), and, coupled with reduced survival of feeding larvae (0.279% versus 0.326% for baseline), the lowest age-1 survivors ratio (average ratio = 0.57) (Figure 3). The difference in results of simulations involving 1-early/1-late toxics episodes, 2-early toxics episodes and temperature episodes is the higher mortality of yolk-sac larvae caused by the late toxics events. Average fraction of yolk-sac larvae surviving was 10.6% for 1-early/1-late toxics episodes, compared to 15.3% for baseline, 14.7% for 2-early toxics episodes, 15.0% for 2 early temperature episodes, and 14.3% for 1-early/1-late temperature episodes. All toxics episodes impose a mortality regardless of when they occur. For 2 of the 5 replicate simulations involving late toxics events, the day-120 spawning peak, which was a major contributor to age-1 survivors, was eliminated resulting in ratios of age-1 survivors of 0.33 and 0.42. For the remaining 3 replicate simulations, the day-120 spawning peak was unaffected by toxics and ratios of age-1 survivors were substantially higher (0.66, 0.70, and 0.71). Similarly, the average age-1 ratios also were generally high for the 2-early and 1-early/1-late temperature episodes simulations (0.86 and 0.72) and 2-early toxics events simulations (0.84), because they could not eliminate the day-120 cohort.

Chronic exposure to toxics applied to all life stages (simulation 10) had the largest effect on age-1 survivors (average ratio = 0.16 of baseline). Increased egg and yolk-sac larva mortality caused most of the reduction; the effect on feeding larvae was less than on eggs and yolk-sac larvae but more than on juveniles. Average age-1 ratios were 0.25 for chronic toxics applied to eggs and yolk-sac larvae only (simulation 7), 0.66 for feeding larvae only (simulation 8), and 0.87 for juveniles only (simulation 9) (Figure 3). Results are consistent with the magnitudes of increased mortalities that

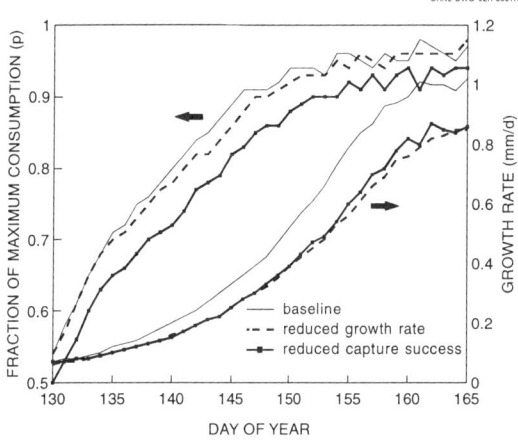

FIGURE 5.—Average growth rate and p values for the day-120 cohort for one of the replicate simulations under baseline, reduced growth rate, and reduced capture success conditions.

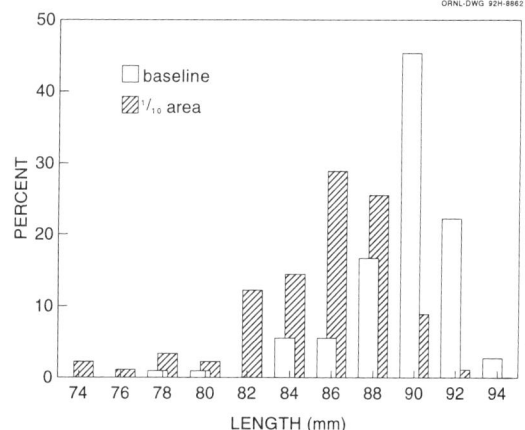

FIGURE 6.—Length frequencies of age-1 survivors under baseline and juvenile habitat 10% of baseline conditions.

were specified for each of the life stages. We added 0.2 to egg and yolk-sac larva mortality and increased the daily probability of dying for feeding larvae and juveniles by 10%. Because feeding larvae have higher probabilities of dying than juveniles (mortality declines with increasing length), the 10% increase in probability of dying for larvae had a bigger consequence than did the 10% increased probability of dying by juveniles. Interestingly, the effects of chronic toxics exposure by life stage are inversely related to durations of the life stages. The least effect was observed on juveniles which have the longest stage duration (approximately 300.4 d); eggs and yolk-sac larvae showed the greatest response and they have the shortest duration (approximately 10.2 d combined).

Increasing mortality of feeding larvae by a factor dependent upon feeding success (simulation 11; K = 0.5 - 0.5 · p) and age (simulation 12; K' = 0.31· $e^{-0.1 \cdot age}$) had effects similar to increasing the daily probability of dying of larvae by 10% (simulation 8; Figure 3). This result was expected because the values of coefficients in the equations for K and K' were selected to result in approximately a 10% increase in larva mortality. Attributes of survivors when mortality depended upon feeding success differed from those when mortality depended upon age or simply increased by 10%. To illustrate, we followed the number surviving, and average values of p, growth rate, and length for the cohort spawned on day 120 (feeding initiated on day 130) for one of the five replicate simulations (Figure 4). The numbers surviving over time in the day-120 cohort declined similarly for the 3 simulations with increased mortality relative to baseline (Figure 4a). When mortality was increased dependent upon feeding success, individuals realizing higher values of p (Figure 4b) were differentially selected to survive. Individuals with higher values of p correspond to those with higher growth rates (Figure 4c) and longer length (Figure 4b). The selective force of increased mortality dependent upon p was slight in magnitude but real in the simulations. The average differences between mean values predicted by increasing mortality based upon p (simulation 11) and the 10% chronic increase in mortality (simulation 8) for days 130 to 165 were 0.013 for p (29 of 36 differences were positive), 0.033 mm/d for growth rate (30 of 36 differences were positive), and 0.35 mm for length (28 of 36 differences were positive). The effect of having increased mortality dependent upon feeding success was detectable in simulations, but the differences between mortality as a function of feeding success, as a function of age, or simply imposed as a 10% increase seem biologically insignificant.

Sublethal toxic effects of reducing growth rate (simulation 13) and reducing prey capture success (simulation 14) produced similar reductions in the number of age-1 survivors (average ratios = 0.63 and 0.60 of baseline, respectively; Figure 3). Directly reducing growth rates or reducing capture success caused similar reductions in growth rates (0.45 and 0.47 mm/d versus 0.50 mm/d for baseline) and, consequently, substantially lower larva survival (S = 0.18% and 0.20% versus 0.33% for baseline).

Reducing growth rates or reducing capture success resulted in almost identical effects on feeding larvae as they grew in simulations. The average growth rate and p value for the cohort spawned on day 120 for one of the replicate simulations shows the expected results of the reduced growth-rate simulation; lower growth rates (by about 10%) and similar p values compared to baseline (Figure 5). Because we reduced growth rates by a percentage, higher growth rates of older (larger) larvae were reduced more. Reduced capture success acted much like directly reducing growth rate, having an effect throughout the larval stage. Intuitively, one might expect reduced capture success to affect smaller (near first feeding) larvae much more than larger (older) larvae. Small larvae encounter fewer prey and have lower p values and would *a priori* seem to be most sensitive to reduced capture success; however, reduced capture success also affected larger larvae because p values remained below 1.0 for all larvae, and larger larvae relied more on larger prey types (copepodites plus adults, and cladocerans). Capturing fewer large prey items translated into the reduced consumption and growth rates of larger larvae in simulations.

Liveable Habitat

Reductions of juvenile habitat to 50% of baseline (simulation 15) had no effect (average ratio of age-1 survivors = 1.0), while a reduction to 10% of baseline (simulation 16) resulted in an average production of 88% of the age-1 baseline survivors (Figure 3). The decreased juvenile survival for 10% of baseline area simulations (S = 13.9% versus 14.7% for baseline) was due to increased density of juveniles reducing food levels and leading to slightly lower growth rates (0.45 versus 0.47 mm/d). Predicted densities of juveniles under the 10% of baseline area, less than $0.01/m^2$, seem reasonable. These densities are comparable to densities reported from July through September in the Potomac River ranging from approximately 0.0002 to about $0.026/m^2$ (MD DNR 1990) and densities averaging about $0.01/m^2$ for 1975 to 1976 (Setzler-Hamilton et al. 1981). The cumulative effect of slightly slower growth rate over the long duration of the juvenile stage (≈ 305 d) results in slightly smaller age-1 survivors (Figure 6; mean of 89.1 mm compared to 92.8 mm for baseline).

Discussion

The major goal of this paper was to demonstrate application of an individual-based model to evaluating environmental quality effects on early life stages of fishes. The most important contribution of this paper is the demonstration of the flexibility of the individual-based modelling approach in handling environmental effects data. Because we used an individual-based model of striped bass, a species for which considerable information from both laboratory and field studies is available, model analyses may provide insights into environmental factors that influence early life survival of striped bass. We will discuss some tentative conclusions resulting from this model analysis, recognizing that our choices of simulations were somewhat subjective and could have biased our results. We will conclude with our perspective for how to improve prediction of population-level consequences of environmental quality effects.

Striped Bass Results

In the simulations, toxics generally had greater effects on striped bass than did temperature. Chronic toxic exposures involving eggs and yolk-sac larvae (simulations 7 and 10) had the biggest impact of all conditions analyzed on the number of age-1 survivors (average ratio of age-1 survivors to baseline less than 0.24), whereas chronic temperature simulations resulted in only a 60% change from baseline in the number of age-1 survivors (1.65 for increased temperature and 0.54 for decreased temperature; Figure 3). Both of the episodic temperature simulations and the 2-early toxic events simulation caused only small reductions in numbers of age-1 survivors (average ratios of age-1 survivors to baseline greater than 0.72). However, episodic toxic simulations that occurred during the latter part of the spawning season and which eliminated the day-120 spawning peak resulted in substantially lower numbers of age-1 survivors (average ratios = 0.33 and 0.42). Toxics had a greater effect than temperature because toxics are assumed fatal, regardless of when the episode occurs, whereas temperature in the simulations is only fatal during the early subperiod when the temperature episode results in ambient temperatures less than 12°C. Temperature episodes could have a greater effect than observed in our simulations in years when early spawning peaks predominate. The early spawning peak used in baseline simulations (day 105) produced 21% of the total eggs yet contributed only 9% of the age-1 survivors.

Reduced liveable habitat for juveniles had the smallest effect on striped bass of the three factors examined (Figure 3). Simulations using 50% of the area of baseline simulations had no effect on the numbers of age-1 survivors (average ratio = 1.0). Crowding juveniles into an area 10% that of baseline did result in decreased and lowered growth rates, and consequently smaller size at age-1 (Figure 6). The differences in predicted growth rates and length due to crowding had only a small effect on mortality (average ratio of age-1 survivors to baseline of 0.86). Our results for young-of-the-year juveniles do not negate Coutant and Benson's (1990) hypothesis that limited available habitat (due to low dissolved oxygen and high temperatures) affects striped bass in Chesapeake Bay. The limiting habitat hypothesis was proposed for fish older than age-0 and for the mainstem of Chesapeake Bay. In our simulations, reduced habitat was imposed on young-of-the-year juveniles, which have higher temperature tolerances than older subadults (Coutant 1985), in a model configured for the Potomac River environment.

In general, simulations indicated that chronic exposures had larger effects then episodic exposures. Chronic exposures to toxics that involved egg and yolk-sac larva stages (simulations 7 and 10) resulted in the lowest numbers of age-1 survivors. Similarly, a chronic decrease in temperature produced the fewest age-1 survivors among all temperature simulations (Figure 3). Episodic events potentially can cause major reductions in numbers of age-1 survivors but they must occur during or shortly after a spawning peak that is potentially a major contributor to recruitment. This was observed in two of the five replicate model simulations in which episodic toxics events eliminated the important day-120 spawning peak, resulting in the lowest ratios of age-1 survivors to baseline (0.33 and 0.42) of all episodic simulations. Houde (1989), using a different modelling approach applied to striped bass and bay anchovy, also concluded that high episodic losses need not be catastrophic and that chronic stresses may be a more serious threat to recruitment potential in all but the most severe episodic events.

The simulations indicated that egg, yolk-sac larva, and feeding-larva stages were more sensitive to variation in environmental quality than was the juvenile stage. Chronic exposures to toxics applied to life stages separately resulted in average ratios of age-1 survivors to baseline of 0.24 for eggs and yolk-sac larvae, 0.66 for feeding larvae, and 0.87 for juveniles (Figure 3). If the amounts by which we changed mortalities are realistic, then these results have important implications for understanding the sensitivities of different life stages to environmental variation. Clearly, if we had increased egg and yolk-sac larva mortality from chronic exposure by less and juvenile probability of dying by more, different conclusions could be drawn concerning the relative sensitivities of life stages. Houde (1987) arbitrarily imposed 25% increases and decreases in mortality rates, growth rates, and stage durations on young-of-the-year striped bass, and computed their effects on numbers of age-1 survivors. He concluded that changes in rates have the smallest effect for eggs and yolk-sac larvae, have a moderate to large effect on feeding larvae, and a small to moderate effect on juveniles. The difference between Houde's analysis and our results is due to the relatively large increases in egg and yolk-sac larva mortalities that we imposed compared to increases in mortality on other life stages. These results point out the importance of determining realistic levels of chronic effects to impose in simulations, even though it is very difficult due to the high uncertainty of exposure levels in nature.

There was little difference in model simulations in the attributes of individuals killed by chronic toxics exposure, whether mortality of feeding larvae was increased equally for all individuals (i.e., 10% increase in probability of dying) or based on the state of the individual (mortality dependent on p values or age). The only consistent difference was observed for mortality based on feeding success (p) which selected faster growing individuals within a cohort as survivors (Figure 4). However, the magnitude of the differential selection was quite small. These results imply that, given the same magnitude of mortality increase, maternal route of exposure (assuming no effects on eggs and yolk-sac larvae, simulation 12) and degree of feeding success acted similarly as mortality increased uniformly for all individuals.

The sublethal effects of reduced growth rate and reduced prey-capture success affected feeding larvae similarly in the simulations. While reducing growth rate would, by definition, affect individuals throughout the larval stage, the effect of reduced capture success on all sizes of larvae was not expected (Figure 5). Much emphasis has been placed on capture success of early stage larvae as a determinant of survival because their success is generally low (Blaxter 1986). Our simulations indicate that effects of reductions in capture success also can be important for larger larvae which realize less than maximum consumption. One reason striped bass larvae in the model do not exceed maximum consumption is that the optimal foraging algorithm prevents individuals from selecting less preferred (smaller) prey when preferred (larger) prey are abundant. Thus, even though larger larvae potentially could encounter and capture enough prey to exceed maximum consumption, not all prey are selected which results in *p* values less than 1.0. The optimal foraging algorithm also depends on prey handling times, which can be affected by sublethal exposures to toxics (Sandheinrich and Atchison 1989). Additional evaluation of the applicability of the optimal foraging algorithm to larval striped bass is warranted.

If there is little density-dependent growth during the larval and juvenile stages, predicted effects of toxics on each life stage can be combined *a posteriori*. The average age-1 ratio was 0.16 for chronic toxics applied to all life stages, which is very close to the product of average age-1 ratios for each life stage (0.24 for eggs and yolk-sac larvae, times 0.66 for feeding larvae, times 0.87 for juveniles = 0.14). If the chronic effects are independent among individuals in a life stage and between life stages, chronic toxics exposure applied to eggs and yolk-sac larvae and feeding larvae will be the same as chronic exposure applied to eggs and yolk-sac larvae separately and feeding larvae separately. Under strong density dependence, the dynamics in one life stage would be affected by the removal of individuals due to chronic exposure to toxics, and, by affecting the numbers entering subsequent life stages, the dynamics in these subsequent life stages. Thus, if density dependence was important, predicted age-1 survivor ratios would be appreciably higher for chronic toxics applied to all life stages compared to the product of age-1 ratios obtained for each life stage separately. Contrasting these results to situations of strong density dependence for striped bass (e.g., Goodyear 1985b) or to other species that exhibit strong density dependence (e.g., Schaaf et al. 1987) would provide insights into the general importance of environmental quality to fishes.

The variability in the number of age-1 survivors resulting from simulations based on changing environmental quality is small compared to inter-annual variability in striped bass recruitment. We fixed the size of female spawners and the temporal pattern of spawning to focus on the variability due to simulations of environmental quality. In model simulations, Cowan et al. (1993) varied factors likely to be important in generating inter-annual variability of age-1 survivors. They varied the size of spawning females, zooplankton densities, densities of congeneric white perch larvae, and the temperature during the spawning and larval development period. Cowan et al. (1993) then varied all of these factors simultaneously to generate the best and worst case combinations (highest and lowest numbers of age-1 survivors). The ratio of predicted age-1 survivors to the number predicted for baseline conditions ranged from 11 for large female spawners to within 0.3 to 3 for the other factors. Best case simulations based on variation of all factors simultaneously produced average to baseline ratios of age-1 survivors of 17; worst case simulations produced average ratios of age-1 survivors of 0.11. Almost all of our environmental quality simulations produced ratios to age-1 survivors between 0.5 and 1.0 (Figure 3); the notable exceptions were the 0.24 for chronic toxics applied to eggs and yolk-sac larvae (Simulation 7) and the 0.16 for chronic toxics applied to all life stages (Simulation 10). Detecting in the field all but the chronic toxics effects applied to eggs and yolk-sac larvae and to all life stages would be difficult, given the greater variability generated by the many other factors that vary from year to year. The relatively small effects of changing environmental quality do not diminish the potential importance of such effects to striped bass recruitment fluctuations. Even a small effect acting on an already depressed year class can cause significant further reductions in recruitment. However, the small magnitude of such effects does make the detection of effects from toxics or other environmental quality factors in any given year difficult.

Perspective

Individual-based modelling offers promise as an alternative to more traditional approaches for assessing environmental quality effects on early life stages of fishes. As demonstrated in this paper, individual-based modelling offers flexibility in the types of environmental quality effects that can be investigated. Using an individual-based model of young-of-the-year striped bass as a case study, we investigated three factors: temperature, toxics, and liveable habitat. Chronic and episodic exposures, and a wide variety of direct effects on individuals of altered development rates and bioenergetics, increased mortality based on age and feeding success, reduced growth rates and reduced prey capture success, and "squeezing" individuals into smaller liveable habitat, were simulated. Few other modelling approaches (e.g., Logan 1986; Goodyear 1985b; see DeAngelis et al. 1990) would permit such easy implementation of the wide variety of environmental quality effects we investigated with the individual-based model of striped bass.

There are five aspects of the present version of the striped bass model that need refinement and additional model development. First, the model focuses on striped bass and represents the rest of the food web in a simplified manner. However, many factors that affect environmental quality for striped bass (e.g., temperature, toxics) also affect the dynamics of other components of the food web which, in turn, can affect striped bass. Bartell (1990) has demonstrated how comparable food-web and population models can produce different predictions of fish dynamics when indirect food-web effects are important. Second, the striped bass model presently simulates only the first year of life. Extending the model to the entire life cycle would permit investigation of long-term consequences of environmental quality effects on early life stages, as well as investigation of environmental quality effects on older than young-of-the-year individuals. Third, much of the information used to configure the striped bass model is based on averaged values of parameters and rates reported in the literature. Inter-individual variability is simulated via differences in prey encounters and captures. There are other, potentially important, processes (e.g., bioenergetics), which express inter-individual variability and which may affect growth and survival. The importance of properly incorporating phenotypic and other sources of inter-individual variability into ecotoxicological assessments (Depledge 1990) and into individual-based models (Chambers 1993) has been emphasized. Fourth, the present model is spatially configured as a single, well mixed compartment. Explicitly representing spatial heterogeneity in model simulations would allow investigation of interactions between spatial location of individuals and their ultimate fate, as well as the consequences of local exposures. Finally, rigorous model corroboration is necessary before the model can be truly used in a predictive mode. The present model is quite complex, and, as with any complex model, individual-based or not, rigorous model testing with field data is severely limited by the availability of adequate empirical information. This lack of rigorous model corroboration does not limit the usefulness of the model, provided proper caution is used in interpreting model results.

The weakest aspect of modelling environmental quality effects on early life stages of fish is determining realistic levels of exposure. Assumed exposure levels are used to define the magnitude of effects to impose in simulations. Conclusions based on the striped bass analysis performed in this paper, especially those involving cross-life-stage and cross-factor comparisons, depend to a large extent on the magnitude of effects imposed. In situ toxicity tests offer one approach for estimating realistic exposure levels (Hall 1991), although to date they have tended to focus more on effects and can be costly to implement. Use of the molecular, biochemical, and histopathological measures of stress have also been proposed as indicators of toxics exposure in nature (see Adams 1990; McCarthy and Shugart 1990). A critical next step is to develop methods that can relate these sub-organismal indices to exposure levels in units that permit comparison to experimental data on effects on individuals.

Many of the effects reported in the literature are specific to a particular study, making it difficult to extrapolate measured effects on individuals to population-level consequences. For example, swimming performance has been proposed as a sensitive indicator of sublethal toxics exposure, and numerous different methods have been used to quantify the effects of a toxics on swimming performance (Little and Finger 1990). However, the lack of standardization of swimming performance measures makes comparisons across studies difficult. Further, rarely are swimming performance measures explicitly related to relevant effects on individuals such as feeding, growth, or predator avoidance (Little and Finger 1990). Laboratory studies on the effects of factors on individuals should include measurement of effects that can be explicitly related to relevant changes in growth or mortality or their constituent subprocesses.

Estimating exposure in nature and measurement of sub-lethal effects that can be explicitly related to growth and mortality processes are priority areas for further research on environmental quality effects on early life stages of fish. Better estimates of exposure and more relevant measures of effects, done in conjunction with further refinement of the individual-based modelling approach, will improve our ability to assess environmental quality effects on fish at the population-level.

Acknowledgments

W. Van Winkle and L. Barnthouse provided useful comments on a previous version of this manuscript. Research sponsored by the Electric Power Research Institute under Contract No. RP2932-2 (DOE No. ERD-87-672) with the U. S. Department of Energy, under Contract No. DE-AC05-84OR21400 with Martin Marietta Energy Systems, Inc. This is Publication 4064 of the Environmental Sciences Division, Oak Ridge National Laboratory.

References

Adams, S. M., editor. 1990. Biological indicators of stress in fish. American Fisheries Society Symposium 8, Bethesda, Maryland.

Atchison, G. J., M. G. Henry, and M. B. Sandheinrich. 1987. Effects of metals on fish behavior: a review. Environmental Biology of Fishes 18:11-25.

Bailey, K. M., and E. D. Houde. 1989. Predation on eggs and larvae of marine fishes and the recruitment problem. Advances in Marine Biology 25:1-83.

Bartell, S. M. 1990. Ecosystem context for estimating stress-induced reductions in fish populations. American Fisheries Society Symposium 8:167-182.

Blaxter, J. H. S. 1986. Development of sense organs and behaviour of teleost larvae with special reference to feeding and predator avoidance. Transactions of the American Fisheries Society 115:98-114.

Boreman, J., and H. M. Austin. 1985. Production and harvest of anadromous striped bass stocks along the Atlantic Coast. Transactions of the American Fisheries Society 114:3-7.

Boynton, W. R., T. T. Polgar, and H. H. Zion. 1981. Importance of juvenile striped bass food habits in the Potomac Estuary. Transactions of the American Fisheries Society 110:56-63.

Brocksen, R. W., and H. T. Bailey. 1973. Respiratory response of juvenile chinook salmon and striped bass exposed to benzene, a water-soluble component of crude oil. Pages 783-791 *in* Proceedings of joint conference on prevention and control of oil spills. American Petroleum Institute., U. S. Environmental Protection Agency, and U. S. Coast Guard, Washington, District of Columbia.

CDFG (California Department of Fish and Game). 1987. Factors affecting striped bass abundance in the Sacramento-San Joaquin River system. Technical Report 20, California Department of Fish and Game, Sacramento.

Cech, J. J., and A. G. Heath. 1992. Sublethal effects of rice pesticides on juvenile fish. Final Report for Contract FG0479, Pesticides Investigations Unit, CDFG, Rancho Cordova, California.

Chambers, R. C. 1993. Phenotypic variability in fish populations and its representation in individual-based models. Transactions of the American Fisheries Society 122:404-414.

CNFRL (Columbia National Fisheries Research Laboratory). 1983. Impacts of contaminants on early life stages of striped bass. U. S. Fish and Wildlife Service, Progress Report 1980-1983. CNFRL, Columbia, Missouri.

Collvin, L. 1985. The effect of copper on growth, food consumption and food conversion of perch, *Perca fluviatilis* L., offered maximal food rations. Aquatic Toxicology 6:105-113.

Cooper, J. C., and T. T. Polgar. 1981. Recognition of year-class dominance in striped bass management. Transactions of the American Fisheries Society 110:180-187.

Coutant, C. C. 1985. Striped bass, temperature, and dissolved oxygen: a speculative hypothesis for environmental risk. Transactions of the American Fisheries Society 114:31-61.

Coutant, C. C. 1987. Poor reproductive success of striped bass from a reservoir with reduced summer habitat. Transactions of the American Fisheries Society 116:154-160.

Coutant, C. C., and D. L. Benson. 1990. Summer habitat suitability for striped bass in Chesapeake Bay: reflections on a population decline. Transactions of the American Fisheries Society 119:757-778.

Cowan, J. H., K. A. Rose, E. S. Rutherford, and E. D. Houde. 1993. Individual based model of young-of-the-year striped bass population dynamics. II. Factors affecting recruitment in the Potomac River, Maryland. Transactions of the American Fisheries Society 122:439-458.

Crowder, L. B., J. A. Rice, T. J. Miller, and E. A. Marschall. 1992. Empirical and theoretical approaches to size-based interactions and recruitment variability in fishes. Pages 237-255 *in* D. L. DeAngelis and L. J. Gross, editors. Individual-based models and approaches in ecology. Routledge, Chapman, and Hall, New York.

Dalton, P. D. 1987. Ecology of bay anchovy (*Anchoa mitchilli*) eggs and larvae in the mid-Chesapeake Bay. Masters Thesis. University of Maryland, College Park.

DeAngelis, D. L., L. W. Barnthouse, W. Van Winkle, and R. G. Otto. 1990. A critical appraisal of population approaches in assessing fish community health. Journal of Great Lakes Research 16:576-590.

DeAngelis, D. L., and L. J. Gross, editors. 1992. Individual-based models and approaches in ecology: populations, communities and ecosystems. Routledge, Chapman, and Hall, New York.

DeAngelis, D. L., K. A. Rose, and M. A. Huston. In press. Individual-oriented approaches to modelling populations and communities. Lecture Notes in Biomathematics, Volume 100.

Depledge, M. H. 1990. New approaches in ecotoxicology: can inter-individual physiological variability be used as a tool to investigate pollution effects. Ambio 19:251-252.

Drummond, R. A., W. A. Spoor, and G. F. Olson. 1973. Some short-term indicators of sublethal effects of copper on brook trout, *Salvelinus fontinalis*. Journal of the Fisheries Research Board of Canada 30:698-701.

Eldridge, M. B., P. Benville, and J. A. Whipple. 1981. Physiologic responses of striped bass *(Morone saxatilis)* embryos and larvae to low, sublethal concentrations of the aromatic hydrocarbon benzene. Proceedings of the Gulf and Caribbean Fisheries Institute 33:52-68.

Eldridge, M. B., J. A. Whipple, and M. J. Bowers. 1982. Bioenergetics and growth of striped bass, *Morone saxatilis*, embryos and larvae. Fishery Bulletin, U. S. 80:461-474.

Fogarty, M. J., M. P. Sissenwine, and E. B. Cohen. 1991. Recruitment variability and the dynamics of exploited marine populations. Trends in Ecology and Evolution 6:241-246.

Goodyear, C. P. 1985a. Relationship between reported commercial landings and abundance of young striped bass in Chesapeake Bay, Maryland. Transactions of the American Fisheries Society 114:92-96.

Goodyear, C. P. 1985b. Toxics materials, fishing, and environmental variation: simulated effects on striped bass populations. Transactions of the American Fisheries Society 114:107-113.

Hall, L. W. 1987. Acidification effects on larval striped bass, *Morone saxatilis*, in Chesapeake Bay tributaries: a review. Water, Air, and Soil Pollution 35:87-96.

Hall, L. W. 1991. A synthesis of water quality and contaminants data on early life stages of striped bass, *Morone saxatilis*. Reviews in Aquatic Sciences 4:261-288.

Hall, L. W., and D. T. Burton, and L. B. Richardson. 1981. Comparison of ozone and chlorine toxicity to the development stages of striped bass, *Morone saxatilis*. Canadian Journal of Fisheries and Aquatic Sciences 38:752-757.

Hewett, S. W., and B. L. Johnson. 1987. A generalized bioenergetics model of fish growth for microcomputers. WIS-SG-87-245, University of Wisconsin Sea Grant Institute, Madison.

Holland, A. F., A. T. Shaughnessy, L. C. Scott, V. A. Dickens, J. Gerritsen, and J. A. Ranasinghe. 1989. Long-term benthic monitoring and assessment program for the Maryland portion of Chesapeake Bay: interpretive report. Power Plant Research Program Report CBRM-LTB/EST-2, Maryland Department of Natural Resources, Annapolis.

Houde, E. D. 1987. Fish early life history dynamics and recruitment variability. American Fisheries Society Symposium 2:17-29.

Houde, E. D. 1989. Subtleties and episodes in the early life of fishes. Journal of Fish Biology 35 (Supplement A):29-38.

Houde, E. D., E. J. Chesney, R. Nyman, and E. Rutherford. 1988. Mortality, growth and growth rate variability of striped bass larvae in Chesapeake subestuaries. Interim Report to Maryland Department of Natural Resources, Contract F112-87-008, Annapolis.

Houde, E. D., and E. S. Rutherford. 1992. Egg production, spawning biomass and factors influencing recruitment of striped bass in the Potomac River and upper Chesapeake Bay. Report to Maryland Department of Natural Resources, Contract Number F145-88-008, Annapolis, Maryland and Chesapeake Biological Laboratory Report [UMCEES]CBL 92-017.

Houde, E. D., and C. E. Zastrow. In press. Ecosystem- and taxon-specific dynamic and energetics properties of fish larval assemblages. Bulletin of Marine Science.

Hughes, R. N. 1980. Optimal foraging theory in the marine context. Oceanography and Marine Biology Annual Reviews 18:423-481.

Huston, M., D. DeAngelis, and W. Post. 1988. New computer models unify ecological theory. BioScience 38:682-691.

Little, E. E., and S. E. Finger. 1990. Swimming behavior as an indicator of sublethal toxicity in fish. Environmental Toxicology and Chemistry 9:13-19.

Logan, D. T. 1986. Use of size-dependent mortality

models to estimate reductions in fish populations resulting from toxicant exposure. Environmental Toxicology and Chemistry 5:769-775.

Mathers, R. A., J. A. Brown, and P. H. Johansen. 1985. The growth and feeding behavior responses of largemouth bass *(Micropterus salmoides)* exposed to PCP. Aquatic Toxicology 6:157-164.

McCarthy, J. F., and L. R. Shugart, editors. 1990. Biomarkers of environmental contamination. Lewis Publishers, CRC Press, Boca Raton, Florida.

McKim, J. M. 1985. Early life stage toxicity tests. Pages 58-95 *in* G. M. Rand and S. R. Petrocelli, editors. Fundamentals of aquatic toxicology, Hemisphere Publishing, New York.

MD DNR (Maryland Department of Natural Resources). 1990. Investigations of striped bass in Chesapeake Bay. Performance Report of Project F-42-R-3 to the U. S. Fish and Wildlife Service. Maryland Department of Natural Resources, Annapolis.

Moore, C. M. 1988. Food habits, population dyanmics, and bioenergetics of four predatory fish species in Smith Mountain Lake, Virginia. Ph.D. Dissertation. Virginia Polytechnic Institute and State University, Blacksburg.

Morgan, M. J., and J. W. Kiceniuk. 1990. Effect of fenitrothion on the foraging behavior of juvenile Atlantic salmon. Environmental Toxicology and Chemistry 9:489-495.

NOAA (National Oceanic and Atmospheric Administration). 1991. Status of the fishery resources off the northeastern United States for 1991. NOAA Technical Memorandum NMFS-F/NEC-86, Northeast Fisheries Science Center, National Marine Fisheries Service, Woods Hole, Massachusetts.

Olson, M. M. 1987. Zooplankton. Pages 38-81 *in* K. L. Heck, editor. Ecological studies in the middle reach of Chesapeake Bay. Lecture Notes on Coastal and Estuarine Studies 23, Springer-Verlag, New York.

Price, K. S., D. A. Flemer, J. L. Taft, G. B. Mackiernan, W. Nehlsen, R. B. Biggs, N. H. Burger, and D. A. Blaylock. 1985. Nutrient enrichment of Chesapeake Bay and its impacts on the habitat of striped bass: a speculative hypothesis. Transactions of the American Fisheries Society 114:97-106.

Rago, P. J., R. M. Dorazio, R. A. Richards, and D. G. Deuel. 1989. Emergency striped bass research study: report for 1988. U. S. Department of the Interior and Department of Commerce, Washington, District of Columbia.

Rose, K. A., and J. H. Cowan, Jr. 1993. Individual-based model of young-of-the-year striped bass population dynamics. I. Model description and baseline simulations. Transactions of the American Fisheries Society 122:415-438.

Rothschild, B. J. 1986. Dynamics of marine fish populations. Harvard University Press, Cambridge, Massachusetts.

Sandheinrich, M. B., and G. J. Atchison. 1989. Sublethal copper effects on bluegill, *Lepomis macrochirus*, foraging behavior. Canadian Journal of Fisheries and Aquatic Sciences 46:1977-1985.

Schaaf, W. E., D. S. Peters, D. S. Vaughan, L. Coston-Clements, and C. W. Krouse. 1987. Fish population responses to chronic and acute pollution: the influence of life history strategies. Estuaries 10:267-275.

Setzler, E. M., W. R. Boynton, K. V. Wood, H. H. Zion, L. Lubbers, N. K. Mountford, P. Frere, L. Tucker, and J. A. Mihursky. 1980. Synopsis of biological data on striped bass, *Morone saxatilis* (Waldbaum). NOAA Technical Report NMFS Circular 433, National Oceanic and Atmospheric Administration, Washington, District of Columbia.

Setzler-Hamilton, E. M., W. R. Boynton, J. A. Mihursky, T. T. Polgar, and K. V. Wood. 1981. Spatial and temporal distribution of striped bass eggs, larvae, and juveniles in the Potomac Estuary. Transactions of the American Fisheries Society 110:121-136.

Setzler-Hamilton, E. M., and L. Hall. 1991. Striped bass. Pages 13-1 to 13.31 *in* S. L. Funderburk, J. A. Mihursky, S. J. Jordan, and D. Riley, editors. Habitat requirements for Chesapeake Bay living resources, 2nd edition. Living Resources Subcommittee, Chesapeake Bay Program, Annapolis, Maryland.

Setzler-Hamilton, E. M., J. A. Whipple, and R. B. MacFarlane. 1988. Striped bass populations in Chesapeake and San Francisco Bays: two environmentally impacted estuaries. Marine Pollution Bulletin 19:466-477.

Sissenwine, M. P. 1984. Why do fish populations vary? Pages 59-94 *in* R. M. May, editor. Exploitation of marine communities. Springer-Verlag, New York.

Stevens, D. E., D. W. Kohlhorst, L. W. Miller, and D. W. Kelly. 1985. The decline of striped bass in the Sacramento-San Joaquin Estuary, California. Transactions of the American Fisheries Society 114:12-30.

Sullivan, B. K., E. Buskey, D. C. Miller, and P. J. Ritacco. 1983. Effects of copper and cadmium on growth, swimming and predator avoidance in *Eurytemora affinis* (Copepoda). Marine Biology 77:299-306.

Tuncer, H. 1988. Growth, survival, and energetics of larval and juvenile striped bass *(Morone saxatilis)* and its white bass hybrid *(M. saxatilis × M. chrysops)*. Masters Thesis. University of Maryland, College Park.

Ulanowicz, R. E., and T. T. Polgar. 1980. Influences of anadromous spawning behavior and optimal environmental conditions upon striped bass *(Morone saxatilis)* year-class success. Canadian Journal of Fisheries and Aquatic Sciences 37:143-154.

Uphoff, J. H. 1989. Environmental effects on survival of eggs, larvae, and juveniles of striped bass in the Choptank River, Maryland. Transactions of the American Fisheries Society 118:251-263.

van der Veer, H. W., and M. J. N. Bergman. 1987. Predation by crustaceans on a newly settled 0-group plaice *Pleuronectes platessa* population in the western Wadden Sea. Marine Ecology Progress Series 35:203-215.

Van Winkle, W., K. A. Rose, and R. C. Chambers. 1993. Individual-based approach to fish population dynamics: an overview. Transactions of the American Fisheries Society 122:397-403.

von Westernhagen, H. 1988. Sublethal effects of pollutants on fish eggs and larvae. Pages 253-346 *in* W. S. Hoar and D. J. Randall, editors. Fish physiology, volume XI. The physiology of developing fish, Part A: eggs and larvae. Academic Press, New York.

Westin, D. T., C. E. Olney, and B. A. Rogers. 1983. Effects of parental and dietary PCBs on survival, growth, and body burdens of larval striped bass. Bulletin of Environmental Contamination and Toxicology 30:50-57.

Westin, D. T., C. E. Olney, and B. A. Rogers. 1985. Effects of parental and dietary organochlorines on survival and body burdens of striped bass larvae. Transactions of the American Fisheries Society 114:125-136.

Zastrow, C. E., E. D. Houde, and E. H. Saunders. 1989. Quality of striped bass *(Morone saxatilis)* eggs in relation to river source and female weight. Rapports et Procès-Verbaux des Réunions Conseil International pour l'Exploration de la Mer 191:34-42.

Management Issues

The Importance of Habitat to the Early Life History of Estuarine Dependent Fishes

DONALD E. HOSS AND GORDON W. THAYER

NOAA, National Marine Fisheries Service, Southeast Fisheries Science Center
Beaufort Laboratory, Beaufort, North Carolina 28516-9722, USA

Abstract.—Habitats for fishes are places where they live and are defined by their physical, biological, and chemical characteristics. During their development, estuarine dependent fishes utilize a wide variety of habitats in the coastal environment as well as within the estuary. Early life-history research has emphasized life-history patterns, abundance, and distribution of fishes in coastal waters and not the habitat they utilize *per se*. Limited data suggest that there are differences in coastal habitat quality that may cause differences in the growth of larvae. Within the estuary, habitats usually have been described by their physical structure, but there is a paucity of data on the actual use of these habitats by larvae and early stage juveniles. We discuss some of the available information for several habitats common to the east and Gulf coasts of the U. S. and conclude that within estuaries the water column and seagrass meadows provide the primary habitats for initial use, but that other habitats are integrally linked through biological, chemical, and physical processes. Because of the coupling between habitats, we recommend that research be addressed more on a basis of landscape, rather than individual habitats.

The habitat of an organism (as defined by Odum 1971) is the place where it lives which frequently changes with developmental stage. Almost 3 decades ago, Hemple (1965) stated that less was known of the transition between larval and juvenile marine fishes than any other life-history stage. Even now we know surprisingly little about the importance of habitat to the larvae of many of our most important species of estuarine fishes. We do know, however, that larvae use a wide variety of habitats during their development, ranging from the continental slope into coastal lagoons, bays, and estuaries up to the freshwater interface.

We also know, without a doubt, that habitat and fishery production are linked, but in most cases these linkages have not been quantified. Many research programs have been, and are being funded to determine the ecological relationships between quality and quantity of habitat and fishery production so that the effects of habitat loss and degradation can be evaluated.

The purpose of this paper is to briefly review the importance of habitat to the larval through early juvenile stages of estuarine dependent fishes, that is, those species utilizing estuarine habitats at some stage of their life cycle.

To determine the effects of habitat loss or degradation on the early life stages, we first need to define the habitat they live in and develop an understanding of ecological factors operating in those habitats. Once done, we can try to determine how natural or man-induced perturbations might affect the habitat and the organisms that use it. Peters and Cross (1992) have discussed the definitions of habitat and have noted, like Ryder and Karr (1989), that it is defined by its physical, biological, and/or chemical characteristics. They conclude that habitat is most frequently associated with structural components that attract individuals, and that environmental properties such as temperature, toxic substances, oxygen content, and light influence its value and use.

In this paper, we will use Odum's basic definition and describe the place where the specific life stage lives by its physical, biological, and chemical characteristics. As much as possible, we will also identify the critical factors operating in that habitat.

Coastal Habitat

On the southeastern and Gulf coasts of the U. S. there is a wide shelf extending out to the edge of the continental slope. Because coastal or barrier islands usually separate coastal water from estuarine waters, the physical and chemical characteristics of the coastal habitat are more stable than estuarine habitats. Except where affected by large riverine inputs (for example, the Mississippi River) water clarity is usually high. In coastal waters, physical processes that set up "fronts" may play a significant role concentrating both larvae and their food.

Along the southeastern and Gulf coasts of the U. S., over 90% of the most important commercial and recreational species of fishes spawn in coastal waters. Because of the lack of distinguishing features for estuaries, most early life studies on estuarine dependent fishes have not emphasized the physical habitat as much as environmental conditions.

Hunter (1983) gave a succinct description of how the relationship between larvae and their environment changes with development. During egg and yolk-sac stages, there is an initial increase in the dispersion of eggs and larvae, a high mortality rate, a decrease in overall biomass, and an almost complete dependence on the immediate planktonic habitat in both a physical and biological sense. As larvae develop into the nektonic or free swimming phase they become progressively more independent of the immediate planktonic habitat, are able to search for food, become more aggregated, and the total biomass begins to increase. In this stage, density-dependent factors in the habitat (such as food) begin to become more important.

Water Column

Atlantic menhaden *(Brevoortia tyrannus)*, Atlantic croaker *(Micropogonias undulatus)*, spot *(Leiostomus xanthurus)*, and summer *(Paralichthys dentatus)*, southern *(Paralichthys lethostigma)*, and gulf *(Paralichthys albigutta)* flounders provide good examples of species that utilize the coastal habitat as larvae and early juveniles. These species spawn offshore to the edge of the Gulf Stream. Pelagic eggs are found in the near surface layer and most of the early larvae occur in the upper portion of the water column. We can describe, at least partially, the physical and chemical characteristics of their habitat. In the laboratory, for example, Hettler (1981, 1983) has demonstrated that pelagic eggs of Atlantic menhaden only float in salinities above 26‰ and that the temperature for hatching and survival is between 15 and 25°C.

Early life-history research of fishes from the coastal habitat has concentrated on life-history, abundance and distribution, feeding, growth, and predation. In the 1920s and 1930s, Hildebrand and Cable (1930) described the development and life-history of 14 species of fishes from the vicinity of Beaufort, North Carolina. They concluded that most of the important food fishes taken in the estuary during the summer months migrate to warmer coastal habitats during the fall and winter to spawn. Subsequent research has confirmed their original hypothesis of offshore fall and winter spawning and estuarine dependence (Dawson 1958; Fore 1970; Powles and Stender 1976; Powles 1981; Judy and Lewis 1983; Shaw et al. 1985).

With the development of sophisticated, depth-discrete samplers, such as the Multiple Opening/Closing Net and Environmental Sensing System (MOCNESS, Wiebe et al. 1976), recent studies have examined vertical distribution and behavior of coastal larvae. Sogard et al. (1987) compared the vertical and horizontal distribution of three species of larvae along three transects in the northern Gulf of Mexico, and reported that only gulf menhaden *(Brevoortia patronus)* showed any consistent patterns in vertical distribution, being more concentrated in surface waters. This has also been reported for the Atlantic menhaden (Checkley et al. 1988). Thus, there appears to be some difference in the use of the coastal habitat in the vertical dimension.

For larvae to survive in the coastal habitat there must be ample food. In general, their diets are known to consist of phytoplankters and zooplankters. Early work by June and Carlson (1971) examined the alimentary tracts of late stage Atlantic menhaden larvae and found the contents to be 99% copepods, while Govoni et al. (1983) noted that early stage gulf menhaden larvae had a diverse diet that included phytoplankton and zooplankton. Spot and Atlantic croaker, two related species taken by Govoni et al. (1983) in the Gulf, although feeding on zooplankton, had dissimilar diets that also differed from the diets of gulf menhaden. This suggests reduced competition and partitioning of the habitat with respect to food.

Differences in habitat "quality" may affect growth of larvae over spatial and temporal scales. Studies on age and growth of estuarine dependent larvae have been conducted in both the Atlantic and Gulf of Mexico (Warlen 1982, 1988; Maillet and Checkley 1989). Warlen (1992) found that the average daily growth rate of wild Atlantic menhaden larvae was about 10% greater than for wild gulf menhaden through their first 60 days. He also found different growth rates for gulf menhaden from different coastal areas in the northern Gulf of Mexico and from the same area at different times. Although far from conclusive, this suggests that there may be natural differences in habitat "quality" causing differences in growth to occur in subunits of the overall habitat. Habitat "quality" in a given area may also differ because of seasonal differences in temperature, salinity, food availability, etc.

Plumes and Fronts

Hydrodynamic and physical processes in the coastal habitat act to establish habitat structure. For example, temperature and salinity establish the pycnocline creating a boundary in the vertical dimension. In the horizontal dimension convergence sets up within frontal zones because of wind and pressure gradients. These water column habitat subdivisions, although difficult to study because of their ephemeral nature, may be very important to the survival of larvae.

Govoni and his colleagues have conducted extensive research around the Mississippi River plume in the Gulf of Mexico. They have looked at the spatial distribution, diet composition, condition of larvae, and mechanisms of accumulation of fish larvae in relation to the plume.

Govoni et al. (1989) reported that ichthyoplankton densities about the plume were often greater at the plume front than inside or outside the plume. The difference was, at times, several orders of magnitude. The increased ichthyoplankton concentration along fronts may have biological relevance. There is some evidence that the zooplankton food of the larvae is more abundant in the mixing zone between plume water and shelf water than it is elsewhere (Ortner et al. 1989) and that the phytoplankton biomass is also increased (Dagg et al. 1988).

For example, larval spot were found to have eaten twice as many food organisms in plume water than larvae in shelf waters (Govoni and Chester 1990). However, when Powell et al. (1990) examined the nutritional conditions of spot larvae associated with the Mississippi River plume, they were unable to demonstrate that larvae have a nutritional advantage when associated with the plume fronts. The number of starving larvae was in fact higher in the plume front than either inside or outside the plume.

Estuarine Habitat

Fish larvae that enter estuaries or are spawned within the estuary have a mosaic of habitats to utilize during their subsequent development: seagrass meadows, salt marshes and their tidal creeks, mangroves, unvegetated bottoms, shell reefs, and the water column itself. We will not deal with several estuarine habitat types, such as shell reefs and drift algae, in this paper because of space constraints.

The functional values of habitats in the estuary have been most frequently described for vegetated habitats. The functional roles of seagrasses (Wood et al. 1969) can be equally applied to other estuarine vegetated habitats (Thayer et al. 1978). There are relatively high levels of organic production and high standing crops of plants. Few organisms feed directly on the living plant, and as a consequence, detritus plays a major role. Leaves, stems, and root systems present surfaces for epibiota which are available as food resources and contribute to the overall primary and secondary production of the system. The above-ground structures promote sedimentation of inorganic and organic matter that are important in nutrient cycling and accretion. The physical structure also provides protection from predation.

Unvegetated sediments also provide physical structure (for example, different granulometries, color), chemical composition, and abundant and diverse primary and secondary food resources. Demersal life stages may actively seek refuge from predators using behavioral modifications such as color pattern changes or surface burrowing. They also may seek benthic unvegetated habitats that provide the nutritional resources, but which are in environmental tolerance zones less conducive to predators (that is, combinations of temperature, salinity, and oxygen conducive to growth and feeding, but where predation pressure is lower). The water column itself is also a habitat that provides abundant and varied food resources and environmental conditions that are conducive to growth and survival (Ryder and Karr 1989).

Seagrasses

Of the three vegetated habitats that we discuss, seagrasses have been longest recognized as a critical nursery area and the most frequently studied regarding fishery use. This has occurred because traditional fishery sampling gear could be used routinely with little modification. Being subtidal, seagrass meadows are available to fishes and researchers most of the time, and this habitat type is found in almost all shallow coastal and estuarine waters. Yet, the nursery value of this habitat is generally ascribed to juvenile and subadult fishes with little emphasis placed on the settling larvae or early stage juveniles (Kenworthy et al. 1988; Zieman and Zieman 1989; and references cited therein).

Seagrass meadows consistently have higher abundances and diversities of organisms than unvegetated areas, and densities of food resources (benthic, epibenthic, and planktonic) tend to be higher within seagrass beds. Habitat heterogeneity, plant biomass, and surface area enhance faunal abundances (Stoner 1980). Fishes foraging within the grassbed canopy are coincidentally protected from larger predators because of grass blade density and surface and because of reduced light penetration (Kenworthy et al. 1988). Additionally, some organisms can orient themselves with the seagrass blades to become camouflaged.

There are few studies dealing with larva settlement and use of seagrass habitats. One might expect, however, that some of the same functions described above for juvenile and subadult fishes would also hold for larvae. Whether larva settlement into habitats is a random or directed process has been the subject of few studies. Bell and Westoby (1986) proposed a model suggesting that larvae settle into the first seagrass bed they encounter and then select micro-sites within the bed; predation pressure is ultimately responsible for abundance and distribution patterns within the bed. Using artificial seagrass habitats with different densities of "leaves," Bell et al. (1987) concluded that abundances of fishes in seagrass beds was not due to larva settlement preferences or post-settlement predation pressure, but rather to the initial availability of larvae to settle indiscriminately into any seagrass shelter. Sogard (1989), using artificial seagrass in New Jersey, has shown that distance from natural eelgrass effects composition of communities. In some instances, fishes and crustaceans apparently traversed expanses of unvegetated habitat to colonize isolated plots of seagrass. This suggests that the "settle-and-stay" (Sogard 1989) hypothesis of Bell and Westoby (1986) may not be the norm. From a limited number of studies such as these, it would appear that we cannot make a definitive statement as to the factors controlling the distribution and abundance of larval fishes within and among seagrass meadows. Predation may play a role, and proximity to other seagrass habitats is important in the composition of the community.

Olney and Boehlert (1988), demonstrated that seagrass beds in the lower Chesapeake Bay were important for fish species brooding eggs (for example, silverstripe halfbeak *Hyporhamphus unifasciatus*) and with demersal adherent eggs (rough silverside *Membras martinica*). On the Pacific coast of the U. S., herring (*Clupea pallasi*) appear to rely on eelgrass blades to attach their eggs. Olney and Boehlert (1988) noted that winter-spring spawners lacked this habitat in Chesapeake Bay, but that seagrass meadows were present and used by larvae of spring-summer spawners: anchovies (*Anchoa* spp.), gobies (*Gobiosoma* spp.), green goby (*Microgobius thalassinus*), sharptail goby (*Gobionellus oceanicus*), northern pipefish (*Syngnathus fuscus*), weakfish (*Cynoscion regalis*), southern kingfish (*Menticirrhus americanus*), red drum (*Sciaenops ocellatus*), silver perch (*Bairdiella chrysoura*), rough silverside (*Membras martinica*), feather blenny (*Hypsoblennius hentz*), and halfbeaks (*Hyporhamphus* spp.). Sampling in New Jersey vegetated and unvegetated habitats, Sogard (1989) suggested that gobies (for example, naked goby, *G. bosc*) initially settled on sand/mud substrates and subsequently migrated to eelgrass habitats. Subsequent sampling (Sogard and Able 1991), however, did not demonstrate such a pattern.

In contrast, the northern regions of North Carolina exhibit almost year-round coverage of seagrasses (eelgrass and shoalgrass) (Thayer et al. 1984), and larval and early juvenile fishes are present in these beds during much of the year. Early developmental stages of fishes (< 25 mm fork or peduncle length) have been collected from high- and low-energy seagrass habitats near Beaufort, North Carolina between February and August 1991 (NMFS, Beaufort Laboratory, unpublished data). Pinfish (*Lagodon rhomboides*), spot, menhaden, and pigfish (*Orthopristis chrysoptera*), dominated collections during February through April, although numerous other species such as speckled worm eel (*Myrophis punctatus*), gulf flounder, summer flounder, spottail pinfish (*Diplodus holbrooki*), bay anchovy (*A. mitchilli*), striped anchovy (*A. hepsetus*), and mojarra (*Eucinostomus* spp.) were also present. During May through August, total densities were lower, but species composition was higher than during winter, with early stages of gobies, blennies, pipefishes, silver perch, sheepshead (*Archosargus probatocephalus*), searobin (*Prionotus* spp.), mojarras, red drum, and spotted seatrout (*C. nebulosus*) larvae or early stage juveniles becoming common.

Use of this habitat by larvae may be only transitory because these fishes may suffer from heavy predation (Olney and Boehlert 1988). This hypothesis was based on presence of lower numbers of larvae during daylight when planktivores tend to feed than at night and the fact that the early life-history stage probably cannot orient itself to the seagrass blades and thus take advantage of the refuge function of the habitat. If, however, visual detection is important, then the decreased light penetration within the seagrass canopy and the presence of seagrass blades, which can interrupt visual reception and actually camouflage small organisms, should play an equally important role for both early and

Mangroves

While the functional characteristics described earlier for vegetated habitats may be intuitively correct, the mangrove habitat has received the least attention, particularly as this relates to its nursery function for fishery organisms. As noted by Thayer et al. (1987), Ley (1992), and Thayer and Sheridan (in press), the limited information stems largely from the lack of suitable collection techniques to address the direct contribution of mangroves to fishery organisms. With the exception of some efforts in Australia (for example, Robertson and Duke 1987, 1990) and India (Krishnamurthy and Jeyaseelan 1981), all of the hypotheses and assumptions on nursery value to fishes is derived from sampling of juvenile and subadults, and there have been few experiments carried out to verify the assumptions.

Thayer and Sheridan (in press) have summarized some of the data on species occurring among prop roots of south Florida mangroves. Forage fishes generally predominate: hardhead silverside *(Atherinomorus stipes)*, silver jenny *(E. gula)*, goldspotted killifish *(Floridichthys carpio)*, spotfin mojarra *(E. argenteus)*, code goby *(G. robustum)*, rough silverside, striped anchovy, and clown goby *(M. gulosus)*, to name a few. Juveniles of commercial and recreational species also are common, but in lower densities: common snook *(Centropomus undecimalis)*, gray snapper *(Lutjanus griseus)*, spotted seatrout, red drum, striped and white mullets *(Mugil cephalus* and *M. curema)*, sheepshead, and great barracuda *(Sphyraena barracuda)*. Salinity patterns appear to affect species composition and abundance (Ley 1992). Average juvenile and adult fish density appeared to be low in upstream areas of Ley's (1992) study sites in Florida Bay which were subject to more variable salinities, and high in downstream areas where salinities tended to be more stable. High densities of fishes and invertebrates in mangrove habitats in part may be the result of protection from predators afforded by both the physical structure of the habitat and the frequently occuring high turbidity.

It is quite possible that U. S. mangrove habitats do not serve an important nursery function for larvae of all but resident species. The life-history strategy of many estuarine dependent fishes is one of offshore spawning, immigration of larvae to estuaries, settlement and growth of juveniles, and emigration of sub-adults to offshore or openwater habitats. Mangrove shorelines frequently are fringed by abundant seagrass habitats. One possible scenario is that larva settlement may occur to a greater extent in the seagrass beds with subsequent movement of early stage juveniles into the mangrove habitat, followed by diel movements between seagrasses and mangroves. In fact, Ley (1992) concluded that paucity of submerged aquatic plants in some areas may result in inadequate intermediate habitat for fishes between planktonic and mangrove life stages.

Thus, we have a poor data base on the direct value to fishery organisms of a habitat that is among the dominant habitats in the tropical and subtropical Americas and which occupies about 200,000 hectares of estuarine and coastal shoreline in the United States. Thayer et al. (in press) recommended research on habitat utilization in several priority areas: development of quantitative sampling methodology for various forest types and the intercalibration of methods for fishery use; comparison of spatial and temporal variation in habitat use by fishes and invertebrates, particularly in relation to critical water levels that permit access; comparing food/feeding ecology and refuge potential in each mangrove habitat; and contrasting these patterns and functions among mangrove, emergent marsh, seagrass, and non-vegetated habitats.

Salt Marshes and Marsh Creeks

Research on the value of *Spartina* and *Juncus* marshes to fishery organisms has dealt primarily with the use of their tidal creeks and with the transfer of energy produced through invertebrates and detritus to fishes (see reviews by Josselyn 1983; Stout 1984; Teal 1986). In the last decade, however, with an increased awareness of the loss of wetlands due to subsidence, sea level rise, and coastal development, there has been a concerted effort to determine the direct use of marsh habitats (that is, the flood marsh surface) by fishes. This includes some experimental studies on functional values of the marsh to growth and survival of fishes.

With the exception of a few studies, there has been little recognition of the potential use by the larval life-history stage. Data suggest that the larvae or juveniles of some species feed extensively on the marsh surface when it is flooded (Weisberg and Lotrich 1982; McIvor et al. 1988; and references cited therein), and that juveniles of many species seek shelter on the marsh surface during high water (Boesch and Turner 1984; McIvor and Odum 1988; Hettler 1989). In light of these findings, one might expect that a large majority of the fish larvae and early juveniles collected in marsh channels have the ability to move onto the marsh surface on flood tides. Boesch and Turner (1984) noted, however, that there have been few experimental tests of the hypothesis that marsh habitats provide protection from predation for fishes.

There are strong indications that larvae of estuarine transient as well as resident species gain access to marshes and may move actively or passively onto salt marsh surfaces during flood tides, and may be present in sizeable numbers. Talbot and Able (1984) found larval killifishes, silversides, and sticklebacks commonly on three New Jersey salt marshes. Kilby (1955) reported larvae and/or early stage juveniles of these species plus several other species of killifishes, live bearers, and mullets in pools in Florida salt marshes, and Subrahmanyam and Drake (1975), while not sampling marsh surfaces *per se*, demonstrated the presence of 11 to 21-mm spot, mojarra, pinfish, anchovy, and flounder in creeks within tidal *Juncus* marshes. Rountree and Able (1992) also demonstrated numerous species present in New Jersey marsh creeks, but most were classified as young-of-the-year or older; there was little evidence of larvae or very early juveniles being present. Rogers and Herke (1985) showed the presence of < 25-mm individuals of sand seatrout *(Cynoscion arenarius)*, spotted seatrout, red drum, gulf menhaden, Atlantic croaker, striped mullet, southern flounder, and sheepshead, to name a few, in interior marshes in Louisiana.

There is little doubt that there are fluxes of material between the marsh surface and adjacent vegetated and unvegetated substrates and the water column in adjacent creeks and the open estuary. In fact, Nixon (1980) has shown that commercial landings of estuarine dependent species are related to the ratio of marsh area to open water area along major areas of the coast with the exception of the Chesapeake Bay. Since 1980, only limited research has been carried out on the direct use of marsh habitats by the early stages of fishes (see previous discussion), and it has emphasized resident and not transient species. It is possible that for the larvae of many species entering an estuary, the water column or seagrass meadows are the major initial habitats of choice (for some demersal species, such as flounders, unvegetated sediments may provide refuge), and that salt marshes or mangroves provide a secondary "choice." Early and later stage juveniles may move readily between habitats, as has been suggested or shown by several investigators (McIvor and Odum 1988; Hettler 1989; Rountree and Able 1992; Rozaz and Reed 1993). It is obvious that studies are required to address life-history stage use of marsh habitats as well as experimental studies on trophic support and refuge potential, not only of resident species, but also of the transient species that do make use of these habitats.

Unvegetated Bottoms

Unvegetated benthic habitats also present structural complexity for settling larvae within the fluctuating environmental conditions of estuaries. Again, however, much of the information available is on juveniles. Some available scientific data indicate, or at least suggest, that unvegetated benthic sediments provide food and refuge for settling larvae, the latter being a function of sediment granulometry and perhaps color. Marliave (1977) investigated substrate preferences and reported that in many instances larvae preferred substrates that often had characteristics preferred by their adults. This was particularly true for grain size, although frequently there were preferences for color within a grain size; color preference may be related to camouflage. This may be a function of ability to bury, as is the case for marbled sole *(Limanda yokohamae)* and Japanese flounder *(P. olivaceus)*, which prefer sand substrates in which they can easily bury (Tanda 1990). Gibson and Robb (1992), however, found that many small (< 30mm) plaice *(Pleuronectes platessa)* failed to bury in fine sediments in the diameter range of 0.062 to 0.125 mm. They suggest that small plaice may be unable to bury completely in any sediment and that this inability may contribute to their vulnerability to predation.

Considerable research on habitat selection by larval and juvenile flatfishes has been conducted (see references in Murchand and Masson 1989; Burke et al. 1991; Gibson and Robb 1992). These studies have examined niche separation in species of flatfishes that may immigrate and settle as larvae in different parts of an estuary. Burke et al. (1991) demonstrated that different habitat preferences for southern and summer flounder were a function of salinity and substrate type. Southern flounder distribution was affected by salinity, being found more frequently between 17 and 24‰, while the summer flounder were unaffected by salinity changes, but were more prevalent on sandy substrates having 99.4 to 99.6% sand. These sediments are more characteristic of downstream sediments than the muddy upper estuary sediments where southern

HABITATS AND POTENTIAL IMPACTS ON FISH

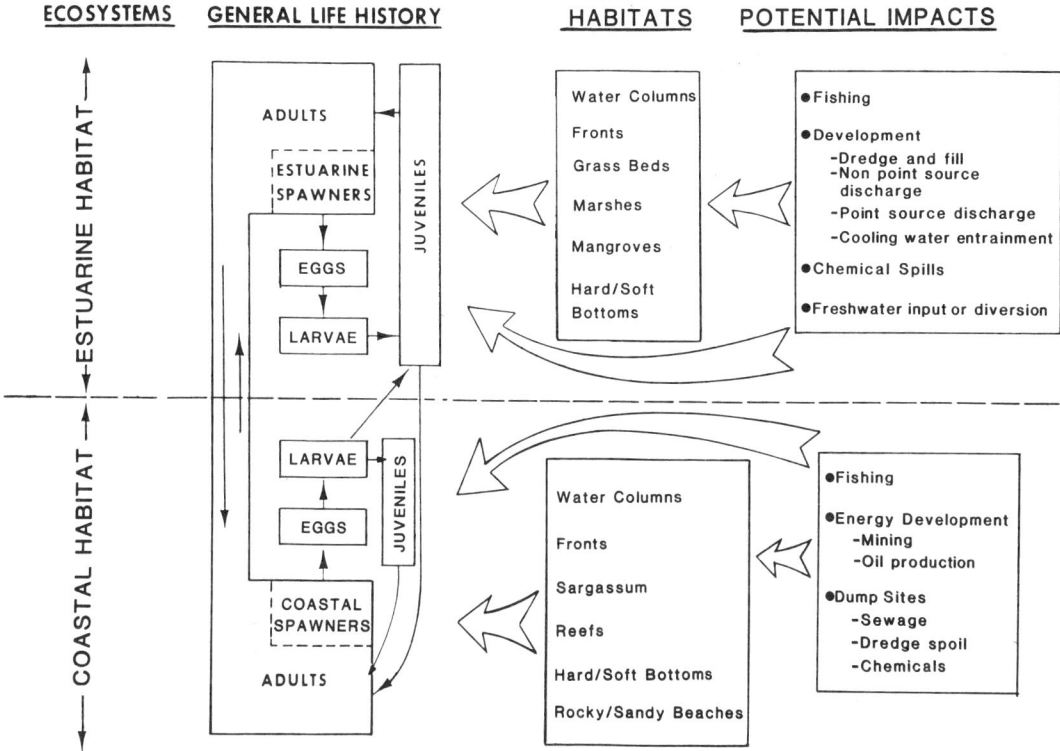

FIGURE 1.—Diagram showing potential environmental modifications to fish habitats and fish life-history stages in coastal and estuarine regions. (See test for discussion).

flounder are prevalent.

Food availability also may play a role in benthic habitat selection by larval flatfishes. Laboratory experiments have shown that summer flounder selected for sand when prey was present in both sand and mud environments, but showed no preference when prey were absent from the substrates (Burke 1991). Edwards and Steele (1968) demonstrated that habitat separation also could occur with plaice, which feed on bivalve siphons, and dabs (*Limanda limanda*), which feed on polychaete tentacles, immediately after settlement.

Summary and Conclusions

Because of the planktonic life style of larvae, changes in the chemical and physical structure of the water column in the coastal zone as well as within estuaries will have the greatest impact (Figure 1). While this is a true statement, it cannot be held inviolate, because many larvae are integrally associated with benthic and vegetated habitats once they have entered the estuary. Additionally, impacts to these structural habitats will have impacts on future fishery yields because of the associations of early juvenile stages with unvegetated and vegetated environments.

In the open coastal area, disturbances to the integrity of the environmentally favorable water column conditions can have major impacts on larval fishes. For example, alterations to the thermal structure can modify the survival of eggs and utilization of yolk reserves of larvae, making larvae more susceptible to mortality through starvation and/or predation (Figure 1). Particulate and dissolved pollutants, including oil spills, can concentrate along fronts and impact both the food resources of larval fishes and the fishes themselves, both of which may concentrate along fronts (Govoni et al. 1989).

Within estuaries, modifications to temperature, salinity, turbidity, and water chemistry conditions (Figure 1), factors that describe the bounds of the water column and structural habitats, vary over shorter distances or within smaller areas than they do in the coastal environment, where the modifications generally are more dispersed. Therefore, the already greater environmental stress that occurs in estuaries can be easily exacerbated by slight alterations in water column characteristics. Reduction in oxygen concentration, as might occur with additions of organic substances, or increased concentrations of toxins, as might occur during a chemical spill or through non-point source pollution, can impose additional stress on behavior and growth of larvae and early stage juveniles. Options for juvenile fishes exposed to stressful conditions entail some energetic cost whether it is tolerance of the condition or movement to a less stressful condition (Miller and Dunn 1980); one might consider this to be even more of a monumental effect on larvae than juveniles because larvae have poorer locomotor capabilities and energy reserves. In either case, only limited data exist.

Modifications of freshwater flow such as channelization can impact both water-column and estuarine bottom habitats (Figure 1) and be both beneficial and detrimental to fishery resources. For example, Rogers et al. (1984) found that low-salinity and freshwater areas were primary zones of recruitment for many species and that peak recruitment and use of these areas appeared to coincide with periods of maximum river flow and influence on the estuary, thereby creating a much larger area of optimum habitat. Zimmerman and Minello (personal communication, NMFS Galveston Laboratory), however, found that freshets had a negative impact on fishery organisms in Texas, when fish moved out of the low-salinity area to follow their preferred benthic food resources which were unable to withstand the change in salinity.

Changes in flow patterns that might occur through dredging operations can alter not only the physical structure of the sediment but also its chemical characteristics. Substrates that were once in low-salinity zones and high in silt-clay content or vice versa can be radically altered in terms of both granular structure and color. These modifications can change habitats from optimal to suboptimal for settling larval flatfishes and their prey.

Intuitively, it seems that physical impacts to vegetated and unvegetated estuarine habitats that either remove or modify the physical and chemical structure of the habitat will have a negative impact on most life-history stages that utilize the habitat for feeding, growth, and/or predator avoidance. Nixon (1980) and Boesch and Turner (1984) among others, have indicated that there are links between fishery yields and various measures of wetland habitat (acres, ratio of marsh to water, area of marsh edge, etc.), and many have linked the physical loss or contamination of habitat with loss of fishery resources. Evidence of direct impacts of contaminants on fishery organisms as well as on their habitats also have been shown.

We know that impacts of chemical and physical modifications of larval fish habitat have occurred and will continue to occur, in the water column in both coastal and estuarine environments. For many species, impacts to unvegetated benthic habitats and to seagrass meadows also will be detrimental to the survival of larvae because of disruption of food resources and/or refuge areas. Modification to salt marsh and mangrove habitats may have less immediate and direct impacts on larval stages of ocean spawned, estuarine dependent fishes. These estuarine habitats, however, do not exist separately from other habitats within the estuary or the watershed. The components of estuarine systems are coupled through water and chemical exchange as well as through movement of fauna, particularly early and later juvenile stages and even adult life forms (Parrish 1989). Therefore, while marsh and mangrove habitats may not play a major direct nursery role for larval stages, they do so for successive stages.

Management, therefore, must be multidimensional and not directed simply at conservation and enhancement of a single habitat type or a single species. Research needs to address habitat issues on a landscape or hydrologic unit basis with such questions as: what is the sequence of habitats that are used by different life-history stages, both temporally and spatially; what is the optimum mix of habitats to produce optimum growth and survival of species in question; are there obligate and facultative habitats for different species; how does geomorphology (size, shape, edge) influence the use and availability of habitats to different life-history stages; what are the costs of sub-lethal stresses (*sensu* Miller and Dunn 1980).

Acknowledgments

Thanks are due to most of the staff of the NMFS Beaufort Laboratory, Estuarine and Coastal Ecology Division. We thank in particular, Drs. Stanley Warlen, John Govoni, Mr. William Hettler and Ms. Valerie Comparetta.

References

Bell, J. D., and M. Westoby. 1986. Variations in seagrass height and density over a wide spatial scale: effects on fish and decapods. Journal of Experimental Marine Biology and Ecology 104:275-295.

Bell, J. D., M. Westoby, and A. S. Steffe. 1987. Fish larvae settling in seagrass: do they discriminate between beds of different leaf density? Journal of Experimental Marine Biology and Ecology 111:133-144.

Boesch, D. F., and R. E. Turner. 1984. Dependence of fishery species on salt marshes: the role of food and refuge. Estuaries 7:460-468.

Burke, J. S. 1991. Influence of abiotic factors and feeding on habitat selection of summer and southern flounder during colonization of nursery grounds. Ph.D. Dissertation. Department of Zoology, North Carolina State University, Raleigh.

Burke, J. S., J. M. Miller, and D. E. Hoss. 1991. Immigration and settlement pattern of *Paralichthys dentatus* and *P. lethostigma* in an estuarine nursery ground, North Carolina, USA. Netherlands Journal of Sea Research 27:393-405.

Checkley, D. M., Jr., S. Raman, G. L. Maillet, and K. M. Mason. 1988. Winter storm effects on the spawning and larval drift of a pelagic fish. Nature 335:346-348.

Dagg, M. J., P. B. Ortner, and F. Al-Yamani. 1988. Winter-time distribution and abundance of copepod nauplii in the northern Gulf of Mexico. Fishery Bulletin, U. S. 86:319-330.

Dawson, C. E. 1958. A study of the biology and life-history of the spot, *Leiostomus xanthurus* Lacépède, with special reference to South Carolina. Contributions from the Bears Bluff Laboratory 28.

Edwards, R., and J. H. Steele. 1968. The ecology of 0-group plaice and common dabs at Loch Ewe. I. Population and food. Journal of Experimental Marine Biology and Ecology 2:215-238.

Fore, P. L. 1970. Eggs and larvae of the gulf menhaden. U. S. Fish and Wildlife Service, Circular 341:11-13.

Gibson, R. N., and L. Robb. 1992. The relationship between body size, sediment grain size and the burying ability of juvenile plaice, *Pleuronectes platessa* L. Journal of Fish Biology 40:771-778.

Govoni, J. J., and A. J. Chester. 1990. Diet composition of larval *Leiostomus xanthurus* in and about the Mississippi River plume. Journal of Plankton Research 12:819-830.

Govoni, J. J., D. E. Hoss, and A. J. Chester. 1983. Comparative feeding of three species of larval fishes in the northern Gulf of Mexico: *Brevoortia patronus, Leiostomus xanthurus*, and *Micropogonias undulatus*. Marine Ecology Progress Series 13:189-199.

Govoni, J. J., D. E. Hoss, and D. R. Colby. 1989. The spatial distribution of larval fishes about the Mississippi River plume. Limnology and Oceanography 34:178-187.

Hemple, G. 1965. On the importance of larval survival for the population dynamics of marine fish. California Cooperative Fisheries Investigations Report 10:13-23.

Hettler, W. F. 1981. Spawning and rearing Atlantic menhaden. Progressive Fish-Culturist 43:80-84.

Hettler, W. F. 1983. Transporting adult and larval gulf menhaden and techniques for spawning in the laboratory. Progressive Fish-Culturist 45:45-48.

Hettler, W. F., Jr. 1989. Nekton use of regularly-flooded saltmarsh cordgrass habitat in North Carolina, USA. Marine Ecology Progress Series 56:111-118.

Hildebrand, S. F., and L. E. Cable. 1930. Development and life history of fourteen teleostean fishes at Beaufort, North Carolina. Bulletin of the U. S. Bureau of Fisheries. 46:383-488.

Hunter, J. R. 1983. On the determinants of stock abundance. Pages 11-16 *in* W. S. Wouster, editor. From year to year: interannual variability of the environment and fisheries of the Gulf of Alaska and the Eastern Bering Sea. Washington Sea Grant Publication, Seattle.

Josselyn, M. 1983. The ecology of San Francisco Bay tidal marshes: a community profile. U. S. Fish and Wildlife Service, Division of Biological Services, Washington, District of Columbia. FWS/OBS-83/23, 102 pp.

Judy, M. H., and R. M. Lewis. 1983. Distribution of eggs and larvae of Atlantic menhaden, *Brevoortia tyrannus*, along the Atlantic coast of the United States. U. S. National Marine Fisheries Service, Special Scientific Report — Fisheries 774.

June, F. C., and F. T. Carlson. 1971. Food of young Atlantic menhaden, *Brevoortia tyrannus*, in relation to metamorphosis. Fishery Bulletin,

U. S. 68:493-512.

Kenworthy, W. J., G. W. Thayer, and M. S. Fonseca. 1988. The utilization of seagrass meadows by fishery organisms. Pages 548-560 *in* D. D. Hook, W. H. McKee, Jr., H. K. Smith, J. Gregory, V. G. Burrell, Jr., M. R. DeVoe, R. E. Sojka, S. Gilbert, R. Banks, L. H. Stolzy, C. Brooks, T. D. Matthews, and T. H. Shear, editors. The ecology and management of wetlands, vol. 1. Ecology of wetlands. Timber Press, Oregon.

Kilby, J. D. 1955. The fishes of two Gulf coastal marsh areas in Florida. Tulane Studies in Zoology 2:175-247.

Krishnamurthy, K., and M. J. P. Jeyaseelan. 1981. The early life history of fishes from Pichavaram mangrove ecosystem of India. Rapports et Procès-Verbaux des Réunions Conseil International pour l'Exploration de la Mer 178:416-423.

Ley, J. A. 1992. Influence of changes in freshwater flow on the use of mangrove prop root habitat by fishes. Ph.D. Dissertation. University of Florida, Tallahassee.

Maillet, G. L., and D. M. Checkley, Jr. 1989. Effects of starvation on the frequency of formation and width of growth increments in sagittae of laboratory-reared Atlantic menhaden *Brevoortia tyrannus* larvae. Fishery Bulletin, U. S. 88:155-165.

Marliave, J. B. 1977. Substratum preferences of settling larvae of marine fishes reared in the laboratory. Journal of Experimental Marine Biology and Ecology 27:47-60.

McIvor, C. C., and W. E. Odum. 1988. Food, predation risk and microhabitat selection in a marsh fish assemblage. Ecology 69:1341-1351.

McIvor, C. C., L. P. Rozas, and W. E. Odum. 1988. Use of marsh surface by fishes in tidal freshwater wetlands. Proceedings of the Symposium on Freshwater Wetlands and Wildlife, Savannah River Ecology Laboratory. U. S. Department of Energy, Office of Technical Information, Washington, District of Columbia.

Miller, J. M., and M. L. Dunn. 1980. Feeding strategies and patterns of movement in juvenile estuarine fishes. Pages 437-448 *in* V. S. Kennedy, editor. Estuarine perspectives. Academic Press, New York.

Murchand, J., and G. Masson. 1989. Process of estuarine colonization by 0-group sole (*Solea solea*): hydrological conditions, behavior, and feeding activity in the Vilaine estuary. Rapports et Procès-Verbaux des Réunions Conseil International pour l'Exploration de la Mer 191:287-295.

Nixon, S. W. 1980. Between coastal marshes and coastal waters — A review of twenty years of speculation and research on the role of salt marshes in estuarine productivity and water chemistry. Pages 437-525 *in* P. Hamilton and K. B. MacDonald, editors. Estuarine and wetland processes. Plenum, New York.

Odum, E. P. 1971. Fundamentals of ecology. W. B. Saunders, Philadelphia.

Olney, J. E., and G. W. Boehlert. 1988. Nearshore ichthyoplankton associated with seagrass beds in the lower Chesapeake Bay. Marine Ecology Progress Series 45:33-43.

Ortner, P. B., L. C. Hill, and S. R. Cummings. 1989. Zooplankton community structure and copepod species composition in the northern Gulf of Mexico. Continental Shelf Research 9:387-402.

Parrish, J. D. 1989. Fish communities of interacting shallow-water habitats in tropical oceanic regions. Marine Ecology Progress Series 58:143-160.

Peters, D. S., and F. A. Cross. 1992. What is coastal fish habitat? Pages 17-22 *in* R.H. Stroud, editor. Stemming the tide of coastal fish habitat loss. Proceedings of the Symposium on Conservation of Coastal Fish Habitat, Baltimore, Maryland. Marine Recreational Fisheries Number 14. National Coalition for Marine Conservation, Inc. Savannah, Georgia.

Powell, A. B., A. J. Chester, J. J. Govoni, and S. M. Warlen. 1990. Nutritional condition of spot larvae associated with the Mississippi River plume. Transactions of the American Fisheries Society 119:957-965.

Powles, H. 1981. Distribution and movements of neustonic young of estuarine dependent (*Mugil* spp., *Pomatomus saltatrix*) and estuarine independent (*Coryphaena* spp.) fishes off the southeastern United States. Rapports et Procès-Verbaux des Réunions, Commission Internationale pour l'Exploration Scientifique de la Mer 178:207-209.

Powles, H., and B. W. Stender. 1976. Observations on composition, seasonality and distribution of ichthyoplankton from MARMAP cruises in the South Atlantic Bight in 1973. South Carolina Marine Resources Center, Technical Report Series 11.

Robertson, A.I., and N. C. Duke. 1987. Mangroves

as nursery sites: comparisons of the abundances and species composition of fish and crustaceans in mangroves and other nearshore habitats in tropical Australia. Marine Biology 96:193-205.

Robertson, A. I., and N. C. Duke. 1990. Recruitment, growth and residence time of fishes in a tropical Australian mangrove system. Estuarine Coastal and Shelf Science 31:723-743.

Rogers, B. D., and W. H. Herke. 1985. Temporal patterns and size characteristics of migrating juvenile fishes and crustaceans in a Louisiana marsh. Louisiana Agricultural Experiment Station Research Report 5.

Rogers, S. G., T. E. Targett, and S. B. Van Sant. 1984. Fish-nursery use in Georgia salt-marsh estuaries: the influence of springtime freshwater conditions. Transactions of the American Fisheries Society 113:595-606.

Rountree, R. A., and K. W. Able. 1992. Fauna of polyhaline subtidal marsh creeks in southern New Jersey: composition, abundance and biomass. Estuaries 15:171-185.

Rozaz, L. P., and D. J. Reed. 1993. Nekton use of marsh-surface habitats in Louisiana deltaic salt marshes undergoing submergence. Marine Ecology Progress Series 96:147-157.

Ryder, R. A., and S. R. Karr. 1989. Environmental priorities: placing habitat in hierarchic perspective. Pages 2-12 in C. D. Levings, L. B. Holt and M. A. Henderson, editors. Proceedings of the National Workshop on Effects of Habitat Alteration on Salmonid Stocks. Canadian Special Publications on Fisheries and Aquatic Science 105.

Shaw, R. F., W. J. Wiseman, Jr., R. E. Turner, L. J. Rouse, Jr., R. E. Condrey, and F. J. Kelly, Jr. 1985. Transport of larval gulf menhaden *Brevoortia patronus* in continental shelf waters of western Louisiana: a hypothesis. Transactions of the American Fisheries Society 114:452-460.

Sogard, S. M. 1989. Colonization of artificial seagrass by fishes and decapod crustaceans: importance of proximity to natural eelgrass. Journal of Experimental Marine Biology and Ecology 133:15-37.

Sogard, S. M., and K. W. Able. 1991. A comparison of eelgrass, sea lettuce, macroalgae, and marsh creeks as habitats for epibenthic fishes and decapods. Estuarine, Coastal and Shelf Science 38:501-519.

Sogard, S. M., D. E. Hoss, and J. G. Govoni. 1987. Density and depth distribution of larval gulf menhaden, *Brevoortia patronus*, Atlantic croaker, *Micropogonias undulatus*, and spot, *Leiostomus xanthurus*, in the northern Gulf of Mexico. Fishery Bulletin, U. S. 85:601-609.

Stoner, A. W. 1980. Perception and choice of substratum by epifaunal amphipods associated with seagrasses. Marine Ecology Progress Series 3:105-111.

Stout, J. P. 1984. The ecology of irregularly flooded salt marshes of the northeastern Gulf of Mexico: a community profile. U. S. Fish and Wildlife Service, Biological Report 85 (7.1).

Subrahmanyam, C. B., and S. H. Drake. 1975. Studies on the animal communities in two north Florida salt marshes. I. Fish communities. Bulletin of Marine Science 25:445-465.

Talbot, C. W., and K. W. Able. 1984. Composition and distribution of larval fishes in New Jersey salt marshes. Estuaries 7:434-443.

Tanda, M. 1990. Studies of burying ability in sand and selection to the grain size for hatchery-reared marbled sole and Japanese flounder. Nihon Suisan Gakkai-shi 56:1543-1548.

Teal, J. M. 1986. The ecology of regularly flooded salt marshes of New England: a community profile. U. S. Fish and Wildlife Service, Biological Report 85 (7.4).

Thayer, G. W., D. R. Colby, and W. F. Hettler, Jr. 1987. Utilization of the red mangrove prop root habitat by fishes in south Florida. Marine Ecology Progress Series 35:25-38.

Thayer, G. W., W. J. Kenworthy, and M. S. Fonseca. 1984. The ecology of eelgrass meadows off the Atlantic coast: a community profile. U. S. Fish and Wildlife Service, Biological Services Program FWS/OBS-84/02.

Thayer, G. W., and P. F. Sheridan. In press. Fish and aquatic invertebrate use of the mangrove prop-root habitat in Florida: a review. Mangrove ecosystems in tropical America: structure, function and management. EPOMEX Serie Cientifica (Univ. Mexico).

Thayer, G. W., H. H. Stuart, W. J. Kenworthy, J. F. Ustach, and A. B. Hall. 1978. Habitat values of salt marshes, mangroves, and seagrasses for aquatic organisms. Pages 235-247 in Wetland functions and values: the state of our understanding. American Water Resources Association, Milwaukee, Wisconsin.

Thayer, G. W., R. R. Twilley, S. C. Snedaker, and P. F. Sheridan. In press. Research information needs on U. S. mangroves: recommendations to the United States National Oceanic and

Atmospheric Administration's Coastal Ocean Program from an Estuarine Habitat Program-funded Workshop. Mangrove ecosystems in tropical America: structure, function and management. EPOMEX Serie Cientifica (Univ. Mexico).

Warlen, S. M. 1982. Age and growth of larvae and spawning time of Atlantic croaker in North Carolina. Proceedings of the Annual Conference of the Southeastern Association of Fish and Wildlife Agencies 34:204-214.

Warlen, S. M. 1988. Age and growth of larval gulf menhaden, *Brevoortia patronus*, in the northern Gulf of Mexico. Fishery Bulletin, U. S. 86:77-90.

Warlen, S. M. 1992. Age, growth and size distribution of larval Atlantic menhaden, *Brevoortia tyrannus*, off North Carolina. Transactions of the American Fisheries Society.

Weisberg, S. B., and V. A. Lotrich. 1982. The importance of an infrequently flooded intertidal marsh surface as an energy source for the mummichog *Fundulus heteroclitus*: an experimental approach. Marine Biology 66:307-310.

Wiebe, P. H., K. H. Burt, S. H. Boyd, and A. W. Morton. 1976. A multiple opening/closing net and environmental sensing system for sampling zooplankton. Journal of Marine Research 34:313-326.

Wood, E. J. F., W. E. Odum, and J. C. Zieman. 1969. Influence of sea grasses on the productivity of coastal lagoons. Pages 495-502 *in* A. A. Castanares, editor. Coastal lagoons. Universidad Nacional Autonoma de Mexico, Cuidad Universitaria, Mexico City.

Zieman, J. C., and R. T. Zieman. 1989. The ecology of the seagrass meadows of the west coast of Florida: a community profile. U. S. Fish and Wildlife Service, Biological Report 85 (7.25).

Certain Decisions with Uncertain Data: Early Life-History Data and Resource Management

PENNY HOWELL

*Connecticut Department Environmental Protection,
Fisheries Division, Post Office Box 719, Old Lyme, Connecticut 06371, USA*

Abstract.—Resource managers are charged with the job of protecting natural populations from overuse by consumers and abuse by non-consumers. Consumption of the adult fish population is a fairly straightforward issue. However, management of pre-recruit members of the population is much more problematic, as management actions are not related to harvest and often involve users of the population's habitat, rather than users of the fish themselves. A current hypothesis which relates the mean size of a population to the size of its spawning area is discussed to clarify the role of early life-history characteristics in determining adult stock size and distribution. In order to superimpose habitat related mortality, or lack of production, onto other sources of mortality, populations of a given species must be bounded in space and the habitat area ranked in order of pollutant exposure. The spatial/temporal structure of the populations and the pollutant load become comparable trials on a relative scale of effects. An example of this approach, which compares winter flounder juvenile length frequencies in five areas, is discussed. The smaller lengths of fish taken from a heavily impacted site is translated into a higher estimated relative mortality.

Resource managers are charged with the job of protecting natural populations from overuse by consumers and abuse by non-consumers. This task is fairly straightforward when the manager deals with consumers; some measure of the population's production rate is matched with some measure of the consumer's consumption rate. If consumption is greater than production, even the most ardent fisherman knows some change must be made. Of course, what that change might be is not necessarily straightforward.

When managers move from effects on the adult population to those on pre-recruit stages, this uncertain situation gets even more tenuous. Managers must then make public policy concerning losses not related to harvest, often necessitating regulation of the population's habitat rather than of the fish themselves. Without the harvest in hand to measure, the first problem is to determine which population, or subpopulation, might be affected; the second problem is to determine what the effect might be.

The effects of habitat modification on local fish stocks can be indirect and gradual. The difficulty in quantifying chronic decline in production or subtle mortality effects has kept discussion of habitat effects out of the realm of public law making. However, in recent decades, the ever greater loss and alteration of nearshore marine habitat, primarily due to human activities, has made assessment of marine habitat loss more compelling. Three quarters of all estuaries nationwide, prime nursery areas for many marine species, were classified as moderately or severely degraded in the National Estuary Study completed in 1970 (Gusey 1978, 1981). Degraded water quality most often occurs from contaminant discharges, industrial coolant water removals, and loss of nearshore shallow habitat due to filling, construction of bulkheads, and sediment dredging. Off the Atlantic coast of the United States, loss of shallow water acreage averaged 4% by state from 1954 to 1968; however, the average loss along the coast from Massachusetts to New Jersey was 6 to 15% (Gusey 1978, 1981). Except for a few limited restoration projects, this loss, measured decades ago, has been permanent.

Habitat effects also can be additive and interactive (Odum 1970) and therefore not clearly related to the resulting size of the stock recruited from the area. However, if the life history of a species is considered in light of its habitat requirements, the relationship between population size and habitat size becomes clearer. A current hypothesis (Sinclair 1988) relates a species' population richness, or number of populations, to the number of geographic settings, referred to as retention areas, within which the species' life cycle is capable of closure. The hypothesis suggests that the absolute or mean

TABLE 1.—Estimates of winter flounder population size and associated habitat area in acres. Population size was estimated by tagging (1) or exploitation (2) methods.

Stock location (reference)	Method	Habitat area (acres)	Estimated population size
Georges Bank (NMFS 1989, Lux 1973)	2	11,580,366	11,954,248
St. Marys Bay, Nova Scotia (Dickie and McCracken 1955)	2	134,400	1,821,039
Narragansett Bay, Rhode Island (Gibson 1989, 1990)	1	84,063	6,385,057
Great South Bay, New York (Lobell 1939, Briggs 1965, Poole 1966)	2	81,920	2,458,171
Barnegat Bay, New Jersey (P. Scarlett personal communication, Gibson 1991)	2	72,593	4,013,937
Peconic Bay, New York (Lobell 1939, Briggs 1965, Poole 1966)	2	14,515	285,300
Moriches Bay, New York (Lobell 1939, Briggs 1965, Poole 1966)	2	8,687	1,513,489
Point Judith Pond, Rhode Island (Grove 1982)	1	1,576	38,700
Ninigret Pond, Rhode Island (Saila 1961b)	1	1,560	137,800
Ninigret Pond, Rhode Island (Worobec 1982)	1	1,560	38,000
Waquoit Bay, Massachusetts (Howe et al. 1976)	1	1,211	51,845
Niantic River, Connecticut (NUSCO 1990)	1	721	62,998
Green Hill Pond, Rhode Island (Saila 1961a)	1	475	11,320
Upper Narragansett Bay, Rhode Island (Gibson 1989, 1990)	1	336	21,936

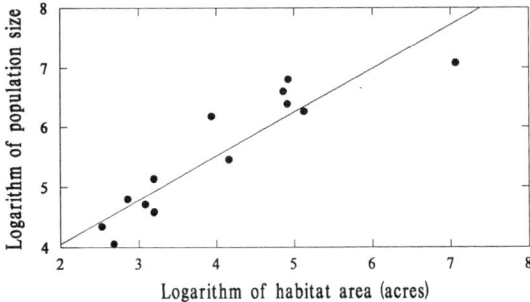

FIGURE 1.—Winter flounder stock size versus habitat area.

size of each population is scaled to the size of these retention areas. Geographic areas whose physical features lead to larva retention, thereby enhancing spawning success and stock cohesion, have been identified for several species from Atlantic herring *(Clupea harengus)* to American eel *(Anguilla rostrata)* (Sinclair 1988).

Winter flounder *(Pleuronectes americanus)* is a species with numerous genetically identifiable populations (Kendall 1912; Perlmutter 1947; Bigelow and Schroeder 1953; Lux et al. 1970; Pierce and Howe 1977; Schenck and Saila 1982). Individual estuaries and offshore banks provide winter spawning grounds for an assemblage of adjacent spawning groups. One can construct a relationship between flounder population size and habitat area using estimates for winter flounder populations from the published literature (Gibson 1991, Table 1). Production area is defined as the inshore embayment, estuary, salt pond, or offshore bank where spawning aggregations occur. Standing stock numbers are either direct adult population estimates from tagging studies on spawning grounds, or indirect adult estimates from known catch divided by estimated exploitation rate for that stock. The logarithm of population size is significantly correlated with the logarithm of habitat area for twelve flounder populations (Figure 1). Although the level of precision of this relationship is only sufficient to make estimates to an order of magnitude, it clearly shows a consistent pattern in population density across many winter flounder habitats. Note that one population, in Ninigret Pond, Rhode Island, is represented twice: its 1,560 acres supported 138,000 winter flounder in 1961, but only 38,000 in 1982, an effect of increased fishing mortality and possibly habitat degradation.

Large flounder populations are associated with large physical structures which promote larva retention, such as bays and offshore banks. Small populations are associated with structures such as coastal ponds. It logically follows that degradation or loss of habitat will reduce successful larva recruitment and ultimate standing stock size. An interesting point is that the relationship between population and area is not constant from small to large population sizes, reflecting the fact that production per unit area decreases with total area. This functional form suggests that changes in habitat area will be most influential over stock size in small populations. The relationship also suggests that management agencies need to be as concerned about the production rate in small areas as in the largest areas.

Given that a population can be identified by the boundaries of its larva-retention area, the next step is to assess the factors in that area that have an effect on larva production. In order to superimpose habitat mortality, or lack of production, onto other sources of mortality, a manager must obtain relative production estimates from habitat areas of varying quality. One approach to this problem is to generate a relative scale of habitat quality by stylizing pollutant effects which occur in a patchy distribution through space and time. The occurrence, or exposure, of a given population can then be similarly bounded in space and time. Finally, the spatial/temporal structure of the populations and the pollutant become comparable trials on a relative scale of effects.

This approach is particularly promising with species that spawn in discrete locations and form discrete populations. Many freshwater and anadromous species are common examples. Marine species are much more problematic because of complex mixing of pollutants and populations within an area, as well as the difficulty in delineating an area as a separate system. However, by examining larval and young-of-year stages of species such as winter flounder, where movements are primarily limited to single nearshore embayments, the problems of bounding the population and the pollutant effect are somewhat simplified.

An Example

As an example of this kind of approach, data from beam trawl catches of winter flounder young of year (YOY) in five embayments along the Connecticut coastline were examined. Of the five embayments, one, New Haven Harbor is the most highly impacted by repeated small scale oil spills and industrial and sewage discharges (Normandeau 1979). It is the busiest commercial harbor in Connecticut, and the third busiest in New England (CT DEP 1992). The other four embayments have impacts from smaller industrial and sewage discharges only. A study of the distribution of benthic invertebrates in New Haven Harbor (Pellegrino 1987) showed that the entire harbor is dominated by few opportunistic and hardy species which reflect deteriorated water quality (Normandeau 1979). The inner harbor had especially low densities and diversity which these authors attributed to high temperatures and low dissolved oxygen, a consequence of sewage eutrophication. Analysis of New Haven Harbor sediments showed widespread contamination by gasoline and No. 2 and No. 6 fuel oil, not seen in sediments taken from eastern harbors (Hoehn and Morris 1977). These data suggest that in New Haven Harbor fish such as juvenile winter flounder are impacted by the combined effects of chronic oil pollution and sewage impacts not seen in the other sites.

Chronic toxicity of petroleum hydrocarbons has been expressed in a variety of degenerative changes in winter flounder exposed to oil in the laboratory. These include reduced food intake and lower condition factor (Fletcher et al. 1981; Khan 1991), altered swimming behavior (Khan 1987), lower juvenile growth rate (Fletcher et al. 1981), and fin and tail necrosis (Khan 1987, 1991). Direct effects on winter flounder reproductive success have been demonstrated in the laboratory (Kuhnhold et al. 1979). Exposure of eggs to low levels (100 ppb) of No. 2 fuel oil significantly reduced viable hatch (19%). Exposure during fertilization and embryonic development delayed hatching 3 to 9 days, a subtle effect that would potentially reduce survival in the wild due to increased predation. Progeny of adults exposed to 10 or 100 ppb fuel oil during gametogenesis showed reduced larva survival and growth. A study of larvae taken from New Haven showed higher early embryo mortality compared to other Long Island Sound sites considered relatively clean (NOAA 1990). Growth retardation was also seen when winter flounder juveniles were chronically exposed to hypoxic conditions in the laboratory (Bejda et al. 1992).

These laboratory data collectively point to pollutant effects which should have significant deleterious effects on an exposed population. However, defensible management policies require that the gap between laboratory results and management recommendations be bridged with site-specific field studies.

Methods

Each embayment was bounded landward by an average spring bottom salinity of 5‰ and seaward by the 6.5 m (18 ft) depth contour. The salinity boundary was a lower lethal limit for winter flounder embryos (Rogers 1976). The depth boundary corresponded generally to the limit of inshore protected habitat. The embayments were divided into 0.5-km grids which were sampled monthly June through September, 1990, and seasonally May-June, July-August, September-October, 1991. For each month or season, a third to a half of the grids were sampled, alternating odd and even grid numbers, using a 1-m beam trawl with 6.4-mm (0.25-inch) bar mesh net towed for 5 min. All juvenile flounder were counted and measured (total length, mm). Catch per tow by 15-mm length intervals was calculated for each sampling period and embayment. The 15-mm length interval approximated the average increase in mean length for all samples from July through October. With the assumption

TABLE 2.—Percentage of young-of-year winter flounder sampled by length interval, 1991. Sites are listed west to east.

Site	Length (mm)					
	20-39	40-59	60-79	80-99	>100	N
SPRING						
Housatonic River	23	65	12	0	0	78
New Haven Harbor	97	3	0	0	0	155
Clinton Harbor	39	53	8	0	0	66
Connecticut River	28	54	19	0	0	208
Thames River	39	59	2	0	0	110
SUMMER						
Housatonic River	0	30	60	10	0	39
New Haven Harbor	27	46	25	1	0	913
Clinton Harbor	7	40	33	20	0	15
Connecticut River	2	29	44	21	4	168
Thames River	1	29	65	5	0	129
FALL						
Housatonic River	0	11	56	33	0	9
New Haven Harbor	0	13	67	33	0	24
Clinton Harbor	0	30	50	10	10	10
Connecticut River	0	5	19	68	8	37
Thames River	0	14	55	27	4	196

TABLE 3.—Percentage of young-of-year winter flounder sampled by length interval, 1990. Sites are listed west to east.

Site	Length (mm)					
	20-39	40-59	60-79	80-99	>100	N
JUNE						
New Haven Harbor	73	28	0	0	0	157
Clinton Harbor	42	45	13	0	0	102
Connecticut River	40	53	6	0	0	47
JULY						
Housatonic River	0	27	66	5	0	41
New Haven Harbor	32	43	25	1	0	162
Clinton Harbor	7	40	31	0	0	29
Connecticut River	14	37	42	6	0	133
Thames River	13	43	39	4	0	67
AUGUST						
Housatonic River	0	27	64	9	0	11
New Haven Harbor	16	46	28	10	0	123
Clinton Harbor	0	25	50	20	5	20
Connecticut River	9	36	48	6	0	33
Thames River	3	27	59	10	1	157
SEPTEMBER						
Housatonic River	10	40	40	10	0	10
New Haven Harbor	10	31	46	12	1	298
Clinton Harbor	4	48	37	11	0	27
Connecticut River	0	17	26	43	13	23
Thames River	7	48	36	7	2	128

that immigration and emigration rates are not related to length, and that length increases linearly with age over this limited size interval, length becomes a surrogate measure of age. Catch-per-tow by length interval for July–October samples from both years were summed for each site and the natural log of abundance was regressed against length intervals from 50 mm to 118 mm. Spring samples were omitted to minimize bias from juveniles newly recruiting to the gear at 20 mm. The slope of abundance over length (age) is a relative measure of mortality over the 4-month time period in the same way as catch curve analysis of abundance by year is for a year class.

Results and Discussion

The resulting comparison of young-of-year flounder length frequencies showed clear differences among the five embayments in 1990 and 1991 (Tables 2 and 3, $P < 0.03$ for chi-square tests by season for 1991). In all sites except New Haven, the majority of age-0 winter flounder sampled in the spring were larger than 40 mm in length. However in New Haven, 97% of the flounder taken in late May 1991 were under 40 mm in length. In 1990, 73% of the flounder taken in June sampling of this harbor were below 40 mm. In summer (July) samples, a third of New Haven flounder were still below 40 mm in both years, while at all other sites less than 15% were below 40 mm in length both years. However, by the fall (September–October, 1991), the majority of flounder at all sites were larger than 60 mm in length.

One explanation for the consistency in fall length frequencies is a substantially higher mortality rate for juveniles not reaching a large enough size to avoid predation. Smaller fish are not only exposed to a wider array of predators, but are also more limited by the number of food sources they can access (Anderson 1988). The combination of these factors would tend to homogenize the size range of surviving juveniles by selecting against the smallest fish. This explanation would support the conclusion that the New Haven population experiences a higher mortality rate due to impaired growth and/or late spawning. Quantification of relative mortality which could be attributed to the smaller size of juveniles taken from New Haven also supports this conclusion. Estimates based on regression of catch-per-tow by length interval resulted in the highest value for the New Haven population (Table 4). A comparable value was obtained for the Housatonic River population; however, data from this site gave a weak correlation ($R^2 = 0.68$, $P = 0.11$), casting doubt on the reliability of the mortality rate estimated from the slope. All other sites showed strong linear relationships ($R^2 > 0.85$, $P < 0.05$).

Although density-dependent growth effects (for example, an inverse relationship between growth and abundance) are probably occurring to some degree, the size frequency seen in the New Haven population is considerably smaller than the Thames River population whose relative abundance (catch-

TABLE 4.—Regression of the natural log of abundance on 15-mm length interval for winter flounder juveniles taken in five embayments from July to October, 1990 to 1991. Sites are listed west to east. The slope of the regression, -Z, is the relative mortality rate, while $e^{-z} = S$ or relative survival rate, over the 4-month time period. The standard error (SE), coefficient of determination (R^2), and probability (P) of a slope not equal to zero are also given for each regression.

Site	Z	SE	R^2	P	S
Housatonic River ($N = 110$, 74 tows)	1.13	0.42	0.11	0.32	0.32
New Haven Harbor ($N = 1,520$, 215 tows)	1.21	0.25	0.88	0.04	0.30
Clinton Harbor ($N = 101$, 63 tows)	0.66	0.05	0.98	0.01	0.52
Connecticut River ($N = 394$, 142 tows)	0.55	0.07	0.95	0.02	0.58
Thames River ($N = 677$, 85 tows)	0.91	0.21	0.86	0.05	0.40

per-tow) is generally much higher. Another possible cause for the smaller size frequency is emigration from the system. However when fish captured in the inner landward half of each gridded area were compared with fish captured in the outer half, there were no differences in the mean or range of lengths at all sites. This overlap in length frequency supports the assumption of minimal movement out of the sampled areas by larger size classes. Circumstantial evidence points to an impaired growth rate caused by the pollutant load of the harbor for the following reasons: (1) the effect was not seen in other sites; (2) it mimics the effects seen in the laboratory with winter flounder exposed to pollutants known to be in the harbor; and (3) no other parsimonious explanation would account for this harbor's smaller fish size in comparison to four other neighboring sites. When the estimates of relative mortality are translated into survival, the New Haven fish show substantially lower survival compared to the eastern sites, suggesting pollutant loadings do have an effect on growth and survival of winter flounder.

Conclusions

Results of this analysis suggest that fisheries managers could use this method to make management decisions concerning the impact of harbor pollutants on its juvenile flounder population. In order to fully equate these juvenile losses to the population as a whole, a measure of relative size for the adult spawning stock is required. Resource managers would then be in a position to weigh the cost of a mitigation program against the cost of yield foregone to the adult stock and the fishery it supports. Usually, as in this case, a complete census of all populations over an extended time period is not available. In lieu of such irrefutable data, management decisions must be made with rough estimates of these numbers. A multi-year tagging study of age-0 recruits was initiated in 1991 to begin the process of estimating the degree of movement within the harbor and, if movement is limited, the size of the New Haven harbor population exposed to harbor pollutants. With the completion of that work, estimated losses due to fishing and non-fishing mortality factors can be compared by means of an early life-history model being developed by a joint effort of several regulatory agencies (Ingham 1992). In the end, decisions based on this model will be only as reliable and as acceptable to the regulatory community as the assumptions inherent in the process of generating the model. And so enters the cloud of uncertainty managers must accept when working with habitat issues.

Acknowledgments

Much of the substance of this paper grew out of the debate and discussion that went into the writing of the Atlantic States Marine Fisheries Commission Interstate Management Plan for Inshore Winter Flounder Stocks, for which the Technical Committee and Management Board provided great insight into effective fisheries management. The development of the relationship between habitat area and population size for this species was one of many contributions by Mark Gibson, of the Rhode Island Department of Environmental Management. I would also like to thank Victor Crecco, of the Connecticut Department of Environmental Protection, for his guidance and helpful review of the New Haven Harbor data.

References

Anderson, J. 1988. A review of size dependent survival during pre-recruit stages in relation to recruitment. Journal Northwest Atlantic Fisheries Science 8:55-68.

Bejda, A., B. Phelan, and A. Studholme. 1992. The effect of dissolved oxygen on the growth of young of the year winter flounder *(Pseudopleuronectes americanus)*. Environmental Biology of Fishes 34:321-327.

Bigelow, H., and W. Schroeder. 1953. Fishes of the Gulf of Maine. U. S. Fish and Wildlife Service. Fishery Bulletin, U. S. 53:1-577.

Briggs, P. 1965. The sport fisheries for winter flounder in several bays of Long Island. N.Y. Fish and Game Journal 12:48-70.

CT DEP (Connecticut Department of Environmental Protection), 1992. Quinnipiac River Fact Sheet, Bureau of Water Management, 2 pp.

Dickie, L., and F. McCracken. 1955. Isopleth diagrams to predict equilibrium yields of a small flounder fishery. Journal of Fisheries Research Board of Canada 12:187-209.

Fletcher, G. Kiceniuk, and U. Williams. 1981. Effects of oiled sediments on mortality, feeding and growth of winter flounder, *Pseudopleuronectes americanus*. Marine Ecology Progress Series 4:91-96.

Gibson, M. 1989. Mortality estimates for Rhode Island winter flounder. Rhode Island Division of Fish and Wildlife Resources Reference Document 89/7.

Gibson, M. 1990. Updated mortality estimates for Rhode Island winter flounder including a retrospective analysis using a relative exploitation approach. Rhode Island Division of Fish and Wildlife Resources Reference Document 90/4.

Gibson, M. 1991. A study of winter flounder population size and production areas along the Atlantic coast: Towards a means for estimating effects of habitat loss. Rhode Island Division of Fish and Wildlife Resources Reference Document 91/2.

Grove, C. 1982. Population biology of the winter flounder, *Pseudopleuronectes americanus* in a New England estuary. M.S. Thesis. University of Rhode Island, Narragansett.

Gusey, W. 1978. The fish and wildlife resources of the Middle Atlantic Bight. Shell Oil Company, Houston, Texas.

Gusey, W. 1981. The fish and wildlife resources of the South Atlantic coast. Shell Oil Company, Houston, Texas 5-23 — 5-25.

Hoehn, T., and J. Morris. 1977. Species abundance, composition, and diversity of marine benthic invertebrates of Connecticut with special consideration for the New Haven oil spill. Connecticut Department of Environmental Protection, Marine Region completion report.

Howe, A., P. Coates, and D. Pierce. 1976. Winter flounder estuarine year-class abundance, mortality, and recruitment. Transactions of the American Fisheries Society 105:647-657.

Ingham, M. 1992. Modelling environmental effects on winter flounder populations. Proceedings of the northeast environmental assessment workshop 2, Northeast Environmental Council, National Marine Fisheries Service, Northeast Region.

Kendall, W. 1912. Notes on a new species of flatfish from off the coast of New England. Bulletin of the U. S. Bureau of Fisheries 30(1910):391-394.

Khan, R. 1987. Effects of chronic exposure to petroleum hydrocarbons on two species of marine fish infected with a hemoprotozoan, *Trypanosoma murmanensis*. Canadian Journal of Zoology 65:2703-2709.

Khan, R. 1991. Influence of concurrent exposure to crude oil and infection with *Trypanosoma murmanensis* (Protozoa: Mastigophora) on mortality in winter flounder, *Pseudopleuronectes americanus*. Canadian Journal of Zoology 69:876-880.

Kuhnhold, W., D. Everich, J. Stegeman, J. Lake, and R. Wolke. 1979. Effects of low levels of hydrocarbons on embryonic, larval and adult winter flounder, *(Pseudopleuronectes americanus)*. Assessment of Ecological Impact of Oil Spills. American Institute of Biological Science 677-711.

Lobell, M. 1939. A biological survey of the salt waters of Long Island, 1938. Report on certain fishes. Winter flounder *(Pseudopleuronectes americanus)*. Supplement 28th Annual Report, New York Conservation Department, Part I: 63-96.

Lux, F. 1973. Age and growth of the winter flounder, *Pseudopleuronectes americanus*, on Georges Bank. Fishery Bulletin, U. S. 71:505-512.

Lux, F., A. Peterson Jr., and R. Hutton. 1970. Geographic variation in fin ray number in winter flounder, *Pseudopleuronectes americanus* (Walbaum), off Massachusetts. Transactions of the American Fisheries Society 99:483-488.

NMFS (National Marine Fisheries Service). 1989. Chart and Statistical Book of the U. S. Northeast Fisheries. Gloucester, Massachusetts: Analytical Services Branch, Research Document NER88.1.

NOAA (National Oceanic and Atmospheric Admininistration). 1990. Final report on a three-year assessment of reproductive success in winter flounder, *Pseudopleuronectes americanus* (Walbaum), in Long Island Sound, with comparisons to Boston Harbor. National Marine Fisheries Service. Milford Laboratory. 112 pp.

Normandeau Associates. 1979. New Haven ecological studies summary report. A. McCuster and W. Bosworth, editors.

NUSCO (Northeast Utilities Service Company), 1990. Monitoring the marine environment of Long Island Sound at Millstone Power Station, Waterford, Connecticut. Annual report, 1990.

Odum, W. 1970. Insidious alteration of the estuarine environment. Transactions of the American Fisheries Society 99:836-847.

Pellegrino, P. 1987. The distribution and abundance of macrobenthos in New Haven Harbor: 1986 (Final Report) submitted to United Illuminating, New Haven, Connecticut.

Perlmutter, A. 1947. The blackback flounder and its fishery in New England and New York. Bulletin of the Bingham Oceanographic Collection Yale University 11(2):1-92.

Pierce, D., and A. Howe. 1977. A further study on winter flounder group identification off Massachusetts. Transactions of the American Fisheries Society 106:131-139.

Poole, J. 1966. Growth and age of winter flounder in four bays of Long Island. New York Fish and Game Journal 13:206-220.

Rogers, C. 1976. Effect of temperature and salinity on the survival of winter flounder embryos. Fishery Bulletin, U. S. 74:52-58.

Saila, S. 1961a. A study of winter flounder movements. Limnology Oceanography 6:292-298.

Saila, S. 1961b. The contribution of estuaries to the offshore winter flounder fishery in Rhode Island. Proceedings of the Gulf and Caribbean Fisheries Institute 14th Annual Session 1961:95-109.

Schenck, R., and S. Saila. 1982. Population identification by biochemical methods with special reference to the winter flounder *Pseudopleuronectes americanus* in the vicinity of Millstone Point, Connecticut. A final report to the Northeast Utilities Service Co. Contract PO 001122.

Sinclair, M. 1988. Marine populations: an essay on population regulation and speciation. Washington Sea Grant. University of Washington Press, Seattle.

Worobec, M. N. 1982. Field analysis of winter flounder *(Pseudopleuronectes americanus)* in a coastal salt pond: abundance, daily ration, and annual consumption. Ph.D. Thesis. University of Rhode Island, Narragansett.

The Potential Role of Larval Fish Culture in Alleviating Population and Habitat Losses

G. JOAN HOLT

The University of Texas at Austin, Marine Science Institute
Post Office Box 1267, Port Aransas, Texas 78373-1267, USA

Abstract.—Water-quality changes and habitat degradation are often detrimental to survival of early life-history stages of fish. Changes in chemical characteristics, such as salinity and dissolved oxygen, and loss of specific habitats required as nursery grounds have been implicated in population declines. The stocking of hatchery-reared young has been proposed as the solution to declining populations and fishery yields. Yet, there is much debate over the success of various enhancement programs in increasing or even maintaining natural stocks due to genetic changes in the target populations and loss of carrying capacity. Progress in larviculture is important for evaluating stock enhancement. Aquaculture production of food and ornamental fishes could replace natural production to some degree, thereby reducing fishing pressure on over-harvested species. Culture of threatened or endangered species to provide critical life-history information and to establish gene banks, as well as the production of test organisms for water-quality and habitat evaluation, depend on progress in larviculture. Further advances are needed in species-specific tagging and marking procedures, methods to identify factors controlling population dynamics, techniques for breeding and rearing for many species, and aquaculture practices such as closed recirculating culture and production of sterile fish. Although increased understanding of biological and ecological components and technological improvement will enhance aquaculture production, a long-term strategy of management and conservation of natural resources is needed.

There is no doubt that fish populations are declining in response to overfishing, habitat degradation, and water-quality changes. The particular vulnerability of early life-history stages to these changes is well documented in other manuscripts in this volume. This paper is a discussion of the role larviculture can fill in easing or diminishing production losses resulting from habitat degradation. Areas where advances in larval fish rearing can contribute are: (1) stock enhancement to augment or mitigate natural stocks; (2) aquaculture of food and ornamental fish to reduce exploitation of natural populations; (3) lab culture of threatened or endangered species as live gene banks; and (4) culture of test organisms for monitoring habitat quality.

Stocking and Enhancement

Fish are stocked in public waters to augment recreational or commercially important fish stocks whose populations have declined due to over-exploitation, habitat loss or degradation, or a combination of the two. In the U.S., state and federal hatcheries release more than 40 million pounds of fish per year (NRC 1992) with the largest effort on Pacific and Atlantic salmon. Historically, this practice was initiated by public hatcheries constructed to mitigate damage from dam construction or other impacts on salmonid populations.

Although artificial propagation efforts with salmonids have been going on for decades, the abundance of most species in recent years has declined in the face of increased fishing pressure, destruction of spawning habitat, and blockage of migratory routes (Waples et al. 1990). Atlantic salmon populations are so reduced and aquaculture and enhancement so well developed that in some places the number of released or escaped hatchery fish is approaching the number of naturally produced salmon (Saunders 1991).

Recent publications have questioned mass releases of hatchery fish as the panacea for salmonid resources and, in fact, have implicated them in the decline of native, wild fish populations (Waples et al. 1990; Hindar et al. 1991; Hutchings 1991; Nehlson et al. 1991; Hilborn 1992). The controversy over the impact of hatchery-reared salmonids on natural populations, the degree to which the fitness of both are affected, and the connection, if any, of enhancement to declining numbers of wild fish can be resolved by monitoring the genetic effect of hatchery releases (Hindar et al. 1991). Genetic markers will be required to identify hatchery fish and resources will be needed for the long-term research needed to answer these questions (Egidius et al. 1991, Saunders 1991).

Marine fish enhancement has been carried out for over a century but with little evidence of beneficial effects (Shelbourne 1964) until released fish were raised to the juvenile stage. Japan releases large numbers of marine fish and shellfish into coastal areas to augment fisheries production (Mito and Fukuhura 1988). Hatchery-reared red sea bream *(Pagrus major)*, a major species for sea ranching in Japan since 1974, have increased in commercial and recreational catches but Kitada (1988) demonstrated that the contribution of hatchery fish is considerable less than estimated by yield-per-recruitment models. He attributed the difference to the survey methods and to high mortality of newly released fish. Norwegian scientist are evaluating the release of juvenile cod as an effective way of enhancing natural populations, by carrying out field experiments to study possible biological, genetic, and behavioral differences between reared and wild fish (Svåsand et al. 1990). Since cod are the basis of an important commercial fishery, tag returns are easily available to evaluate stocking success. Red drum juveniles have been stocked in Texas bays since 1983 to increase populations for the recreational fishery (Gregg and Rutledge 1990). Definitive data are lacking to measure stocking success since released fish cannot be easily tagged. Although attempts to use external tags on the small red drum juveniles (25 mm) have not been successful, current research efforts are focused on the use of other techniques, including genetic marking, to evaluate the impact of hatchery-released fish (Rutledge 1989).

Problems with enhancement can arise from ecological, genetic, or behavioral incompatibilities. Often when fish populations are in trouble, the first response is to replace declining stocks with hatchery-reared young, without a careful examination of the basis for the decline. Or if the causal agent is known, e.g., removal of essential habitat, there may be little attention given to additional environmental quality changes that might intensify the loss. Changes in environmental quality or habitat loss and reduced carrying capacity can prevent successful enhancement attempts.

There is a need for extensive evaluation of past and present stock- enhancement programs with a thorough evaluation of each species, including genetic characterization and habitat evaluation. Inbreeding from using insufficient numbers of broodstock or the use of inappropriate donor stock causes loss of genetic diversity (Waples et al. 1990) and stocking out of season or in new habitats or sites may impact other species through competition or predation.

The stocking decision ultimately depends on knowledge of factors controlling population dynamics and the cost of producing the fish to be stocked. Mathematical models to evaluate the effectiveness of stocking are most sensitive to mortality rate and the ratio of mortality to growth of the stocked fish (Botsford and Hobbs 1984; Polovina 1990; Kitada 1988), parameters for which there are often few data. Genetic marking coupled with full scale hatchery production and stocking may be necessary to determine the effectiveness of an enhancement program (MacCall 1989; Rutledge 1989). The expense of such a project could well exceed the value of the potential increase in fish production.

In summary, to determine whether a fish stocking program is working or should be implemented the following information is needed: (1) data on the mechanisms by which the population is limited, e.g., restricted resources for pelagic larvae, loss of spawning habitat, or overfishing; (2) genetic characterization to determine broodstock selection (number and locality) and heritable markers for monitoring impact on natural populations; (3) biological data concerning requirements for spawning and rearing larvae, and size or stage to release; (4) habitat evaluation for ecological impacts on other species and other life stages, including seasonal timing of releases; and (5) field experiments to test whether large scale releases would be effective.

Aquaculture for Food

Aquaculture not only has the potential to make up for losses from habitat deterioration and recruitment over-fishing but will have to play an ever increasing role in supplying the world's protein requirements. Today aquaculture makes up about 12% of the global fishery production and is predicted to double by the year 2000 and increase to 40% by 2025 when the global population increases to a predicted 8 billion (New 1991). It is generally agreed that the harvest of wild fish stocks is near the maximum and will probably not exceed 100 million tonnes. To maintain the current global consumption of fish and seafood (19.1 kg/person/year) will require an increase in fishery production to over 162 million tonnes by 2025, requiring a quadrupling of aquaculture production in a little over 30 years (New 1991).

Freshwater culture of fish (mainly catfish) and crustaceans (crayfish) represents the fastest growing aquaculture industry in the United States, while marine aquaculture has lagged behind the advancements made in other countries (NRC 1992). Major constraints to marine aquaculture development are the high costs and difficulties of using coastal space, conflicts with other users of the coastal zone, concerns about environmental impacts, and regulatory issues (NRC 1992). Many of these limits could be overcome by the development of technology for high density, water reuse systems for year-round production of marine fish. A major limitation to rapid development in this area is the high cost and unreliable production of fingerlings. Cost reductions would follow from improvements in larva survival and reductions in live food requirements.

Marine fish in general produce large numbers of small eggs with limited nourishment (yolk) and no parental care. Newly hatched larvae are small with rudimentary sensory and motor development. First feeding is critical in that the timing for successful feeding is short, and large numbers (500 to 4,000/L) of small live zooplankton prey are required (Hunter 1981). A major breakthrough in rearing marine larvae occurred in the 1960s when the rotifer *(Brachionus plicatilis)* was identified as suitable food for larvae and mass culture techniques were developed (Nagata and Hirata 1986). Rotifers and/or brine shrimp *(Artemia salina)* are used throughout the world to successfully culture marine larvae, e.g., yellowtail *(Seriola quinqueradiata)*, Japanese flounder *(Paralichthys olivaceus)*, seabass *(Dicentrachus labrax)*, sea bream *(Sparus aurata)*, red drum *(Sciaenops ocellatus)*, striped bass *(Morone saxatilis)*, mahi mahi (dolphin, *Coryphaena hippurus*), and turbot *(Scophthalmus maximus)* (Nagata and Hirata 1986, Sorgeloos et al. 1988).

Although most of the biological concerns about culturing rotifers have been addressed, large scale production is often unpredictable, expensive, and of variable nutritional value. Depending on growing conditions, rotifers and brine shrimp may be missing essential nutrients required for development and growth of larval fish and shellfish (reviewed by Watanabe et al. 1983). High mortalities often occur when larvae are shifted from live prey to inert diets. Many species cannot be weaned until a functional stomach has developed (Person LeRuyet et al. 1993). The production of mass quantities of high quality live prey with increases in size of prey to accommodate the energy needs of growing larvae limits juvenile production in many species. The production of a water-stable, artificial diet that could be digested and assimilated by larvae would be a major breakthrough (Jones et al. 1991)

Larvae of warm water species such as red drum will readily feed on small (< 250 µm), inert particles maintained in the water column by gentle aeration but such non-living diets will not support growth (Holt 1992). The larval digestive system is simple at first feeding, divided into a short foregut, an expanded midgut, and a hindgut. A gall bladder and pancreas are present. Low levels of protease and lipase activity are present in first feeding red drum larvae and increase with age (Holt 1992; Holt and Sun 1991). The inability of first feeding larvae to derive sufficient nutrition from artificial diets may be due to a lack of exogenous enzymes normally supplied by live prey. Jones et al. (1991) noted there is strong evidence to support the idea that prey autolysis is necessary for digestion in carnivorous larvae. Red drum can be grown on a combination of live rotifers and dry compounded diet for 5 days and then strictly weaned to the dry diet on day 8 posthatching (26°C) (Holt 1993). Besides acquiring excellent growth and survival (equivalent to or better than controls fed live prey), early weaning eliminates the brine shrimp feeding stage and reduces cannibalism associated with later weaning. This achievement has led to the testing of specific nutritional requirements of larvae, and is an important step in designing an artificial diet to completely replace the need for live zooplankton cultures.

There are definite problems with marine aquaculture which would impact fishery resources that need to be considered in planning for increased production, including pollution through organic and toxic waste discharge, build-up of suspended solids, reduction in oxygen levels, introduction of pathogens and exotic species, and competitive use of resources. All of these environmental impacts have occurred as a result of intensifying aquaculture production. Many countries are now attempting to rectify impacts by implementing water treatment and water reuse and by moving production away from coastal waters (Eng et al. 1989; Seymour and Bengheim 1991). Onshore closed or semi-closed intensive culture systems have been designed to grow fish in geographic areas where they cannot survive otherwise (e.g., European eel in Denmark and Japan), and to extend seasonal availability of cultured species (sea bream in France and red drum in U. S.). European marine hatcheries for sea bass and sea bream are intensive (100 larvae/L) and have controlled environmental conditions; last year 20 hatcheries produced more than 65 million fry (P. Sorgeloos, keynote address, World Aquaculture Society 1992). The development of very advanced recirculation systems allow northern European farmers to grow eels under severe climatic conditions while preserving the environment by greatly reducing the effluents (Gousset 1990).

Offshore production, such as large, moored net pens off the Irish coast for salmon cultures, and semi-submersible fish farming platforms off the coast of Spain, can help reduce environmental problems typical of inshore sites. Ocean ranching may make important contributions to natural fishery stocks if the concerns raised in the stocking section of this paper are resolved. Finally, problems with potential escape of exotic species into the marine environment can be avoided by basing aquaculture on native species. Studies are necessary to identify native species that will be practical for aquaculture in the region. The task can be simplified by evaluating species closely related to those in culture elsewhere, i.e. native drums, sea basses, penaeids. A commercial aquaculture policy adopted by the Executive Committee of the American Fisheries Society advocates such principals with the aim of protecting both the viability of the aquaculture industry and the integrity of native aquatic communities (Robinette et al. 1991).

Aquaculture of Non-food Fishes

Advances in larviculture can make a major contribution to populations of threatened, endangered, or over-harvested ornamental species and to expansion of the quantity and quality of species used for monitoring and testing. Larvae of many species of concern are poorly known and progress will depend on successful application of current culture practices and the development of new techniques. Valuable early life-history information can be obtained from cultured eggs and larvae, e.g., fecundity, morphological descriptions, stage durations, optimum temperature, time of first feeding, prey preferences, and growth rate. Most of this information is not known and not easily obtained otherwise, but is essential for propagation. An excellent example of the value of larviculture in solving ecological problems, while at the same time forming the bases for a new industry, is captive culture of tropical reef fish.

Coral reefs, like many other tropical resources are being exploited at an alarming rate and collection of fish and invertebrates to supply the growing aquarium trade is a major contributor. In the U. S. alone, 10 to 20 million aquarium enthusiasts keep about 95 million tropical aquarium fish. Other large markets for tropicals include public aquariums and research and education centers. Although the majority of freshwater fishes involved in the trade are cultured, almost all of the marine tropicals are wild-collected (Andrews 1990). Many are taken using methods that are destructive to the coral reef ecosystem, e.g., the use of explosives and sodium cyanide. Cyanide collecting is widespread in the Philippines which supplies 75 to 80% of the marine tropicals throughout the world (Rubec 1986) and has now spread to Indonesia, Malaysia, Thailand, and Haiti (Rubec 1991). Captive rearing would significantly decrease the damage due to harvesting and would reduce the pressure on limited reef resources.

For several years, the spawning and early life history of several species of Gulf of Mexico and Caribbean reef fishes have been investigated in my lab. Procedures include: (1) determining the optimal temperature and photoperiod conditions for controlled spawning; (2) developing rearing systems for newly hatched larvae; (3) testing routinely cultured microzooplankton as potential food for larvae; and (4) identifying the gut contents of wild larvae in order to culture preferred microzooplankton. Three species of fish, cherubfish or pygmy angelfish *(Centropyge argi)*, bluehead wrasse *(Thalassoma bifasciatum)*, and spotfin or Cuban hog fish *(Bodianus pulchellus)*, and the peppermint shrimp (*Lysmata wurdemanni*, Caridea: Hippolytidae) have spawned in captivity. The peppermint shrimp were reared through seven planktonic zoeal stages and settled out as postlarvae beginning on day 35 posthatching, using conventional culture techniques. The fish larvae have not been reared successfully to metamorphosis perhaps because appropriate prey were not available. A study to examine the diets of coral reef larvae revealed that dinoflagellates and tintinnids were preferred food of young, preflexion larvae (Riley and Holt 1993). Since the development of an adequate rearing protocol for larvae is a major problem in raising many small coral reef fishes, this information will be useful for culture of reef fish in captivity. Inadequate knowledge of larva nutrition at first exogenous feeding is the major constraint to mass rearing of these and many other species, e.g., red snapper *(Lutjanus campechanus)*, Atlantic halibut *(Hippoglossus hippoglossus)*. Through improved nutritional quality of first foods and better hygienic procedures, survival rates of an increasingly diverse group of fish will improve.

Conclusions

There is room for much advancement in the use of larviculture to moderate population and habitat losses but recent increased interest in the field assures progress in the near future. Areas where technological and scientific advancements would increase culture success and reduce environmental impacts include: (1) progress in the use and development of tags and marking procedures for identifying stocked fish; (2) thorough evaluation of the habitat and the fishery prior to implementing a stocking program; (3) determination of requirements for rearing larvae of many species, especially reef fishes; (4) development of artificial diets and use of microbes to supply nutrients and improve culture conditions in hatcheries; (5) genetic manipulation of embryos to produce disease resistance, or to produce sterile fish, to reduce impacts on native populations.

Finally, the real need is for a the long-term commitment on the part of the government to fund and to carry thorough studies that will resolve issues identified here. Areas in particular need of dedicated research emphasis are stock-enhancement evaluation and marine aquaculture development. Larviculture can play a prominent role in alleviating losses due to habitat degradation and over-utilization of resources but cannot replace a species, or habitat, or ecosystem once it is gone. It is incumbent upon us to manage our natural resources wisely and to use aquaculture as a tool to assist in that regulation and conservation effort.

References

Andrews, C. 1990. The ornamental fish trade and fish conservation. Journal of Fish Biology 37 (Supplement A):53-59.

Botsford, L. W., and R. C. Hobbs. 1984. Optimal fishery policy with artificial enhancement through stocking: California's white sturgeon as an example. Ecological Modelling 23:293-312.

Egidius, E., L. P. Hansen, B. Jonsson, and G. Naevdal. 1991. Mutual impact of wild and cultured Atlantic salmon in Norway. Journal du Conseil International pour l'Exploration de la Mer. 47:404-410.

Eng, C. T., J. N. Paw, and F. Y. Guarin. 1989. The environmental impact of aquaculture and the effects of pollution on coastal aquaculture development in Southeast Asia. Marine Pollution Bulletin 20:335-343.

Gousset, G. 1990. European eel *(Anguilla* L.) farming technologies in Europe and in Japan: application of a comparative analysis. Aquaculture 87:209-235.

Gregg, B. R., and W. P. Rutledge. 1990. Stocking history of Texas waters through 1989. Texas Parks and Wildlife Management Data Series 27, Austin, Texas.

Hilborn, R. 1992. Hatcheries and the future of salmon in the northwest. Fisheries 17:5-8.

Hindar, K., N. Ryman, and F. Utter. 1991. Genetic effects of cultured fish on natural fish populations. Canadian Journal of Fisheries and Aquatic Sciences 48:945-957.

Holt, G. J. 1993. Feeding larval red drum on microparticulate diets in a closed recirculating water system. Journal of the World Aquaculture Society 24:225-230.

Holt, G. J. 1992. Experimental studies of feeding in larval red drum. Journal of the World Aquaculture Society 23:265-270.

Holt, G. J., and F. Sun. 1991. Lipase activity and total lipid content during early development of red drum *Sciaenops ocellatus*. European Aquaculture Society, Special Publication 15:30-33.

Hunter, J. R. 1981. Feeding ecology and predation of marine fish larvae. Pages 34-77 *in* R. Lasker, editor. Marine fish larvae. Morphology, Ecology and Relation to Fisheries. University of Washington Press, Seattle.

Hutchings, J. A. 1991. The threat of extinction to native populations experiencing spawning intrusion by cultured Atlantic salmon. Aquaculture 98:119-132.

Jones, D. A., M. S. Kamarudin, and L. Le Vay. 1991. The potential for replacement of live feeds in larval culture. European Aquaculture Society, Special Publication 15:141.

Kitada, S. 1988. Evaluation of the restocking effectiveness of sea-farming through the ratio of marked fish in the sampled commercial catch. ACTA Oceanographica Taiwanica 19:71-77.

MacCall, A. D. 1989. Against marine fish hatcheries: ironies of fishery politics in the technological era. CalCOFI Rep. 30:46-48.

Mito, I., and J. Fukuhura. 1988. *in* J. V. Juario and L. V. Benitez, editors. Perspectives in aquaculture development in Southeast Asia and Japan. Aquaculture Department SEAFDEC, Tigbauan, Iloilo, Philippines.

Nagata, W. D., and H. Hirata. 1986. Mariculture in Japan: past, present, and future perspectives. Mini Review of Data File Fishery Research 4:1-38.

Nehlson, W., J. E. Williams, and J. A. Lichatowich. 1991. Pacific salmon at the crossroads: stocks at risk from California, Oregon, Idaho, and Washington. Fisheries 16(2):4-21.

New, M. B. 1991. Turn of the millennium aquaculture. Navigating troubled waters or riding the crest of the wave? World Aquaculture 22(3): 28-49.

NRC (National Research Council). 1992. Report of the Committee on Assessment of Technology and Opportunities for Marine Aquaculture in the United States, Marine Board, Marine Aquaculture — Opportunities for Growth. National Academy Press, Washington, District of Columbia.

Person LeRuyet, J., J. C. Alexandre, L. Thébaud, and C. Mugnier. 1993. Marine fish larvae feeding: formulated diets or live prey? Journal of the World Aquaculture Society 24:211-224.

Polovina, J. J. 1991. Evaluation of hatchery releases of juveniles to enhance rockfish stocks, with application to Pacific ocean perch *Sebastes alutus*. Fishery Bulletin, U. S. 89:129-136.

Riley, C. M., and G. J. Holt. 1993. Gut contents of larval fishes from light trap and plankton net collections at Enmedio Reef near Veracruz, Mexico. Revista de Biologia Tropical 41:53-57.

Robinette, H. R., J. Hynes, N. C. Parker, R. Putz, R. E. Stevens, and R. R. Stickney. 1991. Commercial aquaculture. Fisheries 16(1):18-22.

Rubec, P. J. 1986. The effects of sodium cyanide on coral reef and marine fish in the Philippines. Pages 297-302 *in* J. L. Maclean, L. B. Dizon and L. V. Hosillos, editors. The first asian fisheries forum. Asian Fisheries Society, Manila, Philippines.

Rubec, P. J. 1991. Chronic toxic effects of cyanide on tropical marine fish. Canadian Technical Report Fisheries and Aquatic Sciences 1774(1):243-251.

Rutledge, W. P. 1989. The Texas marine hatchery program — it works! CalCOFI Rep. 30:49-52.

Saunders, R. L. 1991. Potential interaction between cultured and wild Atlantic salmon. Aquaculture 98:51-60.

Seymour, E. A., and A. Bengheim. 1991. Towards a reduction of pollution from intensive aquaculture with reference to the farming of salmonids in Norway. Aquacultural Engineering 10:73-88.

Shelbourne, J. E. 1964. The artificial propagation of marine fish. Advanced Marine Biology 2:1-83.

Sorgeloos, P., P. Léger, and P. Lavens. 1988. Improved larval rearing of European and Asian seabass, seabream, mahi mahi, siganid and milkfish using enrichment diets for *Brachionus* and *Artemia*. World Aquaculture 19(4):78-79.

Svåsand, T., K. E. Jorstad, and T. S. Kristiansen. 1990. Enhancement studies of coastal cod in western Norway. Part I. Recruitment of wild and reared cod/to a local spawning stock. Journal du Conseil International pour l' Exploration de la Mer. 47:5-12.

Waples, R. S., G. A. Winans, F. M. Utter, and C. Mahnken. 1990. Genetic approaches to the management of Pacific salmon. Fisheries 15(5):19-25.

Watanabe, T., C. Kitajima, and S. Fujita. 1983. Nutritional values of live organisms used in Japan for mass propagation of fish: a review. Aquaculture 34:115-143.